THE CHAUVENET PAPERS

A Collection of Prize-Winning Expository Papers in Mathematics

VOLUME II

J. C. ABBOTT, editor

United States Naval Academy

Published and distributed by
THE MATHEMATICAL ASSOCIATION OF AMERICA

CONTENTS

VOLUME I

VOLUME II

CONTENTS

12

CORNELIUS LANCZOS

Cornelius Lanczos was born on February 2, 1893, in Szekesfehervar, Hungary. He studied at the University and the Polytechnicum in Budapest under Fejér and Eötiös where he graduated in 1916. He then went to the University of Szeged and worked with Ortvay for his Ph.D. in the field of Maxwell's equations. He transferred to the University of Freiburg, Germany, and later in 1924 to Frankfurt as assistant of E. Madelung, where he collaborated with Dehn, Hellinger, Seigel, and Szasz on quantum mechanics and integral equations. During the year 1928–29 he was invited by Einstein to Berlin, where he worked with Franck, Kallman, Laue, Nernst, Neumann, Pringsheim, Rubens, Schrödinger, and Wigner at the *Physikalische Technische Reichsanstalt*.

In 1931 he came to the United States and occupied the chair of Mathematical Physics at the University of Purdue until 1946. During the war years 1943–44 he was on the staff of the National Bureau of Standards working on a project on Mathematical Tables. In 1946 he became Senior Research Engineer for Boeing Airplane Company in Seattle. In 1947 he was appointed Walker Ames Lecturer at the University of Washington and in 1949 moved to the National Bureau of Standards for Numerical Analysis at the University of California at Los Angeles. In 1953–54 he was with North American Aviation as a specialist in computing.

In 1954 at the invitation of Prime Minister Eamon de Valera he was appointed Senior Professor in the School of Theoretical Physics of the Dublin Institute for Advanced Studies, where he remained until his retirement in 1968. He died June 26, 1974, at age 81.

He has been Visiting Professor at Oregon State College 1957–58; United States Army Mathematical Research Center, Wisconsin, 1959–60; the Computer Center, Chapel Hill, 1960; University of Michigan, 1962; Science Laboratory, Ford Motor Company, 1964–65; North Carolina State University, 1965 and 1968.

He was a Fellow of the American Association for the Advancement of Science, a member of the American Physical Society, the American Mathematical Society, the Mathematical Association of America, the Society for Industrial and Applied Mathematics, and the Royal Irish Academy. He received the Sc.D. award from Trinity College, Dublin, 1962, the degree of D.Sc. (h.c.) from the National University of Ireland, 1970, the honorary degree of Dr. Nat. Phil. from the Johann Wolfgang Göethe University, Frankfurt, 1972, and the D.Sc. (h.c.) from the University of Lancaster in 1972.

His early mathematical interests were in quantum mechanics and relativity theory, but later he turned to numerical analysis with emphasis on practical techniques of Fourier analysis. He published exact methods of obtaining all eigenvectors and eigenvalues of an arbitrary matrix as well as methods of precision approximation of the Gamma function.

He published eight books: *Variational Principles of Mechanics*, 1949; *Applied Analysis*, 1956; *Linear Differential Operators*, 1961; *Albert Einstein and the Cosmic World Order*, 1965; *Discourse on Fourier Series*, 1966; *Numbers without End*, 1968; *Space through the Ages*, 1970; and *The Einstein Decade: 1905–1915*. In addition, he wrote over one hundred research papers and general articles.

Like most of the Chauvenet Prize winners, Lanczos was devoted to teaching and expository mathematics. In his own words: "Many of the scientific treatises of today are formulated in a half-mystical language, as though to impress the reader with the uncomfortable feeling that he is in the permanent presence of a superman. The present book is conceived in a humble spirit and is written for humble people," [1].

Reference

1. Cornelius Lanczos, The Variational Principles of Mechanics, University of Toronto Press, 1949

LINEAR SYSTEMS IN SELF-ADJOINT FORM

C. LANCZOS, Dublin Institute for Advanced Studies

The principal axis problem of quadratic forms belongs to the best-investigated chapters of analysis. Much attention has been paid to the theory of matrices subjected to arbitrary linear transformations and the normal forms attainable by such transformations; (cf. [1],* p. 58). The relation of the general theory of linear equations to matrix calculus has also found exhaustive treatment; (cf. [2], Chapter IV). It seems, therefore, that the fundamental aspects of this field are practically exhausted. †

It is the purpose of the following discussions to approach the problem of general linear algebraic systems from a somewhat different viewpoint which throws new light on the nature of the principal axis problem by showing that the properties of symmetric matrices are extendable to arbitrary matrices to a surprisingly large degree, without demanding anything but *orthogonal* transformations. In this way the classical theory of linear algebraic forms, developed by Frobenius and Kronecker around the end of the last century (cf. [4], p. 268), which centers around the concept of the "rank" of a matrix, is elucidated from a totally different angle in which a certain *eigenvalue problem* plays the central role.

We formulate the given simultaneous system of n equations in m unknowns as the matrix equation

$$(1) \qquad Ay = b,$$

in which A denotes an arbitrary n-row, m-column matrix—briefly denoted as an $n \times m$ matrix—while y is the unknown and b the given right side, both column vectors of m, respectively, n components. The matrix diagram associated with our system, (if we picture the case $n < m$), looks as follows:

We will replace this diagram by the following extended diagram:‡

---

* Numbers in square brackets refer to the references at the end of the paper.

† A very extensive literature is digested in [3].

‡ The symbol "tilde" ($\sim$) refers to a transposition of rows and columns. If A has complex elements, the transposition shall include a change of i to $-i$ in every element. This generalization is so obvious that we will assume the reality of A and call S a "symmetric" rather than "Hermitian" matrix.

$$
\begin{array}{|c|c|}
\hline
0 & A \\
\hline
\bar{A} & 0 \\
\hline
\end{array}
\;
\begin{array}{|c|}
0 \\
y \\
\end{array}
=
\begin{array}{|c|}
b \\
0 \\
\end{array}
$$

This diagram belongs to the equation $Sw = g$, where the symmetric square matrix $S = \tilde{S}$ is defined as follows:

(3)
$$
S = \;
\begin{array}{|c|c|}
\hline
0 & A \\
\hline
\bar{A} & 0 \\
\hline
\end{array}
\;,
$$

while the column vectors w and g have $n+m$ components, which can be displayed more conveniently by transposing them into row vectors:

(4)
$$
\tilde{w} = \left|\; \underset{0}{\overset{(n)}{\rule{2cm}{0pt}}} \;\right|\; \underset{y}{\overset{(m)}{\rule{2cm}{0pt}}} \;\left|\right.,
$$

$$
\tilde{g} = \left|\; \underset{b}{\rule{2cm}{0pt}} \;\right|\; \underset{0}{\rule{2cm}{0pt}} \;\left|\right..
$$

On the other hand, if these two vectors have the following structure:

(5)
$$
\tilde{w} = \left|\; \underset{x}{\overset{(n)}{\rule{2cm}{0pt}}} \;\right|\; \underset{0}{\overset{(m)}{\rule{2cm}{0pt}}} \;\left|\right.,
$$

$$
\tilde{g} = \left|\; \underset{0}{\rule{2cm}{0pt}} \;\right|\; \underset{c}{\rule{2cm}{0pt}} \;\left|\right.,
$$

we obtain the equation

(6)
$$
Ax = c,
$$

and the number of equations is now *greater* than the number of unknowns. Hence we can without loss of generality assume that $n \leqq m$ and put

(7)
$$
m = n + r \qquad (r \geqq 0).
$$

(In our final results all reference to a preestablished relation between n and m will disappear and the numbers n and m be left completely arbitrary.) A general vector u of the $n+m$-dimensional space associated with the matrix S will be introduced as follows:

$$(8) \qquad \tilde{u} = \left| \frac{(n)}{x} \right| \frac{(m)}{y} \right|.$$

Since S is symmetric, we can transform it into a diagonal matrix D with the help of an orthogonal transformation (the "principal-axis transformation"). The matrix U of this orthogonal transformation is composed of the $n+m$ column vectors u_i which satisfy the following eigenvalue problem:

$$(9) \qquad Su_i = \lambda_i u_i \qquad\qquad (i = 1, 2, \cdots n + m)$$

with the added normalization

$$(10) \qquad |u_i|^2 = 1.$$

As is well known, the eigenvalues λ_i of our problem are all *real*.

In view of the special form of the matrix S our eigenvalue problem separates into the two equations

$$(11) \qquad A y_i = \lambda_i x_i, \quad \tilde{A} x_i = \lambda_i y_i.$$

We will call this system of equations the "shifted eigenvalue problem" because on the right side the vectors x_i and y_i are in a "shifted" position, compared with the usual eigenvalue problem associated with A and $\tilde{A}$.

The eigenvalue problem (11) has a number of interesting properties which we are now going to demonstrate.

1. If we premultiply the second equation by A and substitute on the right side for Ay_i its value from the first equation, we obtain

$$(12) \qquad A\tilde{A}x_i = \lambda_i^2 x_i.$$

Similarly, premultiplying the first equation by $\tilde{A}$ and making use of the second equation we obtain

$$(13) \qquad \tilde{A} A y_i = \lambda_i^2 y_i.$$

We thus see that the vectors x_i and y_i can be defined *separately* in themselves. They are the eigenvectors (principal axes) of the nonnegative symmetric matrices $A\tilde{A}$, respectively, $\tilde{A}A$. The first matrix is an $n \times n$, the second an $m \times m$ symmetric matrix. Hence the vectors x_i and y_i belong to two completely different spaces, the one of n, the other of m dimensions. The vectors x_i, put together columnwise, form an $n \times n$ complete orthogonal matrix X:

$$(14) \qquad \tilde{X}X = X\tilde{X} = I.$$

The same can be done with the vectors y_j, resulting in the complete $m \times m$ orthogonal matrix $\overline{Y}$:

$$(15) \qquad \tilde{\overline{Y}}\overline{Y} = \overline{Y}\tilde{\overline{Y}} = I.$$

2. The sequence of the columns of these two matrices is not arbitrary. The vectors x_i and y_i, although defined independently by the equations (12) and (13), are in actual fact *paired*, since they are also in the relation (11) to each other. To every y_i the corresponding x_i can be found by

$$(16) \qquad x_i = \frac{1}{\lambda_i} A y_i$$

and vice versa, provided that λ_i is not zero. In fact, every solution of the system (11) with nonvanishing λ_i can immediately be extended to a *second* pair of vectors since a simultaneous change of λ_i to $-\lambda_i$ and y_i to $-y_i$ leaves the equations (11) unchanged. Accordingly, all nonvanishing eigenvalues will appear in *pairs* $\pm\lambda_i$ and we can agree that what we will call λ_i, shall be a *positive* number, complemented by the negative eigenvalue $-\lambda_i$. For the time being we want to assume that all the λ_i^2 associated with the eigenvalue problem (12) are different from zero. This provides us with the n positive numbers $\lambda_1, \cdots, \lambda_n$, complemented by the sequence $-\lambda_1, \cdots, -\lambda_n$ of negative eigenvalues.

3. An interesting property of the x_i, y_i vectors is that their length is automatically *equal*. We see from (16) that we get

$$(17) \qquad \bar{x}_i x_i = \frac{1}{\lambda_i^2} \tilde{y}_i \tilde{A} A y_i = \tilde{y}_i y_i.$$

For this reason, if the length of the vector x_i is normalized to 1, the length of the corresponding vector y_i becomes automatically 1.

4. The pairing of the vectors x_i, y_i cannot occur unlimitedly, since we cannot have more than n x-vectors while the number of y-vectors is m. Having obtained the $2n$ pairs (x_i, y_i) and $(x_i, -y_i)$, the remaining y_k-vectors can have no x_k-associates, which is only possible if $\lambda_k = 0$. Hence the eigenvalue zero is always present among the eigenvalues if r is not zero. Since the total number of eigenvalues is $n+m = 2n+r$, we must have r eigenvectors associated with the eigenvalue zero. The corresponding x_k-vectors vanish. We consider these additional y-vectors as columns of a matrix Y_0 which has m rows and r columns. The complete $m \times m$ $\bar{Y}$-matrix is thus composed of the $m \times n$ Y-matrix, associated with the matrix X, and the additional $m \times r$ matrix Y_0, associated with the zero matrix. The complete U-matrix which contains all the principal axes of the matrix S, in the sequence $\lambda = -\lambda_1, \cdots, -\lambda_n; \lambda_1, \cdots, \lambda_n; 0, \cdots, 0$; appears in the form (18) while the diagonal matrix D into which S is transformed if we rotate it into the reference system of the principal axes, becomes (19). Here Λ denotes the $n \times n$ diagonal matrix whose diagonal elements are the n positive numbers $\lambda_1, \cdots, \lambda_n$.

Let us observe that the columns of the matrix U, being composed of the two

$$(18) \qquad U = \begin{array}{c} \\ n \\ \\ m \end{array} \begin{array}{|c|c|c|} \multicolumn{1}{c}{n} & \multicolumn{1}{c}{n} & \multicolumn{1}{c}{m-n} \\ \hline X & X & 0 \\ \hline -Y & Y & \sqrt{2}\,Y_0 \\ \hline \end{array} \; ;$$

$$(19) \qquad D = \begin{array}{|c|c|c|} \multicolumn{3}{c}{n} \\ \hline \begin{array}{c} n \\ -\Lambda \end{array} & & \\ \hline & \begin{array}{c} n \\ \Lambda \end{array} & \\ \hline & & \begin{array}{c} m-n \\ 0 \end{array} \\ \hline \end{array} \, .$$

vectors x_i and y_i, both of the length 1, are normalized to $\sqrt{2}$ instead of 1. For this reason the columns of the matrix Y_0 have to be multiplied by $\sqrt{2}$, in order to uniformize all lengths of the columns of U. The right side of (10) has to be changed accordingly from 1 to 2.

We will now drop the restricting condition that all the λ_i of the eigenvalue problem (12) are positive. Generally only $p \leqq n$ eigenvalues need be different from zero. We thus introduce here a new number p which is characteristic for the matrix A, in addition to the two numbers n and m. This number, which in fact coincides with the "rank" of the matrix A, is here defined by the *number of independent eigensolutions of the system* (11) *which are possible if we demand that the eigenvalue* λ_i *shall be a positive number*. This p can take any value between 1 and the smaller of the two numbers (n, m):

$$(20) \qquad 1 \leqq p \leqq \min (n, m).$$

The case $p = 0$ is excluded if A does not vanish identically because, if all eigenvalues of $A\bar{A}$ are zero, the whole matrix $A\bar{A}$ vanishes, which is only possible if $A = 0$.

In view of this new number p the previous picture of the matrix U and the diagonal matrix D changes to some extent. The multiplicity of the eigenvalue zero has now increased from $m - n$ to $m + n - 2p$. The matrix X, associated with the nonzero eigenvalues, is no longer an $n \times n$ but an $n \times p$ matrix. Moreover, the diagonal matrix Λ which appears in the construction of the matrix D, is no longer an $n \times n$ but a $p \times p$ matrix, composed of the positive numbers $\lambda_1, \cdots, \lambda_p$:

(21) $U =$

	p	p	$n-p$		$m-n$
n	X	X	X_0'	X_0'	0
m	$-Y$	Y	$-Y_0'$	Y_0'	$\sqrt{2}\,Y_0''$

,

(22) $D =$

	p	p	$m+n-2p$
p	$-\Lambda$		
p		Λ	
$m+n-2p$			0

.

We have thus obtained a detailed picture of the structure of the matrices U and D which are associated with the principal axis transformation of the matrix S. The entire principal axis problem is included in the matrix equation*

(23) $SU = UD$

with the added condition

(24) $\tilde{U}U = 2I.$

If now we postmultiply (23) by $\tilde{U}$, we obtain the relation

(25) $2S = UD\tilde{U}.$

We wish to construct S on the basis of this relation, performing the matrix multiplication indicated on the right side of (25). First we obtain the matrix

---

* For this formulation of the principal axis problem cf. [5], p. 93.

UD. We know that multiplication by a diagonal matrix as a second factor means that the columns of the first matrix are in succession multiplied by the successive diagonal elements of the second matrix. This gives

(26)
$$UD = \begin{array}{c|c|c|c} & p & p & n+m-2p \\ \hline n & -X\Lambda & X\Lambda & 0 \\ \hline m & Y\Lambda & Y\Lambda & 0 \end{array} \; .$$

Now we should transpose U and postmultiply by it. We can, however, leave U in its original form (21), if we agree that "row by column multiplication" is changed to "row by row multiplication." We will indicate this kind of multiplication by a little circle ○. First of all we multiply the first n rows of (26) by the first n rows of (21). This gives $-X\Lambda \circ X + X\Lambda \circ X + 0 = 0$. We continue by multiplying the first n rows of (26) by the last m rows of (21). This gives $X\Lambda \circ Y + X\Lambda \circ Y + 0 = 2X\Lambda \circ Y = 2X\Lambda \tilde{Y}$. We now come to the product of the last m rows of (26) with the first n rows of (21): $Y\Lambda \circ X + Y\Lambda \circ X + 0 = 2Y\Lambda \circ X = 2Y\Lambda \tilde{X}$. Finally the last m rows of (26) multiplied by the last m rows of (21) yield $-Y\Lambda \circ Y + Y\Lambda \circ Y + 0 = 0$. The complete result is the following matrix:

(27)
$$2S = \begin{array}{|c|c|} \hline 0 & 2X\Lambda\tilde{Y} \\ \hline 2Y\Lambda\tilde{X} & 0 \\ \hline \end{array} \; .$$

Comparison with the original form (3) of S gives the following fundamental result:

(28) $$A = X\Lambda\tilde{Y},$$
(29) $$\tilde{A} = Y\Lambda\tilde{X}.$$

The second equation contains no new statement since it merely repeats the first relation in transposed form. The equation (28) contains the following fundamental

DECOMPOSITION THEOREM. *An arbitrary nonzero matrix can be written as the product of the $n \times p$ orthogonal matrix X, $(\tilde{X}X = I)$, the $p \times p$ positive diagonal matrix Λ and the transpose of the $m \times p$ orthogonal matrix $Y(\tilde{Y}Y = I)$.*

The remarkable fact about this theorem is that *the principal axes associated with the zero eigenvalue do not participate at all in the formation of the matrix A.* This has a profound effect on the solution problem of the equation $Ay = b$.

The matrices X, Y and Λ which appear in this theorem, are defined by the shifted eigenvalue problem

(30) $$AY = X\Lambda, \quad \tilde{A}X = Y\Lambda,$$

with the added condition that the diagonal elements of Λ shall all be nonzero positive numbers.

Example. Consider the $n \times m$ matrix whose elements are all zero, except the single element a_{ij} which may be given as the complex number c. Show that in this problem $p = 1$, $\lambda_1 = |c|$; all the n elements of the vector x_1 vanish except the element $x_1^{(i)}$; all the m elements of the vector y_1 vanish, except the single element $y_1^{(j)}$. Demonstrate the validity of the relation (28).

Before we continue with the further analysis of our problem, let us recall the two full $n \times n$ and $m \times m$ spaces associated with our eigenvalue problem. We can picture them as follows:

	p	$n-p$			p	$m-p$
n	X	X_0	,	m	Y	Y_0

The matrices X and Y are composed of p mutually orthogonal axes of the two respective spaces. They form a p-dimensional subspace within the full space. Hence it is generally not permissible to convert the relation $\tilde{X}X = I$ to $X\tilde{X} = I$, and the same holds for the matrix Y. The remaining axes, included in the $n \times (n-p)$ orthogonal matrix X_0, respectively the $m \times (m-p)$ orthogonal matrix Y_0, belong to the eigenvalue zero and are thus defined by the two noninterrelated equations

(31) $$\tilde{A}X_0 = 0,$$

respectively,

(32) $$AY_0 = 0.$$

The columns of these two matrices are thus composed of the solutions of the homogeneous equation

(33) $$\tilde{A}x_i^0 = 0 \qquad\qquad (i = 1, \cdots, n - p),$$

respectively,

(34) $$Ay_j^0 = 0 \qquad\qquad (j = 1, \cdots, m - p).$$

In harmony with the orthogonal nature of principal axes we assume that the vectors x_i^0 are mutually orthogonal and their length is 1. The same can be said of the vectors y_j^0. This, however, is not self-evident since the principal axes belonging to a multiple eigenvalue (in this case the eigenvalue zero), are not orthogonal by nature, although they *can* be orthogonalized. We will have use for the vectors ξ_i, which are merely solutions of the homogeneous equation

$$(35) \qquad \qquad \tilde{A}\xi_i = 0 \qquad \qquad (i = 1, \cdots, n - p)$$

without demanding their orthogonalization and the normalization of their length. We merely demand that they shall be linearly independent and that their number shall be $n-p$, in order to span the entire space X_0. Similarly we will consider the $m-p$ linearly independent solutions of the homogeneous equation

$$(36) \qquad \qquad A\eta_j = 0 \qquad \qquad (j = 1, \cdots, m - p)$$

without demanding their orthogonalization and normalization. The vectors ξ_i, taken as column vectors, form the $n \times (n-p)$ matrix Ξ_0, the vectors η_j the $m \times (m-p)$ matrix H_0. These matrices are no longer orthogonal but their orthogonality to the subspaces X and Y remains unchanged:

$$(37) \qquad \qquad \tilde{\Xi}_0 X = 0, \qquad \tilde{H}_0 Y = 0.$$

With this preliminary information we return to the study of the linear equation (1) which could be done by diagonalization in the reference system of the principal axes of S, but we prefer to draw all our conclusions from the decomposition of our matrix A into the product (28). The equation (1) can now be written in the form

$$(38) \qquad \qquad X\Lambda\tilde{Y}y = b.$$

Premultiplication by the matrix $\tilde{\Xi}_0$ gives

$$(39) \qquad \qquad 0 = \tilde{\Xi}_0 b.$$

This equation, if written in the language of vectors, becomes

$$(40) \qquad \qquad (\xi_i \cdot b) = 0 \qquad \qquad (i = 1, \cdots, n - p)$$

and we obtain the following (well-known)

COMPATIBILITY THEOREM. *The equation $Ay=b$ is solvable if and only if the given right side of the equation is orthogonal to every independent solution of the adjoint homogeneous equation $\tilde{A}\xi=0$.*

That the condition (40) is *necessary*, follows from (39). That it is also *sufficient* follows from the fact that if b is perpendicular to the space Ξ_0 (or the equivalent space X_0), it must lie inside the space X, *i.e.* it must have the form

$$(41) \qquad \qquad b = Xb'.$$

But in that case the equation (38)—premultiplying it by $\tilde{X}$—gives at once

(42) $\Lambda \tilde{Y} y = b'.$

This equation is solvable by putting

(43) $y = Y y',$

in which case we obtain

(44) $\Lambda y' = b', \; y' = \Lambda^{-1} b'$

and finally

(45) $y = Y \Lambda^{-1} b' = Y \Lambda^{-1} \tilde{X} b.$

The diagonal matrix Λ contains only nonzero elements in the diagonal and is thus always invertible.

However, the solution (45) is not the *only* solution of our system. An arbitrary vector y, if analyzed in the reference system of the principal axes (Y, Y_0), appears in the following form:

$$y = Y y' + Y_0 y_0'.$$

If we put this expression in (38), we observe that the term with y_0' *drops out completely* from our equation. We thus obtain, as a counterpart of our previous Compatibility Theorem, the following

DEFICIENCY THEOREM: *The equation $Ay = b$ determines uniquely the projection of the vector y into the space Y but leaves its projection into the space Y_0 completely undetermined.*

We will now consider the solution of our system (1) under the following *auxiliary conditions*:

(46) $\tilde{X}_0 b = 0$ (by necessity),

(47) $\tilde{Y}_0 y = 0$ (by choice).

The second condition is not demanded by the original equation. By adding this condition we obtain a definite *particular solution* distinguished by the property that the solution finds its place in a *subspace of smallest capacity, viz.* the space Y.*
This brings us back to the condition (43) and thus to the solution (45).

We know from the general theory of linear operators that the general solution of a linear system of equations is obtainable by adding to any particular solution an arbitrary solution of the homogeneous equation. Applying this principle to our problem we get

(48) $y = y_p + Y_0 \eta,$

where η is an arbitrary column vector of $m - p$ elements while for y_p we can choose the particular solution (45), obtained under the auxiliary condition (47).

---

* This "normalization condition," which makes the solution unique, is equivalent to putting $y = \tilde{A} v$, where the vector v is unrestricted.

We can interpret the deficiency of the given system in the following terms. The equation (1) is not sufficient for the determination of y but it *may* become sufficient by *added information*. The information needed is a statement concerning the projection of y into the space Y_0. Hence we can conceive the equation (1) as *part of a more elaborate system*, the addition taking the form

$$(49) \qquad \tilde{Y}_0 y = b_0,$$

where b_0 is a free column vector of $m - p$ components. In this case the previous solution $y = y_p$ is not more than a *preliminary result*, while the complete solution takes the form (48). But now premultiplication by $\tilde{Y}_0$ yields

$$(50) \qquad \eta = b_0$$

and the complete solution—obtained after complementing the original system (1) by the additional system (49)—becomes

$$(51) \qquad y = y_p + Y_0 b_0 = Y \Lambda^{-1} \tilde{X} b + Y_0 b_0.$$

The result of this analysis may be summarized as follows: The zero-fields X_0, Y_0, associated with the solutions of the homogeneous equations $\tilde{A}x = 0$, $Ay = 0$, do not participate directly in the solution of the linear system (1) but merely decide the *compatibility* and the *deficiency* of the system. The compatibility conditions (46) have to be assumed in order to have a solution at all. The deficiency of the system can be removed by putting the solution in the space Y, assuming that the added information (49) will later provide the missing Y_0-portion of the solution. By this procedure an arbitrarily over-determined (although compatible) or under-determined (and thus deficient) linear system permits a unique solution.

Our solution can be put in the form

$$(52) \qquad y = Gb,$$

where the $m \times n$ matrix G is defined as follows:

$$(53) \qquad G = Y \Lambda^{-1} \tilde{X}.$$

In view of the form (52) of the solution we can conceive the matrix G as the "inverse" of the matrix A. We should thus expect that the product GA has the property of the unit matrix I. It would be a mistake, however, to assume that the product GA must come out as the unit matrix. The product GA does not operate on an *arbitrary* vector b but on a vector which is subject to the condition (46). This means that b is inside the space X and has thus the form (41). Now

$$(54) \qquad AG = X \Lambda \tilde{Y} Y \Lambda^{-1} \tilde{X} = X \tilde{X}$$

and AG operating on b becomes

$$(55) \qquad AGb = X \tilde{X} b = X \tilde{X} X b' = X b' = b,$$

which shows that AG has in fact the property of the unit matrix with respect to all "permissible" vectors b.

On the other hand, let us premultiply (1) by G:

(56) $$GAy = Gb = y.$$

This shows that the product GA must also have the property of the unit matrix, but again operating on a special class of vectors, subject to the condition (47). This condition puts y into the space Y which means that y can be put in the form (43). Now

(57) $$GA = Y\Lambda^{-1}\tilde{X}X\Lambda\tilde{Y} = Y\tilde{Y}$$

and therefore

(58) $$GAy = Y\tilde{Y}y = Y\tilde{Y}Yy' = Yy' = y.$$

Once again the product GA has the property of the unit matrix I with respect to all permissible vectors y.

The matrix (53) has all the properties demanded by E. H. Moore in his "general analysis," (1906); (*cf.* [6]), establishing the "generalized inverse" of a matrix in abstract terms. We can likewise demonstrate that the conditions demanded by R. Penrose (*cf.* [7]; see also R. Rado, [8]) concerning the generalized inverse of a matrix are fulfilled. In our analysis the inverse matrix did not come about by any definitions in terms of matrix equations but by an *explicit method* of solving the linear system (1), based on the properties of an eigenvalue problem.

We can write out the solving matrix more explicitly by substituting for X and Y the constituting column vectors x_i and y_i which appeared in the solution of the eigenvalue problem (11). Let us denote the components of the p vectors x_α by $x_\alpha^{(i)}$ ($i=1, \cdots, n$), the components of the conjugate vectors y_α by $y_\alpha^{(j)}$ ($j=1, \cdots, m$). Then the element g_{ij} of the matrix G comes out as follows:

(59) $$g_{ij} = \sum_{\alpha=1}^{p} \frac{y_\alpha^{(i)} x_\alpha^{(j)}}{\lambda_\alpha},$$

while the element a_{ij} of the original matrix A becomes:

(60) $$a_{ij} = \sum_{\alpha=1}^{p} x_\alpha^{(i)} \lambda_\alpha y_\alpha^{(j)}.$$

More important, however, is another interpretation of the solution (45). We know that the vector y lies inside the space Y which is composed of the p orthogonal vectors $y_1, \cdots, y_p$. Hence y can be analyzed in terms of these vectors:

(61) $$y = \eta_1 y_1 + \eta_2 y_2 + \cdots + \eta_p y_p.$$

On the other hand, the right side b lies inside the space X and can be analyzed in terms of the conjugate orthogonal vectors $x_1, \cdots, x_p$:

(62) $$b = \beta_1 x_1 + \beta_2 x_2 + \cdots + \beta_p x_p$$

where the coefficients β_i are obtainable by projecting b on the axes X_i:

(63) $$\beta_i = (b \cdot x_i).$$

We will call the two conjugate expansions (61) and (62) "co-orthogonal" since they involve two sets of orthogonal vectors which are in a one-to-one correspondence to each other.* Then the linear system (1), under the added auxiliary conditions (46) and (47), establishes the following relation between the coefficients β_i and η_i:

(64) $$\eta_i = \beta_i/\lambda_i.$$

We see that the eigenvalues λ_i play the role of a "transfer function" in going from the right to the left, or from the left to the right.

Since the coefficients β_i are available on the basis of (63), the coefficients η_i become determined on the basis of (64) and the unknown vector y appears in the form of an orthogonal expansion (61), with given coefficients.

Operations in function space. The field of continuous linear operators—*i.e.*, the domain of linear differential or integral equations—can be handled on the basis of matrix operations if we introduce the infinite-dimensional "function space" and the matrices associated with this space; (*cf.* [9], p. 57). The results obtained in the theory of solving arbitrary linear algebraic systems can thus be extended to the realm of linear differential and integral equations. The characteristic feature of our investigation was that we have dealt with an arbitrarily over-determined or under-determined system and yet arrived at a unique solution under the proper auxiliary conditions.

The usual type of boundary-value problems considered in classical analysis are of the so-called "well-posed" type. This means that the given data—the differential equation with a given right side plus the boundary conditions—suffice for a unique solution and that the data can be prescribed freely, without the danger of incompatibility. Such problems realize in the language of matrices the case $n = m = p$: the number of equations is equal to the number of unknowns and the eigenvalue zero is not present (the matrix A is nonsingular).

Our investigation has shown that we can expect a valid and unique solution of a linear system under much more general conditions. The given operator (including the boundary conditions) may or may not comprise all the dimensions of the function space. The classical case usually considered belongs to those operators which comprise the *entire* function space, in both X and Y relations, *i.e.* in relation to the given right side as well as in relation to the unknown function. If we have a problem which is not "well-posed," this merely means that

---

* The expression "bi-orthogonal" would be misleading since it usually refers to two sets of *mutually orthogonal* vectors, while the two expansions (61) and (62) involve two sets of vectors which are orthogonal *within themselves*.

the given operator *omits* certain dimensions of the function space, either with respect to the given right side, or with respect to the unknown, or with respect to both. This, however, is no reason to reject the given boundary value problem. If the omission occurs with respect to the X-space, this means that the given data cannot be given freely but must be contained in a certain subspace of the function space. This condition can be met if we replace the word "prescribed data" by "observed data" because no matter how many surplus data we observe (in addition to the minimum number which would have sufficed for a unique solution), these data cannot be inconsistent since the prescribed mathematical law was in operation throughout our observations. Our operator is restricted to certain dimensions of the function space and cannot lead out of this space, no matter how many observations we perform.

If the given operator omits certain dimensions in the Y relation, this makes our solution incomplete since we obtain no information concerning the missing dimensions. We do get, however, a unique solution in those dimensions which are represented in the operator. We can then add later observations in order to complete our solution with respect to the missing dimensions.

We thus obtain a method for the solution of boundary value problems which can be arbitrarily over-determined (although consistent) or under-determined, and thus far from that "well-posed" type of problems that we expect under the customary conditions. Under these relaxed conditions the "inverse" of the operator does not exist any more in the ordinary sense. But even the "generalized inverse" in the sense of the matrix G which omits the zero-field and avoids the division by zero, need not necessarily exist. In the case of *finite* matrices it cannot happen that the matrix G, defined by (53) and more specifically by (59), should not exist. But in the case of continuous operators the corresponding expansion—called under simplified conditions the "bilinear expansion of the Green's function" (*cf.* [9], p. 360)—becomes an *infinite series* which may or may not converge. In many problems of an unconventional type to which the present theory is applicable, the bilinear expansion becomes in fact meaningless since it has no tendency to converge. Nor does the "inverse bilinear expansion" which corresponds to (60), converge and represent the given operator. This does *not*, however, interfere with the solution of our problem in terms of the two "co-orthogonal expansions" (61) and (62) which remain uniformly convergent even if the expansion of the inverse operator fails, provided that the right side is taken from that restricted subspace of the function space which is allotted to it by the nature of the given operator.

Although the application of the general theory to the field of continuous operators will be discussed in more detail in a separate paper*, it may not be without interest to give an example of the type of unconventional boundary value problems which become solvable by the method here presented.

---

* Proceedings of the International Congress of Mathematicians, Edinburgh, 1958 (to be published in 1960).

Problem. Given a simply-connected domain C of the complex plane, enclosed by a smooth boundary, let it be known that the function $f(z)$ of the complex variable $z = x + iy$ is analytical throughout C including the boundary, and let the value of $f(z)$ be given along the arbitrarily small arc S of the boundary. Find $f(z)$ inside the domain C.

The method of analytical continuation shows that this problem has a unique solution but the theory of analytical functions gives no clue toward a solution which would obtain the value of $f(z)$ at a distant point z directly in terms of the given boundary values along S.

The shifted eigenvalue problem associated with the present problem gives the solution in the following form. Associated with the domain C and the arc S we can define an infinite set of functions

(65) $$f_1(z, z^*), \cdots, f_i(z, z^*), \cdots$$

which exist inside and on the boundary S, S', together with a corresponding set of functions defined along the arc S:

(66) $$g_1(s), \cdots, g_i(s), \cdots.$$

The desired function $f(z)$ can be expanded into the infinite sum

(67) $$f(z) = \sum_{\alpha=1}^{\infty} \gamma_\alpha f_\alpha(z, z^*)$$

where the expansion coefficients γ_i are obtained as follows:

(68) $$\gamma_i = \int_S f(s) g_i(s) ds.$$

None of the functions $f_\alpha(z, z^*)$ are analytical (in view of the dependence on z^*, the "complex conjugate" of z). Nor do these $f_\alpha(z, z^*)$ satisfy the given boundary conditions. In fact, all the $f_\alpha(z, z^*)$ *vanish* along S. And yet, the infinite sum (67) *converges uniformly* to the correct value of $f(z)$ at every point of the domain C (including the outer boundary S') which excludes the arc S.

References

1. H. W. Turnbull and A. C. Aitken, Introduction to the Theory of Canonical Matrices, London, 1932.

2. R. A. Frazer, W. J. Duncan, and A. R. Collar, Elementary Matrices, Cambridge, 1950.

3. C. C. MacDuffee, The Theory of Matrices, New York, 1946.

4. Encyklop. der Math. Wissenschaften, vol. I, Leipzig, 1904.

5. C. Lanczos, Applied Analysis, Englewood Cliffs, N. J., 1956.

6. E. H. Moore, General Analysis, vol. 1, Amer. Philos. Soc., Philadelphia, 1935.

7. R. Penrose, A generalized inverse for matrices, Proc. Cambridge Philos. Soc., vol. 51, 1955, p. 406.

8. R. Rado, Note on generalized inverses of matrices, Proc. Cambridge Philos. Soc., vol. 52, 1956, p. 600.

9. R. Courant and D. Hilbert, Methods of Mathematical Physics, vol. 1, New York, 1953.

13

PHILIP J. DAVIS

Dr. Davis was born on January 2, 1923, in Lawrence, Massachusetts. He received his S.B. from Harvard College in 1943 and his Ph.D. in mathematics from Harvard University in 1950. During the period from 1944 to 1946, he was employed in the Aircraft Loads Division of the National Advisory Committee for Aeronautics at Langley Field, Virginia.

From 1950 to 1952, Dr. Davis was a Post-doctoral Research Assistant at the Harvard School of Engineering. In 1952 he joined the Applied Mathematics Division of the National Bureau of Standards and advanced to Chief of its Numerical Analysis Section in 1958. In 1956–57 he was a Fellow of the John Simon Guggenheim Memorial Foundation.

At various times, Dr. Davis has given courses at Harvard, M.I.T., American University, and the University of Maryland. He has been an Associate Editor of the *Journal of the Society for Industrial and Applied Mathematics* and the *Quarterly of Applied Mathematics*. He is a member of the Editorial Panel of the Monograph Project of the School Mathematics Study Group, and the India Committee on Mathematics Education.

Dr. Davis was the 1960 recipient of the Award in Mathematics of the Washington Academy of Sciences. He was Jeffrey-Williams Lecturer at the Canadian Mathematical Congress and AAAS Chautauqua Lecturer.

Dr. Davis' skills as both a researcher and expositor, in diverse branches of mathematics, is shown in his 41 publications and 2 books. The many who so thoroughly enjoyed his article on the Gamma Function for which he was awarded the 1963 Chauvenet Prize will find equal pleasure in his book *The Lore of Large Numbers*, published in 1961 by Random House and Yale University. In addition he has written the books: *Interpolation and Approximation*, Blaisdell Publishing Co., New York, 1963, and Dover Publications, 1975; *The Mathematics of Matrices*, Blaisdell Publishing Co., New York, 1964; *Numerical Integration*, Blaisdell Publishing Co., Waltham, Mass., 1967 (with Philip Rabinowitz); *3.1416 and All That*, Simon and Schuster, New York, 1969 (with William Chin); *The Schwarz Function and Its Applications*, Carus Mathematical Monograph No. 17, The Mathematical Association of America, 1974; and *Methods of Numerical Integration* (with P. Rabinowitz), Academic Press, New York, 1975 (Japanese translation: 1977).

He recently reiterated his beliefs in the importance of expository writing in the statement: "In these days when the subject of mathematics has grown and has been fragmented to the point where one can no longer communicate intelligently with one's colleagues and when the mathematical sensibility of one generation is not that of its successor, it is of critical importance that first rate expository writing be encouraged and rewarded. It is very necessary, but, alas, it is not sufficient."

In accepting the 1963 Chauvenet Prize, Dr. Davis expressed the hope that his article would serve to interest the undergraduate in pursuing his mathematical studies and to encourage the professional to write in a similar vein. He emphasized the great need for expository articles which make available recent work, and also for articles which have an historical or critical point of view and reveal the creative element in mathematics.

LEONHARD EULER'S INTEGRAL: A HISTORICAL PROFILE OF THE GAMMA FUNCTION

IN MEMORIAM: MILTON ABRAMOWITZ

PHILIP J. DAVIS, National Bureau of Standards, Washington, D. C.

Many people think that mathematical ideas are static. They think that the ideas originated at some time in the historical past and remain unchanged for all future times. There are good reasons for such a feeling. After all, the formula for the area of a circle was πr^2 in Euclid's day and at the present time is still πr^2. But to one who knows mathematics from the inside, the subject has rather the feeling of a living thing. It grows daily by the accretion of new information, it changes daily by regarding itself and the world from new vantage points, it maintains a regulatory balance by consigning to the oblivion of irrelevancy a fraction of its past accomplishments.

The purpose of this essay is to illustrate this process of growth. We select one mathematical object, the gamma function, and show how it grew in concept and in content from the time of Euler to the recent mathematical treatise of Bourbaki, and how, in this growth, it partook of the general development of mathematics over the past two and a quarter centuries. Of the so-called "higher mathematical functions," the gamma function is undoubtedly the most fundamental. It is simple enough for juniors in college to meet but deep enough to have called forth contributions from the finest mathematicians. And it is sufficiently compact to allow its profile to be sketched within the space of a brief essay.

The year 1729 saw the birth of the gamma function in a correspondence between a Swiss mathematician in St. Petersburg and a German mathematician in Moscow. The former: Leonhard Euler (1707–1783), then 22 years of age, but to become a prodigious mathematician, the greatest of the 18th century. The latter: Christian Goldbach (1690–1764), a savant, a man of many talents and in correspondence with the leading thinkers of the day. As a mathematician he was something of a dilettante, yet he was a man who bequeathed to the future a problem in the theory of numbers so easy to state and so difficult to prove that even to this day it remains on the mathematical horizon as a challenge.

The birth of the gamma function was due to the merging of several mathematical streams. The first was that of interpolation theory, a very practical subject largely the product of English mathematicians of the 17th century but which all mathematicians enjoyed dipping into from time to time. The second stream was that of the integral calculus and of the systematic building up of the formulas of indefinite integration, a process which had been going on steadily for many years. A certain ostensibly simple problem of interpolation arose and was bandied about unsuccessfully by Goldbach and by Daniel Bernoulli (1700–1784) and even earlier by James Stirling (1692–1770). The problem was posed to Euler. Euler announced his solution to Goldbach in two letters which were to be the beginning of an extensive correspondence which lasted the duration of Goldbach's life. The first letter dated October 13, 1729 dealt with the interpolation problem, while the second dated January 8, 1730 dealt with integration and tied the two together. Euler wrote Goldbach the merest outline, but within a year he published all the details in an article *De progressionibus transcendentibus seu quarum termini generales algebraice dari nequeunt*. This article can now be found reprinted in Volume I_{14} of Euler's *Opera Omnia*.

Since the interpolation problem is the easier one, let us begin with it. One of the simplest sequences of integers which leads to an interesting theory is 1, $1+2$, $1+2+3$, $1+2+3+4$, $\cdots$. These are the triangular numbers, so called because they represent the number of objects which can be placed in a triangular array of various sizes. Call the nth one T_n. There is a formula for T_n which is learned in school algebra: $T_n = \frac{1}{2}n(n+1)$.

What, precisely, does this formula accomplish? In the first place, it simplifies computation by reducing a large number of additions to three fixed operations: one of addition, one of multiplication, and one of division. Thus, instead of adding the first hundred integers to obtain T_{100}, we can compute $T_{100} = \frac{1}{2}(100)(100+1)$ $= 5050$. Secondly, even though it doesn't make literal sense to ask for, say, the sum of the first $5\frac{1}{2}$ integers, the formula for T_n produces an answer to this. For whatever it is worth, the formula yields $T_{5\frac{1}{2}} = \frac{1}{2}(5\frac{1}{2})(5\frac{1}{2}+1) = 17\frac{7}{8}$. In this way, the formula extends the scope of the original problem to values of the variable other than those for which it was originally defined and solves the problem of interpolating between the known elementary values.

This type of question, one which asks for an extension of meaning, cropped up frequently in the 17th and 18th centuries. Consider, for instance, the algebra of exponents. The quantity a^m is defined initially as the product of m successive a's. This definition has meaning when m is a positive integer, but what would $a^{5\frac{1}{2}}$ be? The product of $5\frac{1}{2}$ successive a's? The mysterious definitions $a^0 = 1$, $a^{m/n} = \sqrt[n]{a^m}$, $a^{-m} = 1/a^m$ which solve this enigma and which are employed so fruitfully in algebra were written down explicitly for the first time by Newton in 1676. They are justified by a utility which derives from the fact that the definition leads to continuous exponential functions and that the law of exponents $a^m \cdot a^n = a^{m+n}$ becomes meaningful for all exponents whether positive integers or not.

Other problems of this type proved harder. Thus, Leibnitz introduced the notation d^n for the nth iterate of the operation of differentiation. Moreover, he identified d^{-1} with $\int$ and d^{-n} with the iterated integral. Then he tried to breathe some sense into the symbol d^n when n is any real value whatever. What, indeed, is the $5\frac{1}{2}$th derivative of a function? This question had to wait almost two centuries for a satisfactory answer.

THE FACTORIALS								
n:	1	2	3	4	5	6	7	8	$\cdots$
$n!$:	1	2	6	24	120	720	5040	40,320	$\cdots$

FIG. 1

INTELLIGENCE TEST

Question: What number should be inserted in the lower line half way between the upper 5 and 6?

Euler's Answer: 287.8852 · · · · . Hadamard's Answer: 280.3002 · · · · .

But to return to our sequence of triangular numbers. If we change the plus signs to multiplication signs we obtain a new sequence: $1, 1 \cdot 2, 1 \cdot 2 \cdot 3, \cdots$. This is the sequence of factorials. The factorials are usually abbreviated $1!, 2!, 3!, \cdots$ and the first five are 1, 2, 6, 24, 120. They grow in size very rapidly. The number 100! if written out in full would have 158 digits. By contrast, $T_{100} = 5050$ has a mere four digits. Factorials are omnipresent in mathematics; one can hardly open a page of mathematical analysis without finding it strewn with them. This being the case, is it possible to obtain an easy formula for computing the factorials? And is it possible to interpolate between the factorials? What should $5\frac{1}{2}!$ be? (See Fig. 1.) This is the interpolation problem which led to the gamma function, the interpolation problem of Stirling, of Bernoulli, and of Goldbach. As we know, these two problems are related, for when one has a formula there is the possibility of inserting intermediate values into it. And now comes the surprising thing. There is no, in fact there can be, no formula for the factorials which is of the simple type found for T_n. This is implicit in the very title Euler chose for his article. Translate the Latin and we have *On transcendental progressions whose general term cannot be expressed algebraically.* The solution to factorial interpolation lay deeper than "mere algebra." Infinite processes were required.

In order to appreciate a little better the problem confronting Euler it is useful to skip ahead a bit and formulate it in an up-to-date fashion: find a reasonably simple function which at the integers $1, 2, 3, \cdots$ takes on the factorial values $1, 2, 6, \cdots$. Now today, a function is a relationship between two sets of numbers wherein to a number of one set is assigned a number of the second set. What is stressed is the relationship and not the nature of the rules which serve to determine the relationship. To help students visualize the function concept in its full generality, mathematics instructors are accustomed to draw a curve full of twists and discontinuities. The more of these the more general the function is supposed to be. Given, then, the points (1,1), (2, 2), (3, 6), (4, 24), $\cdots$ and adopting the point of view wherein "function" is what we have just said, the problem of interpolation is one of finding a curve which passes through the given points. This is ridiculously easy to solve. It can be done in an unlimited number of ways. Merely take a pencil and draw some curve—any curve will do—which passes through the points. Such a curve automatically defines a function which solves the interpolation problem. In this way, too free an attitude as to what constitutes a function solves the problem trivially and would enrich mathematics but little. Euler's task was different. In the early 18th century, a function was more or less synonymous with a formula, and by a formula was meant an expression which could be derived from elementary manipulations with addition, subtraction, multiplication, division, powers, roots, exponentials, logarithms, differentiation, integration, infinite series, *i.e.*, one which came from the ordinary processes of mathematical analysis. Such a formula was called an *expressio analytica*, an analytical expression. Euler's task was to find, if he could, an analytical expression arising naturally from the corpus of mathematics which would yield factorials when a positive integer was inserted, but which would still be meaningful for other values of the variable.

It is difficult to chronicle the exact course of scientific discovery. This is particularly true in mathematics where one traditionally omits from articles and books all accounts of false starts, of the initial years of bungling, and where one may develop one's topic forward or backward or sideways in order to heighten the dramatic effect. As one distinguished mathematician put it, a mathematical result must appear straight from the heavens as a *deus ex machina* for students to verify and accept but not to comprehend. Apparently, Euler, experimenting with infinite products of numbers, chanced to notice that if n is a positive integer,

(1) $$\left[\left(\frac{2}{1}\right)^n \frac{1}{n+1}\right]\left[\left(\frac{3}{2}\right)^n \frac{2}{n+2}\right]\left[\left(\frac{4}{3}\right)^n \frac{3}{n+3}\right]\cdots = n!.$$

Leaving aside all delicate questions as to the convergence of the infinite product, the reader can verify this equation by cancelling out all the common factors which appear in the top and bottom of the left-hand side. Moreover, the left-hand side is defined (at least formally) for all kinds of n other than negative integers. Euler noticed also that when the value $n=\frac{1}{2}$ is inserted, the left-hand side yields (after a bit of manipulation) the famous infinite product of the Englishman John Wallis (1616–1703):

(2) $$\left(\frac{2\cdot 2}{1\cdot 3}\right)\left(\frac{4\cdot 4}{3\cdot 5}\right)\left(\frac{6\cdot 6}{5\cdot 7}\right)\left(\frac{8\cdot 8}{7\cdot 9}\right)\cdots = \pi/2.$$

With this discovery Euler could have stopped. His problem was solved. Indeed, the whole theory of the gamma function can be based on the infinite product (1) which today is written more conventionally as

(3) $$\lim_{m\to\infty} \frac{m!(m+1)^n}{(n+1)(n+2)\cdots(n+m)}.$$

However, he went on. He observed that his product displayed the following curious phenomenon: for some values of n, namely integers, it yielded integers, whereas for another value, namely $n=\frac{1}{2}$, it yielded an expression involving π. Now π meant circles and their quadrature, and quadratures meant integrals, and he was familiar with integrals which exhibited the same phenomenon. It therefore occurred to him to look for a transformation which would allow him to express his product as an integral.

He took up the integral $\int_0^1 x^e(1-x)^n dx$. Special cases of it had already been discussed by Wallis, by Newton, and by Stirling. It was a troublesome integral to handle, for the indefinite integral is not always an elementary function of x. Assuming that n is an integer, but that e is an arbitrary value, Euler expanded $(1-x)^n$ by the binomial theorem, and without difficulty found that

(4)
$$\int_0^1 x^e (1 - x)^n dx = \frac{1 \cdot 2 \cdots n}{(e + 1)(e + 2) \cdots (e + n + 1)}.$$

Euler's idea was now to isolate the $1 \cdot 2 \cdots n$ from the denominator so that he would have an expression for $n!$ as an integral. He proceeds in this way. (Here we follow Euler's own formulation and nomenclature, marking with an * those formulas which occur in the original paper. Euler wrote a plain $\int$ for $\int_0^1$.) He substituted f/g for e and found

(5)
$$\int_0^1 x^{f/g} (1 - x)^n dx = \frac{g^{n+1}}{f + (n + 1)g} \cdot \frac{1 \cdot 2 \cdots \cdot n}{(f + g)(f + 2 \cdot g) \cdots (f + n \cdot g)}.$$

And so,

(6)*
$$\frac{1 \cdot 2 \cdots n}{(f + g)(f + 2 \cdot g) \cdots (f + n \cdot g)} = \frac{f + (n + 1)g}{g^{n+1}} \int x^{f/g} dx (1 - x)^n.$$

He observed that he could isolate the $1 \cdot 2 \cdots n$ if he set $f = 1$ and $g = 0$ in the left-hand member, but that if he did so, he would obtain on the right an indeterminate form which he writes quaintly as

(7)*
$$\int \frac{x^{1/0} dx (1 - x)^n}{0^{n+1}}.$$

He now proceeded to find the value of the expression (7)*. He first made the substitution $x^{g/(f+g)}$ in place of x. This gave him

(8)*
$$\frac{g}{f + g} x^{-f/(g+f)} dx$$

in place of dx and hence, the right-hand member of (6)* becomes

(9)*
$$\frac{f + (n + 1)g}{g^{n+1}} \int \frac{g}{f + g} dx (1 - x^{g/(f+g)})^n.$$

Once again, Euler made a trial setting of $f = 1$, $g = 0$ having presumably reduced this integral first to

(10)
$$\frac{f + (n + 1)g}{(f + g)^{n+1}} \int_0^1 \left(\frac{1 - x^{g/(f+g)}}{g/(f + g)} \right)^n dx,$$

and this yielded the indeterminate

(11)*
$$\int dx \frac{(1 - x^0)^n}{0^n}.$$

He now considered the related expression $(1-x^z)/z$, for vanishing z. He differentiated the numerator and denominator, as he says, by a known (l'Hospital's) rule and obtained

(12)*
$$\frac{-x^z dz\, lx}{dz} \qquad\qquad (lx = \log x),$$

which for $z=0$ produced $-lx$. Thus,

(13)*
$$(1 - x^0)/0 = - lx$$

and

(14)*
$$(1 - x^0)^n/0^n = (-lx)^n.$$

He therefore concluded that

(15)
$$n! = \int_0^1 (-\log x)^n dx.$$

This gave him what he wanted, an expression for $n!$ as an integral wherein values other than positive integers may be substituted. The reader is encouraged to formulate his own criticism of Euler's derivation.

Students in advanced calculus generally meet Euler's integral first in the form

(16)
$$\Gamma(x) = \int_0^\infty e^{-t} t^{x-1} dt, \qquad e = 2.71828 \cdots.$$

This modification of the integral (15) as well as the Greek Γ is due to Adrien Marie Legendre (1752–1833). Legendre calls the integral (4) with which Euler started his derivation the first Eulerian integral and (15) the second Eulerian integral. The first Eulerian integral is currently known as the Beta function and is now conventionally written

(17)
$$B(m, n) = \int_0^1 x^{m-1}(1 - x)^{n-1} dx.$$

With the tools available in advanced calculus, it is readily established (how easily the great achievements of the past seem to be comprehended and duplicated!) that the integral possesses meaning when $x>0$ and thus yields a certain function $\Gamma(x)$ defined for these values. Moreover,

(18)
$$\Gamma(n + 1) = n!$$

whenever n is a positive integer.* It is further established that for all $x>0$

---

* Legendre's notation shifts the argument. Gauss introduced a notation $\pi(x)$ free of this defect. Legendre's notation won out, but continues to plague many people. The notations Γ, π, and ! can all be found today.

(19) $$x\Gamma(x) = \Gamma(x+1).$$

This is the so-called recurrence relation for the gamma function and in the years following Euler it plays, as we shall see, an increasingly important role in its theory. These facts, plus perhaps the relationship between Euler's two types of integrals

(20) $$B(m, n) = \Gamma(m)\Gamma(n)/\Gamma(m+n)$$

and the all important Stirling formula

(21) $$\Gamma(x) \sim e^{-x}x^{x-1/2}\sqrt{(2\pi)},$$

which gives us a relatively simple approximate expression for $\Gamma(x)$ when x is large, are about all that advanced calculus students learn of the gamma function. Chronologically speaking, this puts them at about the year 1750. The play has hardly begun.

Just as the simple desire to extend factorials to values in between the integers led to the discovery of the gamma function, the desire to extend it to negative values and to complex values led to its further development and to a more profound interpretation. Naive questioning, uninhibited play with symbols may have been at the very bottom of it. What is the value of $(-5\frac{1}{2})!$? What is the value of $\sqrt{(-1)}!$? In the early years of the 19th century, the action broadened and moved into the complex plane (the set of all numbers of the form $x+iy$, where $i=\sqrt{(-1)}$) and there it became part of the general development of the theory of functions of a complex variable that was to form one of the major chapters in mathematics. The move to the complex plane was initiated by Karl Friedrich Gauss (1777–1855), who began with Euler's product as his starting point. Many famous names are now involved and not just one stage of action but many stages. It would take too long to record and describe each forward step taken. We shall have to be content with a broader picture.

Three important facts were now known: Euler's integral, Euler's product, and the functional or recurrence relationship $x\Gamma(x)=\Gamma(x+1)$, $x>0$. This last is the generalization of the obvious arithmetic fact that for positive integers, $(n+1)n!=(n+1)!$ It is a particularly useful relationship inasmuch as it enables us by applying it over and over again to reduce the problem of evaluating a factorial of an arbitrary real number whole or otherwise to the problem of evaluating the factorial of an appropriate number lying between 0 and 1. Thus, if we write $n=4\frac{1}{2}$ in the above formula we obtain $(4\frac{1}{2}+1)!=5\frac{1}{2}(4\frac{1}{2})!$ If we could only find out what $(4\frac{1}{2})!$ is, then we would know that $(5\frac{1}{2})!$ is. This process of reduction to lower numbers can be kept up and yields

(22) $$(5\tfrac{1}{2})! = (3/2)(5/2)(7/2)(9/2)(11/2)(1/2)!$$

and since we have $(\frac{1}{2})!=\frac{1}{2}\sqrt{\pi}$ from (1) and (2), we can now compute our answer. Such a device is obviously very important for anyone who must do calculations with the gamma function. Other information is forthcoming from the recurrence relationship. Though the formula $(n+1)n! = (n+1)!$ as a condensation of the arithmetic identity $(n+1) \cdot 1 \cdot 2 \cdots n = 1 \cdot 2 \cdots n \cdot (n+1)$ makes sense only for $n = 1, 2$, etc., blind insertions of other values produce interesting things. Thus, inserting $n = 0$, we obtain $0! = 1$. Inserting successively $n = -5\frac{1}{2}$, $n = -4\frac{1}{2}, \cdots$ and reducing upwards, we discover

$$(23) \qquad (-5\tfrac{1}{2})! = (2/1)(-2/1)(-2/3)(-2/5)(-2/7)(-2/9)(1/2)!$$

Since we already know what $(\tfrac{1}{2})!$ is, we can compute $(-5\tfrac{1}{2})!$ In this way the recurrence relationship enables us to compute the values of factorials of negative numbers.

Turning now to Euler's integral, it can be shown that for values of the variable less than 0, the usual theorems of analysis do not suffice to assign a meaning to the integral, for it is divergent. On the other hand, it is meaningful and yields a value if one substitutes for x any complex number of the form $a+bi$ where $a > 0$. With such substitutions the integral therefore yields a complex-valued function which is defined for all complex numbers in the right-half of the complex plane and which coincides with the ordinary gamma function for real values. Euler's product is even stronger. With the exception of $0, -1, -2, \cdots$ any complex number whatever can be inserted for the variable and the infinite product will converge, yielding a value. And so it appears that we have at our disposal a number of methods, conceptually and operationally different for extending the domain of definition of the gamma function. Do these different methods yield the same result? They do. But why?

The answer is to be found in the notion of an analytic function. This is the focal point of the theory of functions of a complex variable and an outgrowth of the older notion of an analytical expression. As we have hinted, earlier mathematics was vague about this notion, meaning by it a function which arose in a natural way in mathematical analysis. When later it was discovered by J. B. J. Fourier (1768–1830) that functions of wide generality and functions with unpleasant characteristics could be produced by the infinite superposition of ordinary sines and cosines, it became clear that the criterion of "arising in a natural way" would have to be dropped. The discovery simultaneously forced a broadening of the idea of a function and a narrowing of what was meant by an analytic function.

Analytic functions are not so arbitrary in their behavior. On the contrary, they possess strong internal ties. Defined very precisely as functions which possess a complex derivative or equivalently as functions which possess power series expansions $a_0 + a_1(z - z_0) + a_2(z - z_0)^2 + \cdots$ they exhibit the remarkable phenomenon of "action at a distance." This means that the behavior of an analytic function over any interval no matter how small is sufficient to determine completely its behavior everywhere else; its potential range of definition and its values are theoretically obtainable from this information. Analytic functions, moreover, obey the principle of the permanence of functional relationships; if an analytic function satisfies in some portions of its region of definition a certain functional relationship, then it must do so wherever it is defined. Conversely, such a relationship may be employed to extend its definition to unknown regions. Our understanding of the process of analytic continuation, as this phenomenon is known, is based upon the work of Bernhard Riemann (1826–1866) and Karl Weierstrass (1815–1897). The complex-valued function which results from the substitution of complex numbers into Euler's integral is an analytic function. The function which emerges from Euler's product is an analytic function. The recurrence relationship for the gamma function if satisfied in some region must be satisfied in any other region to which the function can be "continued" analytically and indeed may be employed to effect such ex-

THE GAMMA FUNCTION

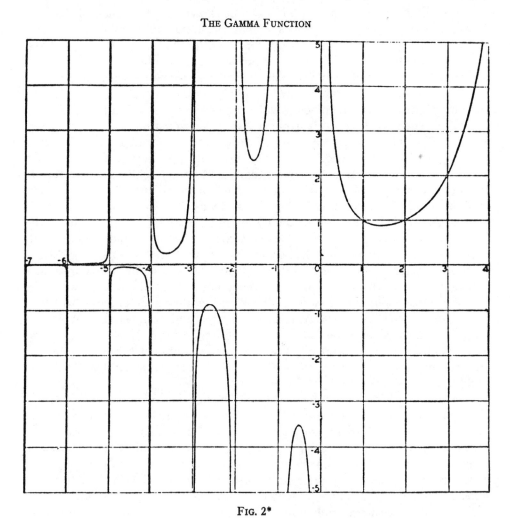

FIG. 2*

* From: H. T. Davis, Tables of the Higher Mathematical Functions, vol. I, Bloomington, Indiana, 1933.

tensions. All portions of the complex plane, with the exception of the values $0, -1, -2, \cdots$ are accessible to the complex gamma function which has become the unique, analytic extension to complex values of Euler's integral (see Fig. 3).

To understand why there should be excluded points observe that $\Gamma(x) = \Gamma(x+1)/x$, and as x approaches 0, we obtain $\Gamma(0) = 1/0$. This is $+\infty$ or $-\infty$ depending whether 0 is approached through positive or negative values. The functional equation (19) then, induces this behavior over and over again at each of the negative integers. The (real) gamma function is comprised of an infinite number of disconnected portions opening up and down alternately. The portions corresponding to negative values are each squeezed into an infinite strip one unit in width, but the major portion which corresponds to positive x and which contains the factorials is of infinite width (see Fig. 2). Thus, there are excluded points for the gamma function at which it exhibits from the ordinary (real variable) point of view a somewhat unpleasant and capricious behavior.

THE ABSOLUTE VALUE OF THE COMPLEX GAMMA FUNCTION, EXHIBITING THE POLES AT THE NEGATIVE INTEGERS

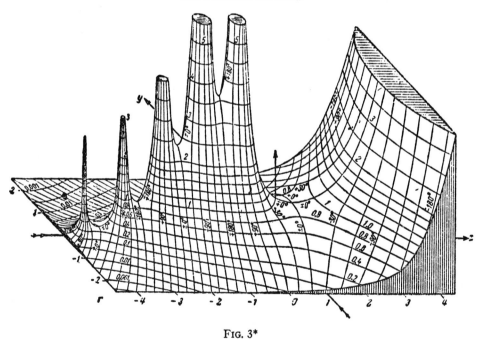

FIG. 3*

But from the complex point of view, these points of singular behavior (singular in the sense of Sherlock Holmes) merit special study and become an important part of the story. In pictures of the complex gamma function they show up as an

* From: E. Jahnke and F. Emde, Tafeln höherer Funktionen, 4th ed., Leipzig, 1948.

infinite row of "stalagmites." each of infinite height (the ones in the figure are truncated out of necessity) which become more and more needlelike as they go out to infinity (see Fig. 3). They are known as poles. Poles are points where the function has an infinite behavior of especially simple type, a behavior which is akin to that of such simple functions as the hyperbola $y = 1/x$ at $x = 0$ or of $y = \tan x$ at $x = \pi/2$. The theory of analytic functions is especially interested in singular behavior, and devotes much space to the study of the singularities. Analytic functions possess many types of singularity but those with only poles are known as meromorphic. There are also functions which are lucky enough to possess no singularities for finite arguments. Such functions form an elite and are known as entire functions. They are akin to polynomials while the meromorphic functions are akin to the ratio of polynomials. The gamma function is meromorphic. Its reciprocal, $1/\Gamma(x)$, has on the contrary no excluded points. There is no trouble anywhere. At the points $0, -1, -2, \cdots$ it merely becomes zero. And the zero value which occurs an infinity of times, is strongly reminiscent of the sine.

In the wake of the extension to the complex many remarkable identities emerge, and though some of them can and were obtained without reference to complex variables, they acquire a far deeper and richer meaning when regarded from the extended point of view. There is the reflection formula of Euler

$$(24) \qquad\qquad \Gamma(z)\Gamma(1 - z) = \pi/\sin \pi z.$$

It is readily shown, using the recurrence relation of the gamma function, that the product $\Gamma(z)\Gamma(1-z)$ is a periodic function of period 2; but despite the fact that $\sin \pi z$ is one of the simplest periodic functions, who could have anticipated the relationship (24)? What, after all, does trigonometry have to do with the sequence 1, 2, 6, 24 which started the whole discussion? Here is a fine example of the delicate patterns which make the mathematics of the period so magical. From the complex point of view, a partial reason for the identity lies in the similarity between zeros of the sine and the poles of the gamma function.

There is the duplication formula

$$(25) \qquad\qquad \Gamma(2z) = (2\pi)^{-1/2} 2^{2z-1/2} \Gamma(z) \Gamma(z + \tfrac{1}{2})$$

discovered by Legendre and extended by Gauss in his researches on the hypergeometric function to the multiplication formula

$$(26) \quad \Gamma(nz) = (2\pi)^{1/2(1-n)} n^{nz-1/2} \Gamma(z) \Gamma\left(z + \frac{1}{n}\right) \Gamma\left(z + \frac{2}{n}\right) \cdot \ \cdots \cdot \Gamma\left(\frac{z + n - 1}{n}\right).$$

There are pretty formulas for the derivatives of the gamma function such as

$$(27) \qquad\qquad d^2 \log \Gamma(z)/dz^2 = \frac{1}{z^2} + \frac{1}{(z + 1)^2} + \frac{1}{(z + 2)^2} + \cdots.$$

This is an example of a type of infinite series out of which G. Mittag-Leffler (1846–1927) later created his theory of partial fraction developments of mero-morphic functions. There is the intimate relationship between the gamma func-tion and the zeta function which has been of fundamental importance in study-ing the distribution of the prime numbers,

$$(28) \qquad \zeta(z) = \zeta(1 - z)\Gamma(1 - z)2^z\pi^{z-1}\sin\tfrac{1}{2}\pi z,$$

where

$$(29) \qquad \zeta(z) = 1 + \frac{1}{2^z} + \frac{1}{3^z} + \cdots.$$

This formula has some interesting history related to it. It was first proved by Riemann in 1859 and was conventionally attributed to him. Yet in 1894 it was discovered that a modified version of the identity appears in some work of Euler which had been done in 1749. Euler did not claim to have proved the formula. However, he "verified" it for integers, for $\tfrac{1}{2}$, and for 3/2. The verification for $\tfrac{1}{2}$ is by direct substitution, but for all the other values, Euler works with divergent infinite series. This was more than 100 years in advance of a firm theory of such series, but with unerring intuition, he proceeded to sum them by what is now called the method of Abel summation. The case 3/2 is even more interesting. There, invoking both divergent series and numerical evaluation, he came out with numerical agreement to 5 decimal places! All this work convinced him of the truth of his identity. Rigorous modern proofs do not require the theory of divergent series, but the notions of analytic continuation are crucial.

In view of the essential unity of the gamma function over the whole complex plane it is theoretically and aesthetically important to have a formula which works for all complex numbers. One such formula was supplied in 1848 by F. W. Newman:

$$(30) \quad 1/\Gamma(z) = ze^{\gamma z}\{(1 + z)e^{-z}\}\{(1 + z/2)e^{-z/2}\} \cdots, \quad \text{where } \gamma = .57721\,56649\cdots.$$

This formula is essentially a factorization of $1/\Gamma(z)$ and is much the same as a factorization of polynomials. It exhibits clearly where the function vanishes. Setting each factor equal to zero we find that $1/\Gamma(z)$ is zero for $z = 0$, $z = -1$, $z = -2, \cdots$. In the hands of Weierstrass, it became the starting point of his particular discussion of the gamma function. Weierstrass was interested in how functions other than polynomials may be factored. A number of isolated factor-izations were then known. Newman's formula (30) and the older factorization of the sine

$$(31) \qquad \sin \pi z = \pi z(1 - z^2)\left(1 - \frac{z^2}{4}\right)\left(1 - \frac{z^2}{9}\right) \cdots$$

are among them. The factorization of polynomials is largely an algebraic matter but the extension to functions such as the sine which have an infinity of roots required the systematic building up of a theory of infinite products. In 1876 Weierstrass succeeded in producing an extensive theory of factorizations which included as special cases these well-known infinite products, as well as certain doubly periodic functions.

In addition to showing the roots of $1/\Gamma(z)$, formula (30) does much more. It shows immediately that the reciprocal of the gamma function is a much less difficult function to deal with than the gamma function itself. It is an entire function, that is, one of those distinguished functions which possesses no singularities whatever for finite arguments. Weierstrass was so struck by the advantages to be gained by starting with $1/\Gamma(z)$ that he introduced a special notation for it. He called $1/\Gamma(u+1)$ the *factorielle* of u and wrote $Fc(u)$.

The theory of functions of a complex variable unifies a hotch-potch of curves and a patchwork of methods. Within this theory, with its highly developed studies of infinite series of various types, was brought to fruition Stirling's unsuccessful attempts at solving the interpolation problem for the factorials. Stirling had done considerable work with infinite series of the form

$$A + Bz + Cz(z-1) + Dz(z-1)(z-2) + \cdots .$$

This series is particularly useful for fitting polynomials to values given at the integers $z = 0, 1, 2, \cdots$. The method of finding the coefficients A, B, C, $\cdots$ was well known. But when an infinite amount of fitting is required, much more than simple formal work is needed, for we are then dealing with a bona fide infinite series whose convergence must be investigated. Starting from the series $1, 2, 6, 24, \cdots$, Stirling found interpolating polynomials via the above series. The resultant infinite series is divergent. The factorials grow too rapidly in size. Stirling realized this and put out the suggestion that if perhaps one started with the logarithms of the factorials instead of the factorials themselves the size might be cut down sufficiently for one to do something. There the matter rested until 1900 when Charles Hermite (1822–1901) wrote down the Stirling series for $\log \Gamma(1+z)$:

$$(32) \quad \log \Gamma(1 + z) = \frac{z(z - 1)}{1 \cdot 2} \log 2 + \frac{z(z - 1)(z - 2)}{1 \cdot 2 \cdot 3} (\log 3 - 2 \log 2) + \cdots$$

and showed that this identity is valid whenever z is a complex number of the form $a + ib$ with $a > 0$. The identity itself could have been written down by Stirling, but the proof would have been another matter. An even simpler starting point is the function $\psi(z) = (d/dz) \log \Gamma(z)$, now known as the digamma or psi function. This leads to the Stirling series

$$\frac{d}{dz} \log \Gamma(z) = -\gamma + (z - 1) - \frac{(z - 1)(z - 2)}{2 \cdot 2!} + \frac{(z - 1)(z - 2)(z - 3)}{3 \cdot 3!} \cdots ,$$

(33)

which in 1847 was proved convergent for $a > 0$ by M. A. Stern, a teacher of Riemann. All these matters are today special cases of the extensive theory of the convergence of interpolation series.

Functions are the building blocks of mathematical analysis. In the 18th and 19th centuries mathematicians devoted much time and loving care to developing the properties and interrelationships between special functions. Powers, roots, algebraic functions, trigonometric functions, exponential functions, logarithmic functions, the gamma function, the beta function, the hypergeometric function, the elliptic functions, the theta function, the Bessel function, the Matheiu function, the Weber function, Struve function, the Airy function, Lamé functions, literally hundreds of special functions were singled out for scrutiny and their main features were drawn. This is an art which is not much cultivated these days. Times have changed and emphasis has shifted. Mathematicians on the whole prefer more abstract fare. Large classes of functions are studied instead of individual ones. Sociology has replaced biography. The field of special functions, as it is now known, is left largely to a small but ardent group of enthusiasts plus those whose work in physics or engineering confronts them directly with the necessity of dealing with such matters.

The early 1950's saw the publication of some very extensive computations of the gamma function in the complex plane. Led off in 1950 by a six-place table computed in England, it was followed in Russia by the publication of a very extensive six-place table. This in turn was followed in 1954 by the publication by the National Bureau of Standards in Washington of a twelve-place table. Other publications of the complex gamma function and related functions have appeared in this country, in England, and in Japan. In the past, the major computations of the gamma function had been confined to real values. Two fine tables, one by Gauss in 1813 and one by Legendre in 1825, seemed to answer the mathematical needs of a century. Modern technology had also caught up with the gamma function. The tables of the 1800's were computed laboriously by hand, and the recent ones by electronic digital computers.

But what touched off this spate of computational activity? Until the initial labors of H. T. Davis of Indiana University in the early 1930's, the complex values of the gamma function had hardly been touched. It was one of those curious turns of events wherein the complex gamma function appeared in the solution of various theoretical problems of atomic and nuclear theory. For instance, the radial wave functions for positive energy states in a Coulomb field leads to a differential equation whose solution involves the complex gamma function. The complex gamma function enters into formulas for the scattering of charged particles, for the nuclear forces between protons, in Fermi's approximate formula for the probability of β-radiation, and in many other places. The importance of these problems to physicists has had the side effect of computational mathematics finally catching up with two and a quarter centuries of theoretical development.

As analysis grew, both creating special functions and delineating wide classes of functions, various classifications were used in order to organize them for purposes of convenient study. The earlier mathematicians organized functions from without, operationally, asking what operations of arithmetic or calculus had to be performed in order to achieve them. Today, there is a much greater tendency to look at functions from within, organically, considering their construction as achieved and asking what geometrical characteristics they possess. In the earlier classification we have at the lowest and most accessible level, powers, roots, and all that could be concocted from them by ordinary algebraic manipulation. These came to be known as algebraic functions. The calculus, with its characteristic operation of taking limits, introduced logarithms and exponentials, the latter encompassing, as Euler showed, the sines and cosines of trigonometry which had been available from earlier periods of discovery. There is an impassable wall between the algebraic functions and the new limit-derived ones. This wall consists in the fact that try as one might to construct, say, a trigonometric function out of the finite material of algebra, one cannot succeed. In more technical language, the algebraic functions are closed with respect to the processes of algebra, and the trigonometric functions are forever beyond its pale. (By way of a simple analogy: the even integers are closed with respect to the operations of addition, subtraction, and multiplication; you cannot produce an odd integer from the set of even integers using these tools.) This led to the concept of transcendental functions. These are functions which are not algebraic. The transcendental functions count among their members, the trigonometric functions, the logarithms, the exponentials, the elliptic functions, in short, practically all the special functions which had been singled out for special study. But such an indescriminate dumping produced too large a class to handle. The transcendentals had to be split further for convenience. A major tool of analysis is the differential equation, expressing the relationship between a function and its rate of growth. It was found that some functions, say the trigonometric functions, although they are transcendental and do not therefore satisfy an algebraic equation, nonetheless satisfy a differential equation whose coefficients are algebraic. The solutions of algebraic differential equations are an extensive though not all-encompassing class of transcendental functions. They count among their members a good many of the special functions which arise in mathematical physics.

Where does the gamma function fit into this? It is not an algebraic function. This was recognized early. It is a transcendental function. But for a long while it was an open question whether the gamma function satisfied an algebraic differential equation. The question was settled negatively in 1887 by O. Hölder (1859–1937). It does not. It is of a higher order of transcendency. It is a so-called transcendentally transcendent function, unreachable by solving algebraic equations, and equally unreachable by solving algebraic differential equations. The subject has interested many people through the years and in 1925 Alexander

Ostrowski, now Professor Emeritus of the University of Basel, Switzerland, gave an alternate proof of Hölder's theorem.

Problems of classification are extremely difficult to handle. Consider, for instance, the following: Can the equation $x^7 + 8x + 1$ be solved with radicals? Is π transcendental? Can $\int dx/\sqrt{(x^3 + 1)}$ be found in terms of specified elementary functions? Can the differential equation $dy/dx = (1/x) + (1/y)$ be resolved with quadratures? The general problems of which these are representatives are even today far from solved and this despite famous theories such as Galois Theory, Lie theory, theory of Abelian integrals which have derived from such simple questions. Each individual problem may be a one-shot affair to be solved by individual methods involving incredible ingenuity.

HADAMARD'S FACTORIAL FUNCTION

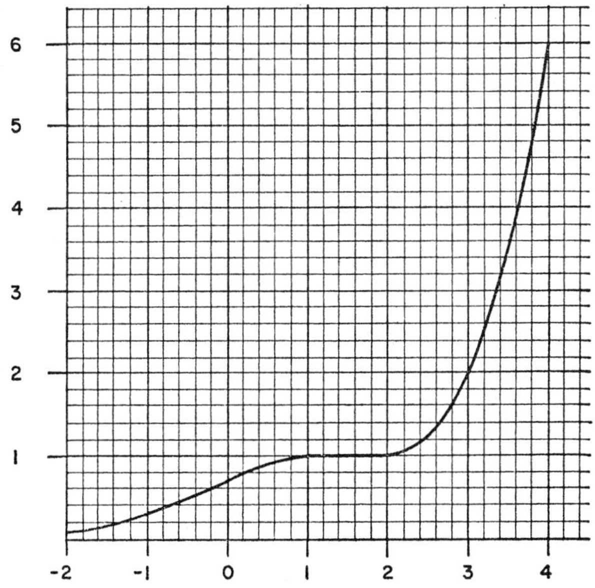

FIG. 4

There are infinitely many functions which produce factorials. The function

$$F(x) = (1/\Gamma(1 - x)) \, (d/dx) \log \{\Gamma((1 - x)/2)/\Gamma(1 - x/2)\}$$

is an entire analytic function which coincides with the gamma function at the positive integers. It satisfies the functional equation $F(x+1) = xF(x) + (1/\Gamma(1-x))$.

We return once again to our interpolation problem. We have shown how, strictly speaking, there are an unlimited number of solutions to this problem. To drive this point home, we might mention a curious solution given in 1894 by Jacques Hadamard (1865–   ). Hadamard found a relatively simple formula involving the gamma function which also produces factorial values at the positive integers. (See Figs. 1 and 4.) But Hadamard's function

$$(34) \qquad y = \frac{1}{\Gamma(1-x)} \frac{d}{dx} \log\left[\Gamma\!\left(\frac{1-x}{2}\right) \bigg/ \Gamma\!\left(1 - \frac{x}{2}\right) \right],$$

in strong contrast to the gamma function itself, possesses no singularities anywhere in the finite complex plane. It is an entire analytic solution to the interpolation problem and hence, from the function theoretic point of view, is a simpler solution. In view of all this ambiguity, why then should Euler's solution be considered the solution par excellence?

From the point of view of integrals, the answer is clear. Euler's integral appears everywhere and is inextricably bound to a host of special functions. Its frequency and simplicity make it fundamental. When the chips are down, it is the very form of the integral and of its modifications which lend it utility and importance. For the interpolatory point of view, we can make no such claim. We must take a deeper look at the gamma function and show that of all the solutions of the interpolation problem, it, in some sense, is the simplest. This is partially a matter of mathematical aesthetics.

A PSEUDOGAMMA FUNCTION

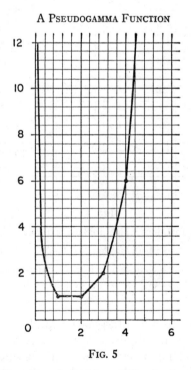

FIG. 5

The function illustrated produces factorials, satisfies the functional equation of the gamma function, and is convex.

We have already observed that Euler's integral satisfies the fundamental recurrence equation, $x\Gamma(x) = \Gamma(x+1)$, and that this equation enables us to compute all the real values of the gamma function from knowledge merely of its values in the interval from 0 to 1. Since the solution to the interpolation problem is not determined uniquely, it makes sense to add to the problem more conditions and to inquire whether the augmented problem then possesses a unique solution. If it does, we will hope that the solution coincides with Euler's. The recurrence relationship is a natural condition to add. If we do so, we find that the gamma function is again not the only function which satisfies this recurrence relation and produces factorials. One may easily construct a "pseudo" gamma function $\Gamma_S(x)$ by defining it between, say, 1 and 2 in any way at all (subject only to $\Gamma_S(1) = 1$, $\Gamma_S(2) = 1$), and allowing the recurrence relationship to extend its values everywhere else.

If, for instance, we let $\Gamma_S(x)$ be 1 everywhere between 1 and 2, the recurrence relation leads us to the function (see Fig. 5).

(35)
$$\begin{aligned}
\Gamma_S(x) &= 1/x & 0 < x &\leq 1; \\
\Gamma_S(x) &= 1, & 1 \leq x &\leq 2; \\
\Gamma_S(x) &= x - 1, & 2 \leq x &\leq 3; \\
\Gamma_S(x) &= (x-1)(x-2), & 3 \leq x &\leq 4; \cdots.
\end{aligned}$$

We might end up with a fairly weird result, depending upon what we start with. Even if we require the final result to be an analytic function, there are ways of doing it. For instance, take any function which is both analytic and periodic with period 1. Call it $p(x)$. Make sure that $p(1) = 1$. The function $1 + \sin 2\pi x$ will do for $p(x)$. Now multiply the ordinary gamma function $\Gamma(x)$ by $p(x)$ and the result $\Gamma(x)p(x)$ will be a "pseudo" gamma function which is analytic, satisfies the recurrence relation, and produces factorials! Thus, we still do not have enough conditions. We must augment the problem again. But what to add?

By the middle of the 19th century it was recognized that Euler's gamma function was the only continuous function which satisfied simultaneously the recurrence relationship, the reflection formula and the multiplication formula. Weierstrass later showed that the gamma function was the only continuous solution of the recurrence relationship for which $\{\Gamma(x+n)\}/\{(n-1)n^x\} \to 1$ for all x. These conditions added to the interpolation problem will serve to produce a unique solution and one which coincides with Euler's. But they appear too heavy and too much like Monday morning quarterbacking. That is to say, the added conditions are hardly "natural" for they are tied in with the deeper analytical properties of the gamma function. The search went on.

Aesthetic conditions were not to be found in the older, analytic considerations, but in a newer, inner, organic approach to function theory which was developing at the turn of the century. Backed up by Cantor's set theory and an emerging theory of topology, the new function theory looked not so much at equations and identities as at the fundamental geometrical properties. The desired condition was found in notions of convexity. A curve is convex if the following is true of it: take any two points on the curve and join them by a straight line; then the portion of the curve between the points lies below the line. A convex curve does not wiggle; it cannot look like a camel's back. At the turn of the century, convexity was in the mathematical air. It was found to be intrinsic to many diverse phenomena. Over the period of a generation, it was sought out, it was generalized, it was abstracted, it was investigated for its own sake, it was applied. Called to attention by the work of H. Brunn in 1887 and of H. Minkowski in 1903 on convex bodies and given an independent interest in 1906 by the work of J. L. W. V. Jensen, the idea of convexity spread and established itself in mean value theory, in potential theory, in topology, and most recently in game theory and linear programming. At the turn of the century then, an application of convexity to the gamma function would have been natural and in order.

The individual curves which make up the gamma function are all convex. A glance at Figure 2 shows this to be true. If, as in the previous paragraph, a pseudogamma function satisfying the recurrence formula were produced by introducing the ripple $1 + \sin 2\pi x$ as a factor, it would no longer be true. It must have occurred to many mathematicians to find out whether the gamma function is the only function which yields the factorial values, satisfies the recurrence relation, and is convex downward for $x > 0$. Unfortunately, this is not true. Figure 5 shows a pseudogamma function which possesses just these properties. It remained until 1922 to discover a correct formulation. But it was not at too far a distance. The gamma function is not only convex, it is also logarithmically convex. That is to say, the graph of $\log \Gamma(x)$ is also convex down for $x > 0$. This fact is implicit in formula (27). Logarithmic convexity is a stronger condition than ordinary convexity for logarithmic convexity implies, but is not implied by, ordinary convexity. Now Harald Bohr and J. Mollerup were able to show the surprising fact that the gamma function is the only function which satisfies the recurrence relationship and is logarithmically convex. The original proof was simplified several years later by Emil Artin, now professor at Princeton University, and the theorem together with Artin's method of proof now constitute the Bohr-Mollerup-Artin theorem. Its precise wording is this:

The Euler gamma function is the only function defined for $x > 0$ which is positive, is 1 at $x = 1$, satisfies the functional equation $x\Gamma(x) = \Gamma(x+1)$, and is logarithmically convex.

This theorem is at once so striking and so satisfying that the contemporary synod of abstractionists who write mathematical canon under the pen name of N. Bourbaki has adopted it as the starting point for its exposition of the gamma function. The proof: one page; the discovery: 193 years.

There is much that we know about the gamma function. Since Euler's day more than 400 major papers relating to it have been written. But a few things remain that we do not know and that we would like to know. Perhaps the hardest of the unsolved problems deal with questions of rationality and transcendentality. Consider, for instance, the number $\gamma = .57721 \cdots$ which appears in formula (30). This is the Euler-Mascheroni constant. Many different expressions can be given for it. Thus,

$$(36) \qquad \gamma = - d\Gamma(x)/dx \big|_{x=1},$$

$$(37) \qquad \gamma = \lim_{n \to \infty} \left(1 + \frac{1}{2} + \frac{1}{3} + \cdots + \frac{1}{n}\right) - \log n.$$

Though the numerical value of γ is known to hundreds of decimal places, it is not known at the time of writing whether γ is or is not a rational number. Another problem of this sort deals with the values of the gamma function itself. Though, curiously enough, the product $\Gamma(1/4)/\sqrt[4]{\pi}$ can be proved to be transcendental, it is not known whether $\Gamma(1/4)$ is even rational.

George Gamow, the distinguished physicist, quotes Laplace as saying that when the known areas of a subject expand, so also do its frontiers. Laplace evidently had in mind the picture of a circle expanding in an infinite plane. Gamow disputes this for physics and has in mind the picture of a circle expanding on a spherical surface. As the circle expands, its boundary first expands, but later contracts. This writer agrees with Gamow as far as mathematics is concerned. Yet the record is this: each generation has found something of interest to say about the gamma function. Perhaps the next generation will also.

The writer wishes to thank Professor C. Truesdell for his helpful comments and criticism and Dr. H. E. Salzer for a number of valuable references.

References

1. E. Artin, Einführung in die Theorie der Gammafunktion, Leipzig, 1931.

2. N. Bourbaki, Éléments de Mathématique, Book IV, Ch. VII, La Fonction Gamma, Paris, 1951.

3. H. T. Davis, Tables of the Higher Mathematical Functions, vol. I, Bloomington, Indiana, 1933.

4. L. Euler, Opera omnia, vol. I_{14}, Leipzig-Berlin, 1924.

5. P. H. Fuss, Ed., Correspondance Mathématique et Physique de Quelques Célèbres Geómètres du XVIII[ieme] Siècle, Tome I, St. Petersbourg, 1843.

6. G. H. Hardy, Divergent Series, Oxford, 1949, Ch. II.

7. F. Lösch and F. Schoblik, Die Fakultät und verwandte Funktionen, Leipzig, 1951.

8. N. Nielsen, Handbuch der Theorie der Gammafunktion, Leipzig, 1906.

9. Table of the Gamma Function for Complex Arguments, National Bureau of Standards, Applied Math. Ser. 34, Washington, 1954. (Introduction by Herbert E. Salzer.)

10. E. T. Whittaker and G. N. Watson, A Course of Modern Analysis, Cambridge, 1947, Ch. 12.

14

LEON A. HENKIN

Professor Henkin was born on April 19, 1921, in Brooklyn, New York. He received his A.B. from Columbia University in 1941 and his M.A. from Princeton University in 1942. After an interval of four years as a mathematician in industry, he returned to Princeton where he obtained his Ph.D. in 1947 under the direction of Alonzo Church. His thesis included a proof of Gödel's completeness theorem for the predicate calculus which has since become the standard proof in almost every presentation of mathematical logic.

Professor Henkin was Fine Instructor and Jewett Fellow at Princeton from 1947 to 1949, having spent four previous years as a mathematician in industry. In addition, he was a Fulbright Scholar in Amsterdam in 1954–55, a Visiting Professor at Dartmouth in 1960–61, a Guggenheim Fellow and member of the Institute for Advanced Study in 1961–62, and a visiting Fellow at All Souls' College, Oxford in 1968–69. He taught at the University of Southern California and has been a member of the faculty of the University of California at Berkeley since 1953 where he has served as chairman twice. He has been Editor for the *Journal of Symbolic Logic* and served a three year term as President for the Association of Symbolic Logic. He has also been a member of the Council of the American Mathematical Society as well as active in CUPM.

Professor Henkin developed the theory of cylindrification algebras which is an algebraic formulation of the theory of quantifiers. His principal work has been in the area of foundations and mathematical logic in which he has published many papers and is a recognized authority.

Professor Henkin's many significant contributions to logic and the foundations of mathematics are contained in his thirty-five publications. His interest in improving mathematical education at all levels is evidenced by his participation in numerous NSF summer institutes and by his book *Retracing Elementary Mathematics*, published in 1962 in cooperation with W. N. Smith, V. J. Varineau and N. J. Walsh by Macmillan Company.

ARE LOGIC AND MATHEMATICS IDENTICAL?*

An old thesis of Russell's is reexamined in the light of subsequent developments in mathematical logic.

LEON A. HENKIN, University of California, Berkeley

It was 24 years ago that I entered Columbia College as a freshman and discovered the subject of logic. I can recall well the particular circumstance which led to this discovery.

One day I was browsing in the library and came across a little volume by Bertrand Russell entitled *Mysticism and Logic*. At that time, barely 16, I fancied myself something of a mystic. Like many young people of that age I was filled with new emotions strongly felt. It was natural that any reflective attention should be largely occupied with these, and that this preoccupation should give a color and poignancy to experience which found sympathetic reflection in the writings of men of mystical bent.

Having heard that Russell was a logician I inferred from the title of his work that his purpose was to contrast mysticism with logic in order to exalt the latter at the expense of the former, and I determined to read the essay in order to refute it. But I discovered something quite different from what I had imagined. Indeed, contrasting aspects of mysticism and logic were delineated by Russell, but his thesis was that each had a proper and important place in the totality of human experience, and his interest was to define these and to exhibit their interdependence rather than to select one as superior to the other. I was disarmed, I was delighted with Russell's lucent and persuasive style, I began avidly to read his other works, and was soon caught up with logical concepts which have continued to occupy at least a portion of my attention ever since.

Bertrand Russell was a great popularizer of ideas, abstract as well as concrete. Probably many of you have been afforded an introduction to mathematical logic through his writings, and perhaps some have even been led to the point of peeping into the formidable *Principia Mathematica* which he wrote with Alfred Whitehead about 1910. You will recall, then, the astonishing contention with which he shocked the mathematical world of that time—namely, that all of mathematics was nothing but logic. Mathematicians were generally puzzled by this radical thesis. Really, very few understood at all what Russell had in mind. Nevertheless, they vehemently opposed the idea.

This is readily understandable when you recall that a companion thesis of Russell's was that logic is purely tautological and has really no content whatever. Mathematicians, being adept at putting 2 and 2 together, quickly inferred that Russell meant to say that all mathematical propositions are completely devoid of content, and from this it was a simple matter to pass to the supposition that he held

---

* The author is professor of mathematics at the University of California, Berkeley, and president of the Association for Symbolic Logic. This article is adapted from an address given 5 September 1961 at the 5th Canadian Mathematical Congress, in Montreal. It is reprinted from the *Proceedings* of the congress, with permission.

all mathematics to be entirely without value. *Aux armes, citoyens du monde mathématique!*

Half a century has elapsed since this gross misinterpretation of Russell's provocative enunciation. These 50 years have seen a great acceleration and broadening of logical research. And so it seems to me appropriate to seek a reassessment of Russell's thesis in the light of subsequent development.

Definitions and Proofs. In order to explain how Russell came to hold the view that all of mathematics is nothing but logic, it is necessary to go back and discuss two important complexes of ideas which had been developed in the decades before Russell came into the field. The first of these was a systematic reduction of all the concepts of mathematics to a small number of them. This process of reduction had indeed been going on for a very long time. As far back as the days of Descartes, for example, we can see at least an imperfect reduction of geometric notions to algebraic ones. Subsequently, with the development of set theory initiated by Georg Cantor, the reduction of the system of real numbers to that of natural numbers marked another great step in this process. But perhaps the most daring of these efforts, the culminating one, was the attempt by a German mathematician, Gotlob Frege, to analyze the notion of natural number still further and reduce it to a concept which he considered to be of a purely logical nature.

Frege's work was almost entirely unnoticed in his own time (the last three decades of the 19th century), but when Bertrand Russell came upon Frege's work he realized its great significance and gave these ideas very wide currency through his own brilliant style of exposition. The ultimate elements into which the notion of natural number was analyzed by Frege and by Russell were entities which they called "propositional functions." To this day there persists a controversy among philosophers as to just what these objects are, but at any rate they are connected with certain linguistic expressions which are like sentences except for containing variables. Just as there is a certain *proposition* associated with (or expressed by) the sentence "U Thant is an astronaut," for example, so there is a *propositional function* associated with the expression "x is an astronaut." Since propositions had long been recognized as constituting one of the most basic portions of the domain of investigation of logicians, and since propositional functions are very closely related to propositions, it was natural to consider these, too, to be a proper part of the subject of logic. It is in this sense that Frege seemed able, by a series of definitions, to arrive at the notion of number, as well as at the other notions under study in various parts of mathematics, starting from purely logical notions.

The second important line of development which preceded Russell, and upon which he drew for his ideas, was the systematic study by mathematical means of the laws of logic which entered into mathematical proofs. This development was initiated by George Boole, working in England in the middle of the 19th century. He discovered that certain of the well-known laws of logic could be formulated with the aid of algebraic symbols such as the plus sign, the multiplication sign, and the equality sign and of variables. For example, Boole used the familiar equation $P \cdot Q = Q \cdot P$ to express the fact that sentences of the form "P and Q" and "Q and

P" must be both true or both false (whatever the sentences P and Q may be), while the generally unfamiliar algebraic equation $-(P \cdot Q) = (-P) + (-Q)$ indicates that the sentence "Not both P and Q" has the same truth value as "Either not P or not Q." Boole demonstrated that through the use of such algebraic notation one can effect a great saving in the effort needed to collate and apply basic laws of logic. Later his work was extended and deepened by the American C. S. Peirce and the German mathematician E. Schröder. And Russell himself, working within this tradition, found it a convenient basis for a systematic development of all mathematics from logic. By combining the symbolic formulation of logical laws with the reduction of mathematical concepts to a logical core, he was able to conceive of a unified development such as was attempted in the *Principia Mathematica*.

From Russell to Gödel. What was the *Principia* like? Well of course the work is still not completed (only three of four projected volumes having appeared); and since Bertrand Russell has most recently seemed to occupy himself with the political effects of certain physical research it may, perhaps, never be completed! Nevertheless, one can see clearly the intended scope of the work. Surprisingly, it reminds one of the present massive undertaking by the Bourbaki group in France. For even though the *Principia* and Bourbaki are very dissimilar in many ways, each attempts to present an encyclopedic account of contemporary mathematical research unified by a coherent point of view.

In the *Principia*, starting from certain axioms expressed in symbolic form which were intended to express basic laws of logic (axioms involving only what Russell conceived to be logical notions), the work systematically proceeds to derive the other laws of logic, to introduce by definition such mathematical notions as the concept of number and of geometric space, and finally to develop the main theorems concerning these concepts as part of a uniform and systemic development.

Viewed in retrospect, the contemporary logician is struck by the willingness of Russell and Whitehead to rest their case on what, for a mathematician, must be considered such flimsy evidence. The world of empirical science, of course, expects to achieve conviction on the basis of empirical evidence, but the quintessence of the mathematician's approach, especially of the mathematical logician's, is the demand always for *proof* before a thesis is accepted. Yet you see that whereas Russell was interested in establishing that in a certain sense all of mathematics could be obtained from his logical axioms and concepts, he never really set out to give a proof of this fact! All he did was to gather the basic ideas that had been developed in a nonformal and unsystematic way by mathematicians before him, and to say, in effect, "You see that I have been able to introduce all this loosely formulated work within the precise framework of my formal system. And it's pretty clear, isn't it, that I have all the tools available to formalize such further work as mathematicians are likely to do?"

In this respect one is reminded of the approach of that first great axiomatizer and geometer, Euclid. Euclid, too, conceived that all propositions of geometry—that is, all the true statements about triangles, circles, and those other figures in which he was interested—could be developed from the simple list of concepts and axioms he gave. But in his case, too, there was never any attempt to prove this fact other than by the empirical process of deriving a large number of geometric propositions from the axioms and then appealing to the good will of the audience, so to speak. "Well," we may imagine him saying, "look how much I have been able to deduce from my axioms. Aren't you pretty well convinced that *all* geometric facts follow from them?"

But of course there were mathematicians and logicians who were *not* convinced. And so the demand for proof was raised.

Actually, the proper formulation of the problem of whether a system of axioms is adequate to establish *all* of the true statements in some domain of investigation requires a mathematically precise formulation of the notion of "true sentence," and it was not until 1935 that Alfred Tarski, in a great pioneering work, made fully evident the form in which semantical notions must be analyzed for mathematical languages. Of course, it is a trivial matter to give the conditions under which any *particular* sentence is true. For example, in the theory of Euclidean geometry the sentence "All triangles have two equal angles" is true if, and only if, all triangles have two equal angles. However, Tarski made it clear that there is no way to utilize this simple technique in order to describe (in a finite number of words) conditions for the truth of all the infinitely many sentences of a language; for this purpose a very different form of definition, structural and recursive in character, is needed.

Even before Tarski's treatment of semantics, indeed as early as 1919, we find the first proof of what we call, in logic, "completeness." The mathematician Emil Post (in his doctoral dissertation published in that year), limiting his attention to a very small fragment of the system created by Whitehead and Russell, was able to show that for any sentence in that fragment which was "true under the intended interpretation of the symbols," one could indeed get a proof by means of the axioms and rules of inference which had been stated for the system. Subsequently, further efforts were made to extend the type of completeness proof which Post initiated, and it was hoped that ultimately the entire system of the *Principia* could be brought within the scope of proofs of this kind.

In 1930, Kurt Gödel contributed greatly to this development and to this hope when he succeeded in proving the completeness of a deductive system based upon a much larger portion of mathematical language than had been treated by Post. Gödel's proof deals with the so-called "first-order predicate logic," which treats of mathematical sentences containing variables of only one type. When such a sentence is interpreted as referring to some mathematical model, its variables are interpreted as ranging over the elements of the model; in particular, there are no variables ranging over sets of model elements, or over the integers (unless these happen to be the elements of the particular model). Now Gödel shows that if we have any system of axioms of this special kind, then whenever a sentence is true in every model satisfying these axioms there must be a proof of finite length, leading from the axioms to this sentence, each line of the proof following from preceding lines by one of several explicitly listed rules of logic. This result of Gödel's is among the most basic and useful theorems we have in the whole subject of mathematical logic.

But the very next year, in 1931, the hope of further extension of this kind of completeness proof was definitely dashed by Gödel himself in what is certainly the deepest and most famous of all works in mathematical logic. Gödel was able to demonstrate that the system of *Principia Mathematica*, taken as a whole, was *incomplete*. That is, he showed explicitly how to construct a certain sentence, about natural numbers, which mathematicians could recognize as being true under the intended interpretation of the symbolism but which could not be proved from the axioms by the rules of inference which were part of that system.

Now, of course, if Gödel had done nothing more than this, one might simply conclude that Russell and Whitehead had been somewhat careless in formulating their axioms, that they had left out this true but unprovable sentence from among the axioms, and one might hope that by adding it as a new axiom a stronger system which was complete would be achieved. But Gödel's proof shows that this stronger system, too, would contain a sentence which is true but not provable; that, indeed, if this system were further strengthened, by the addition of this new true but unprovable sentence as an axiom, the resulting system would again be incomplete. And indeed, if a whole infinite sequence of sentences were to be obtained by successive applications of Gödel's method, and added simultaneously to the original axioms of *Principia*, the same process could still be applied to find *another* true sentence still unprovable.

Actually, Gödel described a very wide class of formal deductive systems to which his method applies. And most students of the subject have been convinced that any formal system of axioms and rules of inference which it would be reasonable to consider as a basis for a development of mathematics would fall in this class, and hence would suffer a form of incompleteness. From this viewpoint it appears that one of the basic elements on which Russell rested his thesis that all mathematics could be reduced to logic must be withdrawn and reconsidered.

Consistency and Decision Problems. I have been talking about completeness, which has to do with the adequacy of a formal system of axioms and rules of inference for proving true sentences. But I must mention, also, a second aim of the Russell-Whitehead *Principia* which also fared ill in the subsequent development of mathematical logic. Russell and Whitehead were very much concerned with the question of *consistency*. While they hoped to have a complete system, one containing proofs for all correct statements, they were also concerned that their system should *not* contain proofs of incorrect results. In particular, in a consistent system such as they sought, it would not be possible to prove both a sentence and its negation.

To understand their concern with the question of consistency it is necessary to recall the rude wakening which mathematicians sustained in 1897 in connection with Cantor's theory of transfinite numbers. For centuries before the time of Cantor mathematicians simply assumed that anyone who was properly educated in

358   LEON A. HENKIN

their subject could distinguish a correct proof from an incorrect one. Those who
had trouble in making this distinction were simply "weeded out" in the course of
their training and were turned from mathematics to lesser fields of study. And no
one took up seriously the question of setting forth, in explicit and mathematical
terms, exactly what was meant by a correct proof.

Now when Cantor began his development of set theory he concerned himself
with both cardinal and ordinal numbers of transfinite type. (These numbers can be
used for infinite sets in very much the same way that we use ordinary numbers for
counting and ordering finite sets.) Many of the properties of transfinite numbers
are identical to those of ordinary numbers, and in particular Cantor showed that
given any ordinal number b, we can obtain a larger number, $b+1$. However, in
1897 an Italian mathematician, C. Burali-Forti, demonstrated that there must be a
largest ordinal number, by considering the set of all ordinal numbers in their
natural order. Mathematicians were unable to find any point, either in the argument of Cantor or in that of Burali-Forti, which they intuitively felt rested on
incorrect reasoning. Gradually it was realized that mathematicians had a genuine
paradox on their hands, and that they would have to grapple at last with the
question of just what was meant by a correct proof. Later, Russell himself
produced an even simpler paradox in the intuitive theory of sets, based upon the
set of all those sets which are not elements of themselves.

This background sketch will make clear why it was that Russell and Whitehead
were concerned that no paradox should be demonstrable in their own system. And
yet they themselves never attempted a *proof* that their system was consistent! The
only evidence they adduced was that a large number of theorems had been
obtained within their system without encountering paradox, and that all attempts
to reproduce within the system of *Principia Mathematica* the Burali-Forti paradox,
and such other paradoxes as were shown, had failed.

As with the question of completeness, mathematicians were not satisfied with
an answer in this form, and there arose a demand that an actual proof of the
consistency be given for the system of *Principia* (and for other systems then
considered). The great and illustrious name of David Hilbert was associated with
these efforts to achieve consistency proofs for various portions of mathematics, and
under his stimulus and direction important advances were made toward this goal,
both by himself and by his students. But as with the efforts to prove completeness,
Hilbert's program came to founder upon the brilliant ideas of Kurt Gödel.

Indeed, in that same 1931 paper to which I have previously referred, Gödel was
able to show that the questions of consistency and completeness were very closely
linked to one another. He was able to show that *if* a system such as the *Principia*
were truly consistent, then in fact it would not be possible to produce a sound
proof of this fact! Now this result itself sounds paradoxical. Nevertheless, when
expressed with the technical apparatus which Gödel developed, it is in fact a
precisely established and clearly meaningful mathematical result which has per-
suaded most, though admittedly not all, logicians that Hilbert's search for a
consistency proof must remain unfulfilled.

I should like finally to mention a third respect in which the original aim of mathematical logicians was frustrated. The questions of consistency and completeness clearly concerned the authors of *Principia Mathematica*, but the question of decision procedures seems not to have been treated to any serious extent by Russell and Whitehead. Nevertheless, this is an area of study which interested logicians as far back as the time of Leibniz. Indeed, Leibniz himself had a great dream: He dreamt that it might be possible to devise a systematic procedure for answering questions—not only mathematical questions but even questions of empirical science. Such a procedure was to obviate the need for inspiration and replace this with the automatic carrying out of routine procedure. Had Leibniz been conversant with today's high-speed computing machines he might have formulated his idea by asserting the possibility that one could write a program of such breadth and inclusiveness that any scientific question whatever could be placed on tape and, after the machine had been set to work on it for some finite length of time, a definitive reply would be forthcoming.

Logic after 1936. Leibniz's idea lay dormant for a long time, but it was natural to revive it in connection with the formal deductive systems which were developed by mathematical logicians in the early part of this century. Since these logicians had been interested in formulating mathematical ideas within a symbolic calculus and then manipulating the symbols according to predetermined rules in order to obtain further information about these mathematical concepts, it seemed natural to raise the question of whether one could not devise purely automatic rules of computation which would enable one to reach a decision as to the truth or falsity of any given sentence of the calculus. And while the area of empirical science was pretty well excluded from the consideration of 20th-century logicians seeking such decision procedures, it was perhaps not beyond the hope of some that a system as inclusive as that of the *Principia* could some day be brought within the scope of such a procedure.

Efforts to find decision procedures for various fragments of the *Principia* were vigorous and many. The doctoral dissertation of Post, for example, contained some efforts in this direction, and further work was produced during the succeeding 15 years by logicians of many countries. Then in 1936 Alonzo Church, making use of the newly developed notion of recursive function, was able to demonstrate that for a certain fragment of mathematical language, in fact for that very first-order predicate logic which Gödel, in 1930, had showed to be complete, no decision procedure was possible. And so with decision procedures, as with proofs of completeness and consistency, efforts to establish a close rapport between logic and mathematics came to an unhappy end.

Well, I have brought you down to the year 1936. Probably most mathematicians have heard at least something of the development which I have sketched here. But somehow the education in logic of most mathematicians seems to have been terminated at about that point. The impression is fairly widespread that, with the discoveries of Gödel and Church, the ambitious program of mathematical logicians in effect ground to a halt, and that since then further work in logic has been a sort of helpless faltering by people, unwilling to accept the cruel facts of life, who are still seeking somehow to buttress the advancing frontiers of mathematical research by finding a nonexistent consistency proof.

And yet this image is very far indeed from reality. For in 1936, just at the time when, many suppose, the demise of mathematical logic had been completed, an international scholarly society known as the Association for Symbolic Logic was founded and began publication of the *Journal of Symbolic Logic*. In the ensuing 25 years this has greatly expanded to accommodate a growing volume of research. And at present there are four journals devoted exclusively to publishing material dealing with mathematical logic, while many articles on logic appear in a variety of mathematical journals of a less specialized nature.

In the space remaining I should like to mention very briefly some of the developments in mathematical logic since 1936.

Sets and Decision Methods. I have found it convenient for this exposition to divide research in mathematical logic into seven principal areas. And first I shall mention the area dealing with the foundations of the theory of sets.

To explain the connection of this field with logic it should be mentioned that those objects which Russell and Whitehead had called "propositional functions" are, in fact, largely indistinguishable from what are now called "sets" and "relations" by mathematicians. From a philosophical point of view there is perhaps still room for distinguishing these concepts from one another. But since, in fact, the treatment of propositional functions in *Principia Mathematica* is extensional (so that two functions which are true of exactly the same objects are never distinguished), for mathematical purposes this system is identical to one which treats of sets and relations.

Among systems of set theory which have been put forth by logicians as a basis for the development of mathematics, the principal ones are the theory of types used by Whitehead and Russell themselves, subsequently amplified by L. Chwistek and F. Ramsey, and an alternative line of development initiated by E. Zermelo, to which important contributions were subsequently made by A. Fraenkel and T. Skolem. Still another system, having certain characteristics in common with each of these two principal forms, was advanced and has been studied by W. Quine and, to some extent, by J. B. Rosser. Of these systems the Zermelo-type system has probably received most attention, along with an important variant form suggested and developed by J. von Neumann, P. Bernays, and Gödel.

Among the significant efforts expended on these systems were those directed toward establishing the status of propositions such as the axiom of choice and the continuum hypothesis of Cantor. Here the names of Gödel and A. Mostowski are especially prominent.

Gödel showed that a strong form of the axiom of choice and the generalized continuum hypothesis are simultaneously consistent with the more elementary axioms of set theory—under the assumption that the latter are consistent by themselves. Mostowski showed that the axiom of choice is independent of the more elementary axioms of set theory, provided that a form of these elementary axioms is selected which does not exclude the existence of nondenumerably many "*Urelemente*" (objects which are not sets). The independence of the axiom of choice from systems of axioms such as that used in Gödel's consistency proof, and the independence of the continuum hypothesis in any known system of set theory, remain open questions.

More recently the direction of research in the area of foundations of set theory seems to have shifted from that of formulating specific axiom systems and deriving theorems within them to consideration of the totality of different realizations of such axiom systems. It is perhaps J. Shepherdson who should be given credit for the decisive step in this shift of emphasis, although his work clearly owes much to Gödel's. Subsequent work by Tarski, R. Vaught, and R. Montague has carried this development much further.

An important tool in their work is the concept of the *rank* of a set, which may be defined inductively as the least ordinal number exceeding the rank of all elements of the set. This notion may be used to classify models of set theory according to the least ordinal number which is not the rank of some set of the model. Recently there have been some very interesting contributions by Azriel Lévy to these studies. His efforts have been directed toward successively strengthening the axioms of set theory so as to penetrate increasingly far into the realm of the transfinite.

A second area that I would delineate in contemporary logical research is that dealing with the decision problem. While it is true that the work of Church made it clear that there could be no *universal* decision procedure for mathematics, there has remained a strong interest in finding decision procedures for more modest portions of mathematical theory. Of special interest here is Tarski's decision method for elementary algebra and geometry, and an important extension of it which was made by Abraham Robinson. Wanda Szmielew has also given an important decision procedure—namely, one for the so-called "elementary theory" of Abelian groups. By contrast, the elementary theory of *all* groups was shown by Tarski to admit of no decision procedure. In fact, Szmielew and Tarski considered exactly the same set of sentences—roughly, all of those sentences which can be built up by the use of the group operation symbol, and variables ranging over the group elements, with the aid of the equality sign, as well as the usual logical connectives and quantifiers. If we ask whether any given sentence of this kind is true for all *Abelian* groups, it is possible to answer the question in an automatic way by using the method of Szmielew. But if we are interested in which of these sentences are true for *all* groups, then Tarski's proof shows that it is impossible to devise a machine method to separate the true from the false ones.

A result closely related to Tarski's is that of P. Novikov and W. Boone concerning the nonexistence of a decision method which would enable one to solve the word problem for the theory of groups, a problem for which a solution had long been sought by algebraists. In fact it is a simple matter to show that the

Novikov-Boone result is equivalent to the nonexistence of a decision method for a certain *subset* of the sentences making up the elementary theory of groups—namely, all those sentences having a special, very simple, form. Hence, this result is stronger than Tarski's.

Recursive Functions. Now the key concept whose development was needed before negative solutions to decision problems could be achieved was the concept of a recursive function. Intuitively speaking this is simply a function from natural numbers to natural numbers which has the property that there is an automatic method for computing its value for any given argument. A satisfactory and explicit mathematical definition of this class of functions was first formulated by J. Herbrand and Gödel. But it remained for S. C. Kleene to develop the concept to such an extent that it now underlies a very large and important part of logical research.

Much of the work with recursive functions has been along the line of classifying sets and functions, a classification similar to that involving projective and analytic sets in descriptive set theory. Kleene himself, his students Addison and Spector, and other logicians, including Post, Mostowski, J. Shoenfield, and G. Kreisel, have contributed largely to this development. Also to be mentioned are the applications which initially Kleene, and subsequently others, have attempted to make of the concept of recursive function by way of explicating the notion of "constructive" mathematical processes. In this connection several attempts have been made to link the notion of recursive function with the mathematical viewpoint known as intuitionism, a radical reinterpretation of mathematical language which was advanced by L. Brouwer and developed by A. Heyting.

Algebra, Logic, and Models. A fourth area of logical research deals with material which has recently been described as algebraic logic. This is actually a development which can be traced back to the very early work of Boole and Schröder. However, interest in the subject has shifted away from the formulation and derivation of algebraic equations which express laws of logic to the consideration of abstract structures which are defined by means of such equations. Thus, the theory of Boolean algebras, of relation algebras, of cylindric and polyadic algebras have all successively received attention; M. Stone, Tarski, and P. Halmos are closely associated with the central development here. The algebraic structures studied in this domain may be associated in a natural way with mathematical theories, and this association permits the use of very strong algebraic methods in the metamathematical analysis of these theories.

A fifth area of modern logical research concerns the so-called theory of models. Here effort is directed toward correlating mathematical properties possessed by a class of structures defined by means of given mathematical sentences with the structural properties of those sentences themselves.

A very early example is Garrett Birkhoff's result that, for a class of structures to be definable by means of a set of equational identities, it is necessary and sufficient that it be closed under formation of substructures, direct products, and homomorphic images. Characterizations of a similar nature were given for classes definable by universal elementary sentences (Tarski) and by any elementary sentences (J. Keisler).

A related type of result is R. Lyndon's theorem that any elementary sentence whose truth is preserved under passage from a model of the sentence to a homomorphic image of that model must be equivalent to a sentence which does not contain negation signs. In a different direction, E. Beth has shown that if a given set symbol or relation symbol is not definable in terms of the other symbols of an elementary axiom system, then there must exist two distinct models of these axioms which are alike in all respects except for the interpretation of the given symbol. (This proves the completeness of A. Padoa's method of demonstrating nondefinability.) A logical interpolation theorem of W. Craig's provides a close link for the results of Lyndon and Beth.

A sixth area which can be discerned in recent work on logic concerns the theory of proof. This is perhaps the oldest and most basic portion of logic, a search for systematic rules of proof, or deduction, by means of which the consequences of any propositions could be identified. In recent work, however, logicians have begun to depart in radical ways from the type of systems for which rules of proof were originally sought. For example, several attempts have been made to provide rules of proof for languages containing infinitely long formulas, such as sentences with infinitely many disjunctions, conjunctions, and quantified variables. Tarski, Scott, C. Karp, W. Hanf, and others have participated in such efforts. Curiously enough, while this direction of research seems at first very far removed from ordinary mathematics, one of the important results was used by Tarski to solve a problem, concerning the existence of measures on certain very large spaces, which had remained unsolved for many years.

The last area of logical research I should like to bring to your attention is a kind of converse study to what we have called algebraic logic. In the latter we are interested in applying methods of algebra to a system of logic. But there are also studies in which results and methods of logic are used to establish theorems of modern algebra. The first to have made such applications seems to have been the Russian mathematician A. Malcev, who in 1941 indicated how the completeness theorem for first-order logic could be used to obtain a result on groups. Subsequently the same technique was used by Tarski to construct various non-Archimedean ordered fields. Perhaps the best-known name in this area is that of Abraham Robinson, who formerly was associated with the University of Toronto in Canada. Among his contributions was the application of logical methods and results to improve a solution, given in 1926 by E. Artin, to Hilbert's 17th problem (17th of the famous list of problems presented in his address to the International Congress of Mathematicians in 1900). Robinson showed that when a real polynomial which takes only nonnegative values is represented as a sum of squares of rational functions, the number of terms needed for the representation depends only on the degree and number of variables of the given polynomial, and that it is independent of the particular coefficients.

Russell's Thesis in Perspective. I hope that this very brief sketch of some of the areas of contemporary logical research will give some idea of the ways in which logicians have reacted to the theorems of Gödel and Church which, in the period 1931 to 1936, dealt so harshly with earlier hopes. Speaking generally, one could describe this reaction as compounded of an acceptance of the impossibility of realizing the original hopes for mathematical logic, a relativization of the original program of seeking completeness and consistency proofs and decision methods, an incorporation of the new methods and constructs which appeared in the impossibility proofs, and the development of quite new interests suggested by generalization of early results.

Now with this background, let us return to Russell's thesis that all of mathematics can be reduced to logic. I would say that if logic is understood clearly to contain the theory of sets (and this seems to be a fair account of what Russell had in mind), then most mathematicians would accept without question the thesis that the basic concepts of all mathematics can be expressed in terms of logic. They would agree, too, that the theorems of all branches of mathematics can be derived from principles of set theory, although they would recognize that no fixed system of axioms for set theory is adequate to comprehend all of those principles which would be regarded as "mathematically correct."

But perhaps of greater significance is the consensus of mathematicians that there is much more to their field than is indicated by such a reduction of mathematics to logic and set theory. The fact that *certain* concepts are selected for investigation, from among all logically possible notions definable in set theory, is of the essence. A true understanding of mathematics must involve an explanation of which set-theory notions have "mathematical content," and this question is manifestly not reducible to a problem of logic, however broadly conceived.

Logic, rather than being all of mathematics, seems to be but one branch. But it is a vigorous and growing branch, and there is reason to hope that it may in time provide an element of unity to oppose the fragmentation which seems to beset contemporary mathematics—and indeed every branch of scholarship.

References

Some of the early work of Boole, Frege, and Cantor has been made more available by relatively recent translations into English. Examples are the works cited. The *Principia Mathematica* of Russell and Whitehead, while largely devoted to a formidable formalism, contains many very readable sections of an introductory or summary character. An account of the fundamental theorems of Gödel and Church may be found in Kleene. In describing logical research since 1936 my aim has been not to give a complete history but only to indicate the range of activity by mentioning some of the most actively cultivated areas. Accordingly, I have referred to only a very small sample of the literature, selecting (where possible) works which are largely self-contained. With reference to the following list, for an example of an axiomatic theory of sets and an important contribution toward its metatheory, see Gödel. For an account of the decision problem, see Tarski (1951) for positive solutions and Tarski *et al.* (1953) for negative. The theory of recursive functions and their uses is well described in Kleene. Some historical remarks on algebraic logic, as well as detailed results, may be found in Henkin and Tarski. A very recent and important contribution to the theory of models is that of Keisler, where reference to earlier works may be found. For an account of recent work on infinitely long formulas, see Henkin; an application of such work to a problem on the existence of certain measures appears in Tarski (in press). Applications of logic to algebra are described in Robinson.

G. Boole, An Investigation of the Laws of Thought, Dover, New York, new ed., 1951.

G. Cantor, Contributions to the Founding of the Theory of Transfinite Numbers, Open Court, London, 1915 (P. Jourdain, trans.).

G. Frege, The Foundations of Arithmetic, Philosophical Library, New York, 1950 (J. L. Austin, trans.).

K. Gödel, "The Consistency of the Axiom of Choice," Annals of Mathematics Study No. 3, Princeton Univ. Press, Princeton, N.J., 1940.

L. Henkin, "Some remarks on infinitely long formulas," in Infinitistic Methods, Pergamon, New York, 1961.

————, and A. Tarski, "Cylindric algebras," in "Lattice Theory" (Proc. Symposium in Pure Mathematics, Providence, 1961), American Mathematical Society, in press, vol. 2.

H. J. Keisler, Indagationes Mathematicae 23, 477 (1961).

S. C. Kleene, Introduction to Metamathematics, Van Nostrand, New York, 1952.

A. Robinson, Complete Theories, North-Holland, Amsterdam, 1956.

B. Russell and A. N. Whitehead, Principia Mathematica, Cambridge Univ. Press, Cambridge, England (1910, 1912, 1913, respectively), vols. 1-3.

A. Tarski, A Decision Method for Elementary Algebra and Geometry, Univ. of California Press, Berkeley, 1951.

————, "Some problems and results relevant to the foundations of set theory," in "Proceedings of the International Congress for Logic, Methodology, and Philosophy of Science, Stanford, 1960," Stanford Univ. Press, in press.

————, A. Mostowski, R. Robinson, Undecidable Theories, North-Holland, Amsterdam, 1953.

15

JACK K. HALE AND JOSEPH P. LASALLE

Professor Hale was born on October 3, 1928, in Dudley, Kentucky. He received his A.B. from Berea College in Kentucky in 1949 and his Ph.D. from Purdue University in 1953. After a one-year instructorship at Purdue, he was associated successively with the Sandia Corporation in 1954–57, Remington-Rand Univac in 1957–58, and RIAS in 1958–64. In September 1964, he returned to academic life as a Professor in the Division of Applied Mathematics at Brown University. During the period 1973–76 he served as Chairman of the Division of Applied Mathematics.

Professor Hale also served as associate editor for the *Journal of Differential Equations*, *Applicable Mathematics*, *Journal of Nonlinear Analysis*, and the *Proceedings of the Royal Society of Edinburgh*. He was a member of the American Mathematical Society, Sigma Xi, and Fellow of the American Society of Mechanics.

He has written approximately one hundred papers including the book *Theory of Functional Differential Equations*, most of this work being in the field of differential equations, both linear and nonlinear.

Professor LaSalle was born on May 28, 1916, in State College, Pennsylvania. He received his B.S. from Louisiana State University in 1937 and his Ph.D. at the California Institute of Technology in 1941. After serving as Henry Laws Fellow at the California Institute of Technology in 1939–41, he served successively as Instructor at the University of Texas in 1941–42, at the M.I.T. Radar School in 1942–43, as Visiting Research Associate at Princeton University in 1943–44, as Research Associate at Cornell University in 1944–46, as Assistant Professor at Notre Dame University in 1946–48, advancing there to a full professorship in 1951. In 1958, Professor LaSalle left Notre Dame University to become Associate Director of the Mathematics Center at RIAS. In September 1964, he returned to academic life as Professor of Applied Mathematics and Director of the Center for Dynamical Systems at Brown University.

Professor LaSalle was President of SIAM in 1963. He also is an AAAS Fellow and has served as Editor of the *Journal of Mathematical Analysis and Applications*, Editor of the *SIAM Journal on Control*, and Editor-in-Chief of the *Journal of Differential Equations*. During the period 1950–52, he served as scientific advisor to the commander of the U.S. Naval Force in Germany.

Professors Hale and LaSalle have made many important contributions to the field of applied mathematics and, in particular, differential equations, which are contained in a large number of papers in both domestic and foreign publications.

In accepting the 1965 Chauvenet Prize, Professors Hale and LaSalle expressed themselves greatly pleased by the honor they had received. They emphasized that they wrote the paper for which they were awarded the Chauvenet Prize because of their strong feeling concerning the value of expository papers.

DIFFERENTIAL EQUATIONS: LINEARITY VS. NONLINEARITY*

JACK K. HALE† and JOSEPH P. LASALLE†

FOREMOST AMONG THE MATHEMATICAL CHALLENGES in modern science and technology is the field of nonlinear differential equations. They are becoming increasingly important in fields as diverse as economics and space flight, ichthyology and astronomy. Since many of the equations defy complete solution, a good starting point in understanding their value and significance comes with an examination of the geometric point of view, which can be used to study solutions qualitatively. A logical next step is to review some of the basic properties and peculiarities of linear systems, after which it is easy to see that almost none of these properties hold for nonlinear systems—which can approximate nature more closely. Some of Liapunov's simple geometric ideas are invaluable in discussing the problem of stable performance, and this in turn helps to illustrate the modern theory of automatic control. The contrast between linear and nonlinear systems is striking, and this will be illustrated by simple examples. Much has been learned recently about differential equations, but there are still many major unsolved problems.

1. INTRODUCTION

In trying to analyze some process or system that occurs in nature or is built by man, it is almost always necessary to approximate the system by a mathematical model, where the state of the system at each instant of time is given as the solution of a set of mathematical relations. Differential equations are particularly useful models, that this field of mathematics has become increasingly important to mechanical systems, biological systems, economic systems, electronic circuits, and automatic controls. Differential equations are, as Lefschetz has said, "the cornerstone of applied mathematics."

The intricate history of differential equations began around 1690 with Newton and Leibniz, and since then the theory of differential equations has challenged most of the world's greatest mathematicians, providing a tremendous stimulus for much of the modern development of mathematics.

Up to the time when Cauchy proved in the early 1800's that differential equations do actually define functions, mathematicians were primarily concerned with finding explicit solutions to the differential equations which arose when they applied Newton's laws of motion to elementary problems in mechanics and physics. The few solutions which they found had profound scientific and technological implications, for they provided the basis for the development of mechanics and much of classical physics—and spurred the industrial revolution.

Ideally one would like to find all explicit solutions of every differential equation, but it soon became apparent that this was an impossible task. Of

* An article entitled "Analyzing Nonlinearity," which appeared in the June 1963 issue of International Science and Technology, was based in part on this paper.

Received by the editors, April 10, 1963.

† Research Institute for Advanced Studies (RIAS) Baltimore, Md.

necessity the emphasis turned to qualitative properties of the family of functions which was defined by the solutions of each differential equation. Real progress in this direction began near the end of the last century with the work of two of the greatest men in differential equations—Henri Poincaré of France and Alexander Mikhailovich Liapunov of Russia. Most of the modern theory of differential equations can be traced to these two men. Liapunov's work founded the great school of differential equationists which still flourishes in the Soviet Union, while much of the American effort can be traced to some of Poincaré's disciples. Poincaré dealt primarily with the geometric properties of solutions, together with some techniques for the computation of special solutions. On the other hand, Liapunov generally sought as much information as possible about the stability properties of solutions without actually knowing explicit solutions. The work of Poincaré, besides contributing to the theory of differential equations, stimulated many of the abstract developments in today's mathematics. On the other hand, Liapunov's contributions have led to methods which yield quantitative as well as qualitative information about stability and automatic control.

At this point one might well ask why, with the advent of high speed computers, there are any problems left. The answer is quite simple. If the information that is desired concerns only one particular solution, then computers are adequate. Frequently, however, what is wanted is general but qualitative information about all of the solutions, as well as those properties of the solutions which remain unchanged even when the equations themselves are perturbed. Moreover, in problems of design the engineer has an infinity of systems to choose from and he wants to identify those which have some special property or properties. For example, he may be looking for a system in which all solutions tend in time to have the same behavior, regardless of the initial state of the system, or the question may be to classify all those systems which have the same qualitative behavior.

This is not to say that the differential equations of today can satisfy engineering requirements perfectly—or that their use eliminates all problems. The equations still represent an idealization, and in most cases the equations themselves cannot be known precisely. Furthermore, one is usually unable to predict the exact perturbations which will affect the actual system. It is necessary therefore to discuss the sensitivity of the model to slight changes. Any system which is to perform a specified task must maintain stable performance under a broad range of perturbations.

There are certain systems of differential equations for which it is possible to find general solutions, and these include the so-called linear systems. The general solution of a linear system can be obtained by finding only a finite number of specific solutions. This fact explains the simplicity and also the limitations of linear systems. Almost no systems are completely linear, and linearity is another approximation to reality. One purpose of this article is to point out some of the limitations of a linear approximation and also to point out some of the new and interesting phenomena that occur when the differential equations are nonlinear. It is not simply that nature imposes nonlinearity on us. The fact of the matter is that man can use nonlinearity to improve the performance of the instruments and machines that serve him.

2. GEOMETRIC ANALYSIS

A geometric point of view is helpful in obtaining qualitative information about the solutions of a differential equation. We can demonstrate this by starting with an old law—Newton's second law of motion, whose statement in mathematical terms is a differential equation. Take the simplest situation of a particle of mass m moving along a straight line and subject to a force (Fig. 1). We select a point 0 on this line from which to measure distance, so that a positive distance will mean the particle is to the right of 0 and negative distance will mean it is to the left. The distance of the particle from 0 at time t we will call $x(t)$ or simply x. The derivative dx/dt is the velocity of the particle and the derivative of velocity d^2x/dt^2 is its acceleration. For purposes of simplicity we can adopt the convenient notation of Newton, with $\dot{x} = dx/dt$ and $\ddot{x} = d^2x/dt^2$. The force may well depend on the position and velocity of the particle, and we denote this force and the dependence by $f(x, \dot{x})$. Newton's second law of motion then states that

$$m\ddot{x} = f(x, \dot{x});$$

the mass times the acceleration is equal to the force applied. Since we need not be concerned at the moment with units of measurement, let us take the mass to be unity ($m = 1$). The differential equation for the motion of the particle is then

$$\ddot{x} = f(x, \dot{x}). \tag{1}$$

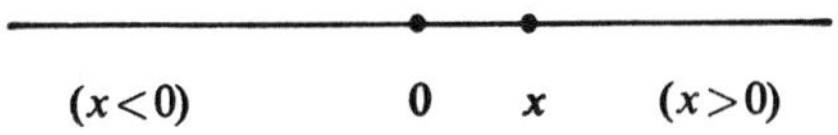

$$(x<0) \qquad \underset{0}{\bullet} \quad x \qquad (x>0)$$

FIG. 1.

Neglecting the effects of the atmosphere, Newton's law for a freely falling body of unit mass near the earth is

$$\ddot{x} = -g, \tag{2}$$

where g is the acceleration of gravity. Near the earth, the gravitational acceleration g may be assumed to be constant. If we know the initial state of the system, which means we know the initial distance x_0 of the particle from 0 and its initial velocity y_0, then at time t

$$\dot{x} = -gt + y_0$$

$$x = -\tfrac{1}{2}gt^2 + y_0 t + x_0, \tag{3}$$

This is the solution of the differential equation and is the law of freely falling bodies partially verified by Galileo from the Tower of Pisa.

To understand the geometric point of view of Poincaré and Liapunov we introduce a new symbol y for the velocity $\dot{x}$. Then $\dot{y} = \ddot{x}$, and the equation (1) involving a second derivative (a second order equation) can be expressed as two equations involving first derivatives (two first order equations):

$$\dot{x} = y$$

$$\dot{y} = f(x,y);$$
(4)

or, to consider the more general situation:

$$\dot{x} = p(x,y)$$

$$\dot{y} = q(x,y).$$
(5)

In the physical problem the numbers x, y describe the "state" of the system—the position and velocity of the particle. Thus if we introduce coordinates (Fig. 2) each point (x,y) in the plane represents a state of the system. This plane is sometimes called the "phase plane" or "state space" of the system. We can now look upon the differential equations (5) as defining a flow in the plane. When we are given the differential equations, we are given the velocity of the flow at each point of the plane (Fig. 2). The horizontal or x-component of the velocity of the flow is $p(x,y)$ and the vertical or y-component of this velocity is $q(x,y)$. This velocity of the flow, which is now something quite different from the physical velocity of the system, can be represented by a vector (bold-face type) $\mathbf{v} = (p(x,y), q(x,y))$. This vector describes at each point (x,y) how—in magnitude and direction—the state of system is changing. Starting at a given initial point the flow of this point defines a curved path, which is the curve defined by a solution of the differential equation.

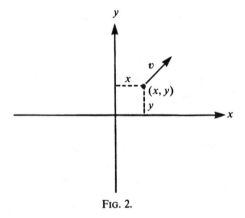

FIG. 2.

Since this picture is important to further understanding, let us look at several examples. With $\dot{y} = x$, the equations for the freely falling body (eq. 2) are

$$\dot{x} = y$$

$$\dot{y} = -g.$$

Here the horizontal component of the flow is the height y of the point above the horizontal axis, and the vertical component is always negative and is constant. The solution (3) describes the flow precisely. Starting initially at the point (x_0, y_0), the solution is

$$y = -gt + y_0$$
$$x = -\frac{1}{2} gt^2 + y_0 t + x_0.$$

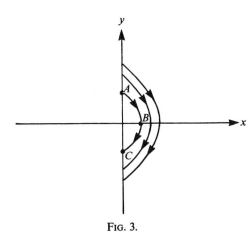

FIG. 3.

The path defined by this solution is a parabola (Fig. 3). The arrows in Figure 4 show the direction of the flow with increasing time. Now remember that this flow describes the changing state of the system. Let the initial state be the point A of Figure 3. The particle—let us assume it is a ball—starts on the earth $(x=0)$ and is given an initial velocity upwards (y_0 is positive). The flow in the phase plane describing this motion is then along the parabola from A towards the point B. Until it reaches B the ball is rising (x is increasing) and its speed y is decreasing. At B the particle has reached its maximum height, it stops instantaneously ($y=0$), and then starts to drop. Its state now follows the path from B to C. The velocity y of the ball becomes negative (it is falling) and its distance x from the ground is decreasing. The point C corresponds to the ball hitting the ground.

Let us also look quickly at a second example that will be of importance to us later on. It has to do with the motion of a mass attached to an elastic spring (Fig. 4), the so-called simple harmonic oscillator. The equations of motion are

$$\ddot{x} + x = 0, \tag{6}$$

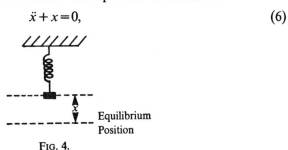

Equilibrium
Position

FIG. 4.

where again since we need not be concerned with units of measurement we may take the mass and the coefficient of elasticity of the spring to be one. Here x measures the distance of the mass from its equilibrium position. The equivalent pair of first order equations is

$$\dot{x} = y$$
$$\dot{y} = -x.$$

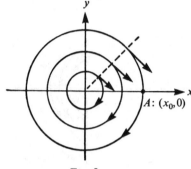

FIG. 5.

At each point the flow is perpendicular (Fig. 5) to the radial line from the origin, and so we know that the flow is circular in a clockwise direction about the origin. This we may see directly from the differential equation, since $\frac{d}{dt}(x^2 + y^2) = 2x\dot{x} + 2y\dot{y} = 2xy - 2xy = 0$. The distance of a point in the flow from the origin does not change. For this particular system this is also the statement of the law of conservation of energy. Thus, if the initial state of the system corresponds to the point A, the circular flow describes the oscillation of the mass. At A it has its maximum displacement and the distance x_0 is the amplitude of the oscillation. Since the speed of the flow is $\sqrt{\dot{x}^2 + \dot{y}^2} = \sqrt{x^2 + y^2} = r$, the radius of the circle, the period of the oscillation (the time to make one revolution of the circle) is 2π. The period does not depend on the amplitude of the oscillation. The solutions of this differential equation give a good approximation of the small oscillations of a pendulum; and the fact that the period does not change with the amplitude of the oscillation was what Galileo observed as a young boy while watching a lamp swing in the cathedral of Pisa.

This geometric picture of the flow in state space is basic to the geometric or qualitative theory of differential equations. In the geometric theory one gives up the futile attempt of finding general solutions of all differential equations and attempts instead to obtain as much information as possible about the flow and the nature of the solutions defined by the flow without explicitly solving the equations. This was illustrated for the simple harmonic oscillator. Although the differential equations of motion could in this case have easily been solved (it is a linear problem), we obtained all of our information about the oscillations without exhibiting the solutions.

3. PROPERTIES OF LINEAR SYSTEMS

To understand some of the differences between linear and nonlinear differential equations, it is first necessary to discuss in some detail equation (5) when $p(x,y)$, $q(x,y)$ are linear in x and y; that is, the equations

$$\dot{x} = ax + by$$

$$\dot{y} = cx + dy \tag{7}$$

where a, b, c, d are real numbers.

The general solution of this linear system is obtained by knowledge of only two special solutions which are easy to compute. The other main implication of linearity is that the behavior of the solutions near the solution $x = y = 0$ yields the behavior of the solutions in the entire phase plane. Two further properties of linear systems are that all solutions are defined for all values of time and that there can never be isolated periodic solutions (see the example of the simple harmonic oscillator in the previous section). A few remarks will also be made about solutions which become unbounded in time (phenomena of resonance) when the simple harmonic oscillator is subjected to external forces. Even though the qualitative properties above are discussed only for the second order system (7), they are also true for n^{th} order systems.

In the case of second order systems we notice first of all that if (x_1, y_1) and (x_2, y_2) are any two solutions of (7), then for any constants e_1, e_2, the pair $(e_1 x_1 + e_2 x_2, e_1 y_1 + e_2 y_2)$ is also a solution. This property is generally referred to as the principle of superposition. One can verify rather easily that *the principle of superposition is the distinguishing characteristic of linear systems*; that is, if the principle of superposition holds for equation (5) for any functions $p(x,y)$, $q(x,y)$ which are continuous, then $p(x,y)$, $q(x,y)$ must be linear.

This principle makes it possible to find explicit solutions for linear systems. In fact, suppose $(x_1(t), y_1(t))$, $(x_2(t), y_2(t))$ are any two solutions of (7) which have the property that, for any constants u, v, there are constants e_1, e_2 such that the system of equations

$$x_1(0)e_1 + x_2(0)e_2 = u$$

$$y_1(0)e_1 + y_2(0)e_2 = v$$

is satisfied. Then if $(x(t), y(t))$ is a solution of (7) with $x(0) = u$, $y(0) = v$ then $x(t) = e_1 x_1(t) + e_2 x_2(t)$, $y(t) = e_1 y_1(t) + e_2 y_2(t)$ for all values of t, since this is true at $t = 0$ and there is only one solution of the differential equation passing through any given point. Consequently, to solve the linear equation (7) it is only necessary to exhibit two solutions with the above property. If λ is a root of the equation

$$\lambda^2 - \lambda(a+d) + ad - bc = 0, \tag{8}$$

then $x = e^{\lambda t}u$, $y = e^{\lambda t}v$ is a solution of (7) for some constants u, v. Furthermore, if the two roots are λ_1, λ_2 and $\lambda_1 \neq \lambda_2$, then the two solutions needed are of this form. If $\lambda_1 = \lambda_2$, one may need to consider functions $te^{\lambda t}$. In any case, the solutions are defined for all values of t. We continue the discussion for the case in which system (7) is of the form

$$\dot{x} = y$$

$$\dot{y} = -x - 2\rho y, \quad \rho^2 - 1 \neq 0.$$

(9)

$\rho < 0, \rho^2 < 1$ $\rho > 0, \rho^2 < 1$ $\rho = 0$

(a)   (b)   (c)

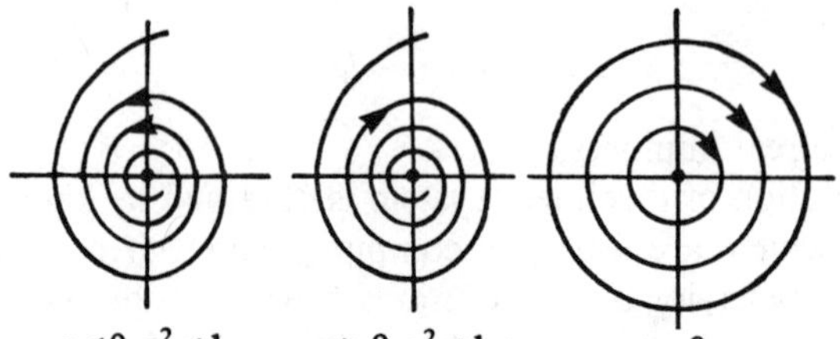

FIG. 6.

Under these hypotheses equation (8) has two distinct roots whose real parts are the same sign as ρ and for $\rho^2 < 1$ an imaginary part equal to $\sqrt{1 - \rho^2}$.

For this situation, we reach the following conclusions:

(I) *If $\rho < 0$, then all solutions of (9) approach zero as t increases indefinitely*;

(II) *If $\rho > 0$, then all solutions of (9) become unbounded as time increases indefinitely*;

(III) (*the simple harmonic oscillator*) *If $\rho = 0$, then all solutions of (9) are periodic with the same period, 2π.*

The paths of the solutions for these three cases are shown in Fig. 6 for $\rho^2 < 1$. The word *all* should be emphasized, since this is a specific characteristic of linear systems. To be precise, the behavior of the paths of a linear system in a local neighborhood of the solution $x = 0$, $y = 0$ yields the behavior of the paths in the entire phase plane. *Linear systems are by nature provincial—global behavior can be predicted from local behavior.*

If an equation (7) is an idealization of some physical system, then the constants a, b, c, d are determined by the special characteristics of the physical system itself. What happens if the system is subjected to external forces? The new differential equations then have the form

$$\dot{x} = ax + by + f(t)$$

$$\dot{y} = cx + dy + g(t)$$

(10)

where $f(t)$, $g(t)$ are some given functions of time. We have already encountered one such equation in the discussion of a freely falling body near the earth; namely, equation (10) with $a = 0$, $b = 1$, $c = 0$, $d = 0$, $f(t) = 0$, $g(t) = g$, the acceleration due to gravity.

As another example consider the forced harmonic oscillator

$$\ddot{x} + x = a \cos \beta t, \quad \alpha \neq 0, \quad \beta > 0,$$

or

$$\dot{x} = y$$

$$\dot{y} = -x + \alpha \cos \beta t$$

(11)

which has a particular solution

$$x_0 = \frac{\alpha}{1-\beta^2} \cos \beta t, \quad y_0 = -\frac{\alpha\beta}{1-\beta^2} \sin \beta t$$

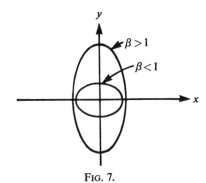

Fig. 7.

provided that $\beta^2 \neq 1$. Since $x_0^2 + y_0^2/\beta^2 = \alpha^2/(1-\beta^2)$ the path in the phase plane is shown in Fig. 7 and is an ellipse whose semiaxes are $\alpha/(1-\beta^2)$, $\beta\alpha/(1-\beta^2)$. For any value of α, no matter how small, the amplitude of this oscillation can be made as large as desired by choosing β close to 1. For $\beta = 1$, the above solution is meaningless and (11) actually has an unbounded solution. This phenomenon is known as *resonance*.

In the case of an idealized pendulum, one can understand resonance intuitively. Suppose the pendulum is oscillating without the influence of an external force and suddenly one begins to apply a very small force at the point of maximum positive displacement from the origin in such a way as to assist the motion in its return to the origin (Fig. 8). Of course, if the model of the system is taken as linear, then the time the pendulum takes to return to the point of maximum positive displacement does not depend on the amplitude of the oscillation, and the applied force will be

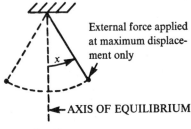

Fig. 8.

periodic of period 2π. If the model is truly linear, the amplitude of the motion will continually increase in time no matter how small the applied force; and eventually the pendulum will rotate around its point of support.

In the case of no damping in (9), that is, $\rho=0$, we have seen that some types of bounded inputs or forcing functions to the system may yield unbounded outputs (resonance). On the other hand, a linear system (9) with damping ($\rho<0$) always yields bounded outputs for all bounded inputs.

4. SOME ASPECTS OF NONLINEAR SYSTEMS

The principle of superposition cannot hold for nonlinear sytems. This implies that in general the solutions of a nonlinear system cannot be represented in a simple manner in terms of a special set of solutions, and thus the qualitative discussion of the solutions becomes even more important. In particular, it can be shown that the solutions of nonlinear systems generally have completely different local and global properties. Furthermore, all solutions need not be defined for all values of time, and there may be isolated periodic solutions (closed curves in phase space). Also, with a more realistic approximation to the harmonic oscillator, the phenomenon of resonance cannot occur. Even though none of the properties of linear systems are preserved, *some of the local properties of nonlinear systems can be determined by a linear analysis.*

An *equilibrium* or *rest point* of a differential system is a constant solution. For system (5), all equilibrium points are the constant pairs (c,d) such that $p(c,d)=0$, $q(c,d)=0$. At an equilibrium point, there is no flow in the phase space.

To make a very simple comparison between linear and nonlinear systems, consider the linear first order equation

$$\dot{y}=-y \tag{12}$$

and the nonlinear first order equation

$$\dot{y}=-y(1-y). \tag{13}$$

The general solution of the linear equation (12) is given by $y=ae^{-t}$ where a is an arbitrary constant and is equal to the initial state of the system at time zero. On the other hand, if the initial state of system (13) at time zero is a, then by direct calculation one observes that the solution of equation (13) is

$$y=\frac{a}{1-a+ae^{-t}}e^{-t}.$$

The curves in the phase space y and the phase-time space (y,t) for the two equations for various values of a are shown in Fig. 9. As is observed from this figure, new phenomena occur even for the simplest nonlinear equations. For nonlinear systems there can be more than one equilibrium point, each of which is isolated; and some solutions of a nonlinear system may become unbounded in a finite interval of time. Also, the behavior of the solutions of (13) with initial values greater than one is completely different from those with initial values less than one. None of these situations occurs in linear systems.

It is possible for linear analysis to yield some information about (13). We first study the solutions near the equilibrium point $y = 0$. Since the equation is $\dot{y} = -y + y^2$ it would seem reasonable (and it was proved by Liapunov) that the equation $\dot{y} = -y$ is a good approximation near $y = 0$. An inspection of the curves in Figures 9a and 9b shows this is actually the case. However, an analysis of $\dot{y} = -y(1-y)$ which is based on $\dot{y} = -y$ for all initial values is erroneous, since all of its solutions do not approach zero and some are actually unbounded. To analyze the behavior of (13) near $y = 1$, let $y = 1 + z$ and study the behavior near $z = 0$. The new equation for z is

$$\dot{z} = z + z^2.$$

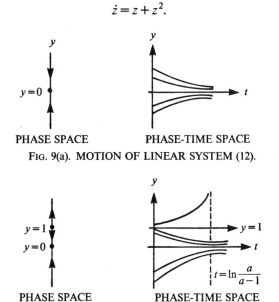

PHASE SPACE   PHASE-TIME SPACE

FIG. 9(a). MOTION OF LINEAR SYSTEM (12).

PHASE SPACE   PHASE-TIME SPACE

FIG. 9(b). MOTION OF NONLINEAR SYSTEM (13).

Near $z = 0$, it again seems reasonable to take the linear approximation $\dot{z} = z$ whose solutions are $z = be^t$, where b is the initial state at zero. The linear analysis then yields the result that solutions of (13) near $y = 1$ tend to diverge from $y = 1$, which is actually the case. On the other hand, the linear analysis would also say that all solutions are defined for all values of t, whereas we have seen this is not the case. Thus we see that *linear approximation is often useful but has its limitations*.

As indicated above, it is impossible for linear homogeneous equations (7) to have isolated periodic solutions; but this is clearly not the case for nonlinear equations. Consider the second order equation

$$\dot{x} = y$$
$$\dot{y} = -x + 2\rho(1 - x^2)y \tag{14}$$

where ρ is a positive constant.

Equation (14) is called van der Pol's equation. He studied it in connection with his work on the triode oscillator; and for large values of ρ, he also suggested that this equation explains some irregularities in the heart beat. The same equation in a different form was studied by Lord Rayleigh in his investigations of the theory of sound.

To understand (14), we first observe that $x=0$, $y=0$, is an equilibrium point and analyze the behavior of nearby solutions by linearizing the equations. These linear equations are (9) with $\rho>0$. Thus, all solutions of the linear equations leave the equilibrium point as t increases. Liapunov has shown that the same result is true for the solutions of (14) when the initial values are sufficiently close to the equilibrium point but different from it. Near the rest point and for $\rho^2<1$, the solution curves of (14) in the phase plane are shown in Fig. 6a.

On the other hand, the term $2\rho(1-x^2)y$ in (14) represents a frictional force and for large values of x, this frictional force actually has a damping effect. As a result of this, one can show that the solutions of (14) are bounded and there actually is a curve C in the (x,y)-plane across which the solution curves move from the outside inward. By this linear analysis, one can also construct a curve C_1 which lies inside C such that the solution curves move from the inside of C_1 to the outside. This is depicted in Fig. 10 where the arrows designate the direction of the motion along solution curves. Since there are no equilibrium points in the region between C_1 and C, intuition leads to the conjecture that there must be a closed solution curve (yielding a periodic motion) in this region. This is precisely the case and is a consequence of a nontrivial result proved around the turn of the century by Bendixson. By a more detailed argument, one can actually show there is only one closed solution curve of (14), and all other solution curves except $x=0$, $y=0$, approach this curve as t increases indefinitely (see Fig. 11).

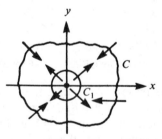

FIG. 10.

Oscillations of the above type are called self-sustained, since they occur without the influence of any external forces but arise simply from the internal structure of the system and the manner in which energy is transferred from one state to another. The point to be made is that such *self-sustained oscillations can only be explained by a nonlinear theory*.

The motion of a clock or watch can be explained by an investigation of self-sustained oscillations, and many attempts have even been made to apply the concept in explaining the interaction of various biological species. For example, suppose there are two isolated species of fishes, of which one species feeds on the

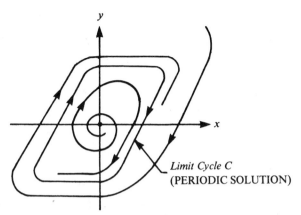

FIG. 11.

other—which in turn feeds on vegetation which is available in unlimited quantity. For this situation, it is possible to construct two first-order differential equations (whose solution gives the quantities of each species at time t) in such a way as to give very realistic information. The differential equations have a self-sustained oscillation (that is, the quantity of each species varies periodically in time); and for every initial state (except the equilibrium states) the number of each species approaches this particular periodic behavior.

A further contrast between linearity and nonlinearity is that nonlinearity can prevent resonance in the sense described in the section on linear systems. In our discussion of the pendulum, the linear equation (6) was said to be a good approximation for small amplitudes. A more realistic approximation is the equation

$$\dot{x} = y$$
$$\dot{y} = -x + \tfrac{1}{6}x^3. \tag{15}$$

Observing that $\frac{d}{dt}(x^2 + x^4/12 + y^2) = 0$, we see that the solution curves lie on the curves $x^2 + x^4/12 + y^2 = $ constant. In a neighborhood of $x = y = 0$, the paths are closed and thus the solutions are periodic, but now the period varies with the path, in contrast to linear systems. In our intuitive discussion of resonance on p. 375, the constancy of the period was the main element for the plausibility of resonance. But, if the period varies with amplitude, then a periodic disturbance will become out of phase with the free motion and the forcing function should be a hindrance to increasing amplitude. This can be shown to be the case for systems without damping in which the period of the oscillations varies with amplitude (in which case the system must be nonlinear), small periodic disturbances do not cause instability. Here is a situation where stability is a consequence of nonlinearity, even though no frictional forces are present. It is typical of the advantages which may sometimes be obtained by deliberately introducing nonlinearities into systems.

5. STABILITY

Stability is one of those concepts that we all understand even though we might not be able to define it precisely. We speak of a stable personality or a stable economy, and we recognize the relative stability in both voltage and frequency of the electric power supplied to our homes. Basically we have a system of some type which operates in some specified way under certain conditions. If these conditions change slightly, how does this affect the operation of the system? If the effect is small, the system is stable. If not, the system is unstable. Take the simple example of a ball rolling along a curve in a plane. The points A, B, C and D of Figure 12 are the rest positions (equilibrium states) of the ball. Positions A and C are clearly unstable, since the slightest perturbation will send the ball rolling down the hill. The positions B and D are stable, since a slight perturbation causes the ball to oscillate about the rest position but remain close to it. This is a particular instance of a quite general physical principle. In mechanics, potential energy at a point is the amount of work expended in moving a unit mass to that point from some arbitrary but fixed point. The principle then states that *a minimum of potential energy corresponds to a stable rest position. If the rest position is not a minimum it is unstable*. This is so easily comprehended that physics teachers seldom bother to prove it.

This law was enunciated by Lagrange around 1800, but it was Liapunov some 90 years later who was the first to appreciate fully this principle and its extensions. Liapunov's extension is called his "direct" or "second" method for the study of stability. The method is said to be direct because his criteria for stability and instability can be applied as soon as the differential equations of the system are known and requires no specific knowledge of their solutions.

But let us now examine these concepts more precisely in terms of our fundamental system

$$\dot{x} = p(x,y)$$
$$\dot{y} = q(x,y). \tag{16}$$

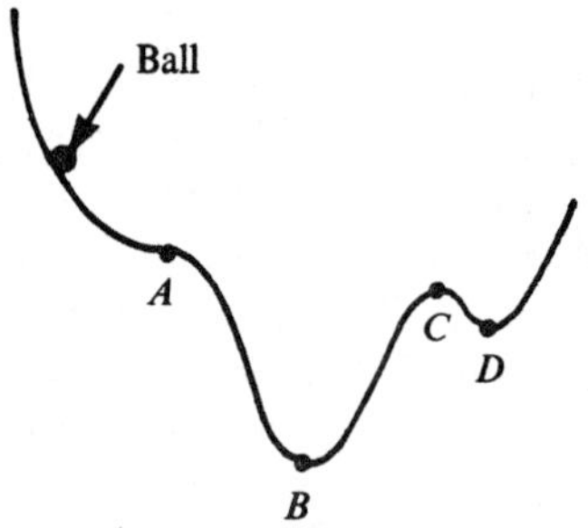

FIG. 12.

Here (x,y) represents the deviation from some desired state or performance of the system. The origin 0 then corresponds to the desired behavior and is an equilibrium state of the system: $p(0,0)=0$ and $q(0,0)=0$. There is no flow at this point, and if the system is initially at 0 it remains for all time at that point. However, such a statement concerns the mathematical model and neglects reality, and this is precisely why we are interested in stability. Stability is concerned with the question of what happens when the system is perturbed from this equilibrium state.

Let C_1 be an arbitrary circle about the origin. Then if for each such C_1 we can find a concentric circle C_2, which is so located that solutions which start inside C_2 remain inside C_1 (Fig. 13), we say that the origin is *stable*. If this is not the case we say that it is *unstable*. If there is also another circle C_0 about a stable origin which is such that solutions starting inside C_0 tend to *return* to the origin (Fig. 14), we say that the origin is *asymptotically stable*; and if *all* solutions tend to return to the origin, the origin is *asymptotically stable in the large*. For example, in Figure 6c the origin is stable but not asymptotically stable. In Figure 6a the origin is asymptotically stable and in Figure 6b the origin is unstable.

Consider once again van der Pol's equation (14) which can be written as

$$\ddot{x} - 2\rho(1 - x^2)\dot{x} + x = 0.$$

Letting $y = \dot{x} - 2\rho(x - x^3/3)$, $\epsilon = -2\rho$, we obtain the two first-order equations

$$\dot{x} = y + \epsilon(\tfrac{1}{3}x^3 - x)$$

$$\dot{y} = -x. \tag{17}$$

In contrast to the previous discussion, we will assume $\epsilon = -2\rho$ is positive ($\epsilon > 0$).

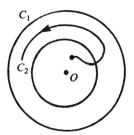

FIG. 13.

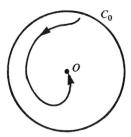

FIG. 14.

This is equivalent to reversing the flow in the phase plane, and the picture of the flow is now that of Figure 15. The origin is asymptotically stable, since every solution inside the limit cycle tends toward the origin. However, if the system were to be perturbed to a state outside the limit cycle this is no longer the case, and it can actually be shown that there are solutions outside the limit cycle that go to infinity in finite time. The region inside the limit cycle is called the *region of asymptotic stability*, and it is the size and shape of this region that determines just how stable the system is.

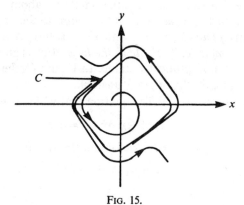

FIG. 15.

It so happens in this case it is possible to decide the asymptotic stability of the origin by omitting the nonlinear term $x^3/3$ and examining only the linear approximation. Since the linear approximation is asymptotically stable it is always asymptotically stable in the large. The region of asymptotic stability for the linear system is always the whole space, and the linear approximation does not and cannot give any information as to how stable the system is. *The extent of the stability of a system is determined by its nonlinearities.* A decided advantage of Liapunov's method is that it takes into account the nonlinearities of the system and permits an estimation of the region of asymptotic stability without calculating any solutions. Even in van der Pol's equation, the calculation of the limit cycle C (which determines the region of asymptotic stability) is extremely complicated, but fortunately the existence of a rather large region of asymptotic stability can be established without knowing C.

Liapunov's method, like Lagrange's principle, is extremely simple. In essence Liapunov's method consists of selecting a suitable function $V(x,y)$ in phase space such that (a) it has a minimum at the equilibrium point being investigated, and (b) the contours (surfaces along which the function is a constant) of the function surround the equilibrium point. Then if the flow of solutions in phase space can be shown to cross these contours from the outside towards the inside the equilibrium point is stable, at least for perturbations that keep the state of the system within the largest contour for which the flow is always inward.

In carrying out Liapunov's method we actually check whether the Liapunov function, $V(x,y)$, decreases along solution-flow lines. Since we required that $V(x,y)$ have a minimum at the equilibrium point, the fact that it decreases along solutions means these solutions are crossing the contours of $V(x,y)$ in the desired direction. Remember, we do this without solving the differential equation.

Of course, nothing comes for free in this life—the trick is to be able to find a suitable Liapunov function. We chose the coordinate system in the example of van der Pol's equation that follows so that $V(x,y)$ would have a particularly simple form; don't be deceived thereby, it is not always that easy.

For van der Pol's equation we can take

$$V(x,y) = \tfrac{1}{2}x^2 + \tfrac{1}{2}y^2.$$

At the origin $V(0,0)=0$ and everywhere else $V(x,y)$ is greater than zero. Computing the rate of change $dV/dt = \dot{V}$ of V along solutions we obtain from equation (17)

$$\dot{V} = x\dot{x} + y\dot{y} = -\tfrac{1}{3}\epsilon x^2(3 - x^2).$$

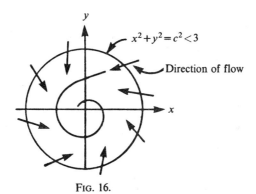

FIG. 16.

Within the circle of radius $x^2+y^2<3$, x^2 is less than 3 and $\dot{V} \le 0$. Actually within this circle $\dot{V}<0$ if $x \ne 0$. This tells us that within any circle $x^2+y^2=c^2<3$ the solutions cross the circle as shown in Figure 16. This means that every solution starting inside $x^2+y^2=3$ tends to the origin as t tends to infinity. The origin is asymptotically stable, and we have the additional information that independently of ϵ (all $\epsilon>0$) the region of asymptotic stability in this coordinate system is always at least as large as this circle. This also gives us information about the limit cycle C. In fact, for all non-zero ϵ, the limit cycle lies outside this circle of radius $\sqrt{3}$.

The Liapunov method which we have discussed is the only general method available at present for the study of stability and has been widely used, particularly in the Soviet Union, to solve practical problems of stability. Success has been achieved in studying the stability of motion of rigid bodies, in investigating the stability of a large class of nonlinear control systems, and in determining the stability of nuclear reactors. The method and its extensions also play an important theoretical role in the study of differential equations.

In theory it should always be possible to decide stability by this method, but there remains the practical difficulty of constructing suitable Liapunov functions—functions of the type V used above. There are some methods which guide one in the construction of Liapunov functions but effective use of the method requires ingenuity and experience. There are, as yet, no general schemes for using computers to decide the stability of nonlinear systems.

6. THEORY OF AUTOMATIC CONTROL

Man has been aware of the feedback principle of automatic control through most of recorded history, and man himself embodies the principle. The Babylonians, some 2000 years before the birth of Christ, recognized a primitive form of feedback in their opening and closing of ditches to control the moisture content of the soil. A more sophisticated feedback was used in highly accurate Arabian water clocks developed near the beginning of the Christian era. It was a float-operated valve, similar to that still used today in our bathrooms, to control the water level in a tank. A classic example of a conscious application of feedback to achieve automatic control is the flyball-governor which Watt used in 1788 to regulate the speed of a steam engine under a varying load. As the engine's speed deviated from the desired value, the change in speed itself was used to regulate the throttle and correct the error. Through the development of electronics and more sophisticated mechanical devices, we now have automatic control systems of great complexity, yet the basic principle of feedback control remains the same—whether it be used for the automatic control of an industrial process, a phase of our economy, a nuclear reactor, or the attitude of a space vehicle.

Our interest is in the fact that this principle and its application can be subjected to mathematical analysis. This was first demonstrated by a British physicist, Maxwell, in 1868 and independently by a Russian engineer, Vyshnegradskii, around 1876. They concerned themselves mainly with small errors, that is, small deviations from the performance desired, and they approximated the actual system by a linear one. They observed that an improper use of feedback could enhance the error instead of diminishing it and that the proper use of feedback involves an analysis of stability. Does the error tend to zero or not?

This same procedure of linear approximation and analysis of stability was the mathematical foundation for Minorsky's design of an automatic steering device for the battleship New Mexico in the early 1920's. The behavior he predicted on the basis of his simple mathematical model was then confirmed by observed fact. In one of those strange quirks of history, however, the U. S. Navy did not adopt automatic steering at that time for the simple reason that it might deprive helmsman of practice. The logic is similar to that of the British Admiralty when they first rejected the idea of using steam engines on warships on the grounds that sparks from the engines would burn holes in the sails.

Even during World War II, in spite of all the sophisticated developments in control, the theory was restricted for the most part to linear approximations, linear feedback, and a linear analysis of stability. Even though the analysis was made in

this limited fashion, optimization of certain performance criteria was quite successful. However, in the past decade there has been an explosive growth in the
development of a theory of control in the United States and the Soviet Union. A
complete bibliography on almost any one aspect of the subject may amount to 300
or 400 papers. Fortunately, it is possible to point out a few basic aspects of the
subject and to illustrate the theory simply by comparing different solutions of a
single not-too-complex problem.

The problem is that of moving a body through a viscous fluid with a force
(thrust) F which can drive or impede the motion of the body. The equations of
motion are

$$\ddot{x} + \beta\dot{x} = F, \tag{18}$$

where β is positive and is the coefficient of the viscosity of the fluid. Let us suppose
that what we want to control is the speed $\dot{x}$ of the body; that is, we want $\dot{x} = a$
where a is a given constant. Our ability to control the motion of the body lies in
our ability to select F. We are interested only in velocity, so with $\dot{x} = y$, we have the
first order equation

$$\dot{y} = \beta y = F.$$

First let us look at what the engineer calls "open loop" control as opposed to
"closed loop" or feedback control. In open loop control no information concerning
the deviation from desired performance is used to adjust the control force F. There
is no connection between the output y and the input F. The control loop is open.
Thus, since we are not using information on whether the body is going too fast or
too slow, the best that we can do is to make F a constant. In this case the solution
for the velocity is

$$y = y_0 e^{-\beta t} + \frac{F}{\beta}(1 - e^{-\beta t}),$$

where y_0 is the initial velocity of the body. Thus, independently of the initial
velocity, y approaches F/β exponentially as t approaches infinity. We want this
terminal velocity to be a and select $F = \beta a$. This is the desired result but the
approach to a will be quite slow if β is small. As we shall see, this open loop
control also has a more serious disadvantage.

Now let us compare this with the feedback control $F = -\delta(y - a) + \beta a$, where δ
is some positive constant. The control force F now depends linearly upon the
difference between the actual velocity of the body and that desired. This is linear
feedback control. The differential equation of the controlled system is

$$\dot{y} + \beta t = -\delta(y - a) + \beta a,$$

or

$$\dot{y} + (\beta + \delta)y = (\beta + \delta)a.$$

The feedback has the effect of changing the coefficient of viscosity and now

$$y = y_0 e^{-(\beta + \delta)t} + a(1 - e^{-(\beta + \delta)t}).$$

Again, independently of the initial velocity, y approaches a but at a faster rate. Note also that for a small error in control $((y - a)$ small) the control force being used is about the same as before.

Although this in itself illustrates some gain in using feedback control, a more important advantage is the increased stability of operation achieved by feedback. Suppose that we do not know the value of β exactly or that we want a system which will operate over a range of values of β. To make matters simple, suppose that when we originally designed the system we thought that the coefficient of viscosity was β but now the body is moving in a fluid whose coefficient of viscosity is $\frac{1}{2}\beta$ (it has become warmer). Under open loop control the equation of motion is in this circumstance

$$\dot{y} + \tfrac{1}{2}\,\beta y = \beta a,$$

and with feedback control

$$\dot{y} + \tfrac{1}{2}\,\beta y = -\delta(y - a) + \beta a.$$

Under open loop control the velocity y now approaches $2a$, an error of 100 percent. Under feedback control y approaches

$$\frac{\beta + \delta}{\tfrac{1}{2}\,\beta + \delta}\,a,$$

and by making δ large we can make the control insensitive to changes in viscosity. For example, if $\delta = 10\beta$, y approaches $\frac{22}{21}a$, which is an error of less than five percent. *Feedback control makes the system relatively insensitive to changes in environment.*

But there is still a more important feature to feedback control. In a sense feedback control enables a system to adapt to its environment. The linear feedback of the above example will now be adjusted by means of a nonlinear feedback in such a way that the system adapts to changes in viscosity. This suggests that *nonlinearity is essential to adaptation.*

Assume now that we have no information about β except that it is constant. We also no longer assume that β is positive (if β is negative, then the system we are controlling is unstable). The controlled system will now be of the following form:

$$\dot{y} + \beta y = (b - \beta_0)(y - a) + ba$$
$$\dot{b} = g(y - a).$$

The linear feedback control, which is the right-hand side of the first equation, was selected by asking ourselves what would work if we knew β and afterwards replacing β by an adjustable parameter b. The second differential equation determines the adjustment of b, and this adjustment is to depend only upon the observation of the error $y - a$. The function g remains to be selected, and β_0 is any positive number which we are also free to select. Thus the control assumes no knowledge of the constant β. What we want is to have y approach the desired velocity a for all initial velocities y_0 and any initial value of b. The function g, which determines the design of the mechanism for adjusting b, can be selected using an extension of Liapunov's method. We want y to approach a and we may suspect that b will approach the unknown constant β. This then suggests that we take as a Liapunov function

$$V = \tfrac{1}{2}(y-a)^2 + \tfrac{1}{2}(b-\beta)^2.$$

Using the differential equations of our system, we obtain after a simple computation the rate of change of V along solutions to be

$$\dot{V} = (y-a)\dot{y} + (b-\beta)\dot{b} = -\beta_0(y-a)^2 + (b-\beta)(g + y(y-a)).$$

Thus, if we take $g = -y(y-a)$,

$$\dot{V} = -\beta_0(y-a)^2.$$

This means that V is always decreasing as long as there is an error in control, and it can be shown that y approaches a and that b does in fact approach β. The choice of β_0 is still open, but the appearance of β_0 in the equation for $\dot{V}$ tells us that the larger the value of β_0 the faster y approaches a. Thus, regardless of the value of β, this control system always reduces the error to zero and the feedback adjusts itself to adapt to the environment.

This leads to the most recent advancement in control theory—the study of optimal processes of control. The problem is, relative to some performance criterion, to select within a given class of available controls the best possible control. The limitations of space make it impossible to carry out the solution in detail in this article, but it is feasible to illustrate the nature of the problem and the nature of the solution that can be derived from the modern theory of optimal control. We are still dealing with a body moving in a viscous fluid, but now let us assume that we want to bring it to rest at the point $(0,0)$ in the phase plane in minimum time. Realistically, assume that our power source is limited, placing constraints on the force F available for control. We might assume, for example, that F must lie between -1 and 1 ($-1 \leqslant F \leqslant 1$). The equations of motion are as before

$$\ddot{x} + \dot{x} = F$$

and the equivalent system is

$$\dot{x} = y$$
$$\dot{y} = -y + F, \qquad -1 \leqslant F \leqslant 1.$$

The desired state of the system is $x = 0$, $y = 0$, and we want to bring the system to this state in the shortest possible time. There is available, as we have said, a theory for solving this problem; and the optimal control law is shown in Figure 17. The phase plane is divided into two parts by a curve C whose equation is

$$x = -y + \log_e(1+y), \quad y \geqslant 0$$
$$x = -y - \log_e(1-y), \quad y \leqslant 0.$$

This curve is called the "switching curve." At a state (x,y) above this switching curve optimal control is obtained by taking $F = -1$ and below the curve $F = 1$. In

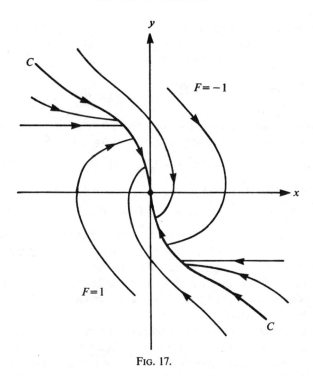

FIG. 17.

other words, F changes sign at the switching curve. This optimal control law is unique. Any other control law satisfying the constraint $-1 \leqslant F \leqslant 1$ takes longer to bring the system to rest at $x = 0$, $y = 0$.

The control law here is said to be "bang-bang." It jumps rapidly from -1 to 1 and at all times uses the maximum control available, which is intuitively what one expects. The control, although made up of linear pieces, is highly nonlinear on the whole and does what no linear control could do—it reduces the error to zero in finite time. Under linear control the error can only tend to zero exponentially. *Nonlinear control can achieve performance far beyond that possible by linear control.*

7. FUTURE TRENDS

Most of the important questions in second order systems have now been answered. This is not surprising since the geometry of integral curves in two dimensions (in a plane) is not too complicated. More precisely, the only possible solution curves which can be limits of the other solution curves (that is, approached as time increases indefinitely or decreases indefinitely) are points of equilibrium or closed curves. As soon as third order systems are considered, however, the geometry becomes vastly more complicated, and it is possible to have limit curves which "fill up" a torus (the surface of a doughnut). Such limit curves still possess oscillatory properties, and one of the most important problems today, as it has also been for the last 75 years, is to characterize the possible solution curves for an nth order system which can be limits of other solution curves. Much intelligent experimentation by engineers is still needed to discover the type of oscillatory phenomena that can occur in practice for higher order systems of differential equations, and this should lead to interesting new classes of nonlinear differential equations which will have to be investigated. For cases in which the nonlinearities are small, specialized techniques are already available for systems of any order.

One of the most important remaining problems which concerns stability of an equilibrium point is that of obtaining general methods or algorithms for the construction of Liapunov functions. A solution of this problem could lead to an efficient use of computers in deciding questions of stability. Another important area involves characterising those properties of the geometric structure in the phase space of all of the solution curves of an nth order system which remain unchanged even when the differential equations themselves are subject to arbitrary but small perturbations. This problem is completely solved for second order equations, but very little is known for higher order systems.

Among astronomical problems, it has still not been proved that a system involving the interaction of n planets is stable. Even if it is, however, it is important to know whether it would still be stable under relatively small perturbations. For example, what would happen if a planet exploded? This problem has recently been solved for periodic disturbances, and is attracting the attention of some of the best young mathematicians in the world.

Since the current views of automatic control are somewhat different from those of the past, there are of course many interesting and unsolved problems in this area. Given a performance criterion and a specified law of control, it is not always clear that an optimum control law exists. Pontryagin's maximum principle gives necessary conditions for the existence of an optimum, but sufficient conditions have only been given in very particular situations. After the existence has been established, efficient methods must be devised for the computation of the optimal control laws. Computational methods exist for systems where the control enters linearly, but the computation for nonlinear systems still must proceed by local linearization. The solution of many more specific examples of automatic control will certainly lead to great strides in the entire area.

Finally, it is difficult to see how a system can be made to be truly adaptive without having the control law depend upon a certain portion of the past history of the system. Insertion of such hereditary dependence in the problem takes one beyond the realm of differential equations. Much research is now being conducted on this extended type of equation with the initial aim being to understand the essential differences between these and ordinary differential equations. It is quite clear that the applications of equations with hereditary dependence will reach beyond automatic control, and, historically, they were being discussed long before the conception of adaptive control.

In a way, this is typical of the entire history of differential equations—particularly those involving nonlinearity. Specific practical application has usually stimulated the initial interest in classes of differential equations. As soon, however, as significant progress is made in any given area, a whole host of applications becomes apparent.

RECOMMENDED READING

Good elementary introductions to the modern theory of differential equations are a book by Birkhoff and Rota, *Ordinary Differential Equations*, Ginn, Boston, 1962, and a book by Walter Leighton, *Ordinary Differential Equations*, Wadsworth, Belmont, California, 1963. For a discussion of the theory of nonlinear oscillations which is not too sophisticated mathematically and contains many physical applications, the reader should consult Minorsky, *Nonlinear Oscillations*, Van Nostrand, Princeton, N.J., 1962. A book by LaSalle and Lefschetz, *Stability by Liapunov's Direct Method*, Academic Press, New York, 1961, gives a self-contained and rather elementary introduction to the subject of stability as well as some applications to control problems.

A well-written treatment of oscillations of all sorts is contained in J. J. Stoker's *Nonlinear Vibrations*, Interscience, New York, N.Y., 1950. Russian contributions to the field are summarized succinctly and critically in a chapter by one of the co-authors, LaSalle and Solomon Lefschetz, that forms Part 6 of *Recent Soviet Contributions to Mathematics*, Macmillan, New York, N.Y., 1962, which LaSalle and Lefschetz also edited.

A unified treatment of oscillations is given in *Oscillations in Nonlinear Systems* by Hale, McGraw-Hill, New York, N.Y., 1963.

The only book presently available in English on modern control theory is Pontryagin, Boltjanskii, Gamkrelidze, and Mishchenko, *The Mathematical Theory of Optimal Processes*, Interscience, New York, N.Y., 1962. More elementary books on this subject will probably appear in the near future.

More advanced books on the theory of differential equations are S. Lefschetz, *Differential Equations: Geometric Theory*, 2nd Edition, Interscience, New York, N.Y., 1963, and E. A. Coddington and N. Levinson, *Theory of Ordinary Differential Equations*, McGraw-Hill, New York, N.Y., 1955.

16

GUIDO L. WEISS

Professor Weiss was born on December 29, 1928, in Trieste, Italy. He received his Ph.B. degree in 1949, his M.S. in 1951, and his Ph.D. in 1956, all from the University of Chicago. During 1953 and part of 1954, he was employed as a mathematician in the Advisory Board on Simulation (a project at the University of Chicago). In 1955, Professor Weiss was appointed as an instructor at De Paul University; he was promoted to Assistant Professor in 1956 and to Associate Professor in 1959. From May to September 1960 he was a Visiting Professor at the University of Buenos Aires. During 1960–61 he spent a year as National Science Foundation Postdoctoral Fellow at the Institute Poincaré at the Sorbonne. In 1961, he was appointed Associate Professor at Washington University and was promoted to Professor in 1963. He spent the Summer Quarter of 1963 as Visiting Professor at the University of Chicago. During the 1964–65 academic year, he was associated as Senior Postdoctoral National Science Foundation Fellow with the University of Geneva.

Professor Weiss' many substantial contributions in the field of analysis, and specifically to the theory of harmonic functions and harmonic analysis, are contained in his numerous papers which have appeared in both domestic and foreign publications.

In accepting the 1967 Chauvenet Prize, Professor Weiss expressed his strong feeling that the writing of expository papers should be encouraged and that, for this reason, he was very happy to write the chapter on "Harmonic Analysis" in *MAA Studies in Mathematics*. He found it most encouraging that, while never expecting this, he had received such a high honor for writing this article.

HARMONIC ANALYSIS

GUIDO L. WEISS, Washington University, St. Louis

1. INTRODUCTORY REMARKS CONCERNING FOURIER SERIES

Classical harmonic analysis deals mainly with the study of Fourier series and integrals. It occupies a central position in that branch of mathematics known as analysis; in fact, it has been described [10, p. xi] as "the meeting ground" of the theory of functions of a real variable and that of analytic functions of a complex variable. Consequently it arises, in a natural way, in several different contexts. Moreover, many basic notions and results in mathematics have been developed by mathematicians working in harmonic analysis. The modern concept of function was first introduced by Dirichlet while studying the convergence of Fourier series; more recently, the theory of distributions (generalized functions) was developed in close connection with the study of Fourier transforms. The Riemann and, later, the Lebesgue integrals were originally introduced in works dealing with harmonic analysis. Infinite cardinal and ordinal numbers, probably the most original and striking notions of modern mathematics, were developed by Cantor in his attempts to solve a delicate real-variable problem involving trigonometric series.

It is our purpose to present some of the main aspects of classical and, to a lesser extent, modern harmonic analysis. The development of the former uses principally the theories of functions of a real variable and of a complex variable while the latter draws heavily from the ideas of abstract functional analysis. Consequently, we shall assume the reader to be acquainted with the material usually presented in a first course in Lebesgue integration or measure theory and to have an elementary knowledge of analytic function theory; moreover, we will require a minimal knowledge of functional analysis.*

Suppose that we are given a real- or complex-valued function f, defined on the real line, periodic of period 1 (that is, $f(x)=f(x+1)$ for all x) and (Lebesgue) integrable when restricted to the interval $(0,1)$.† Its *Fourier transform* is then the function $\hat{f}$, defined on the integers, whose value at k (the kth *Fourier coefficient*) is

$$\hat{f}(k) = \int_0^1 f(t)e^{-2\pi i k t}\, dt, \quad k = 0, \pm 1, \pm 2, \pm 3, \ldots .$$

The *Fourier series* of f is the series

$$\sum_{k=-\infty}^{\infty} \hat{f}(k)e^{2\pi i k x} \tag{1.1}$$

---

* The reader should be acquainted with the elementary properties of Banach spaces and Hilbert spaces, as well as those of linear operators acting on these spaces. He is well advised to glance over the article on functional analysis in Volume 1 of the *MAA Studies in Mathematics*.

† It follows that f must be integrable over any finite subinterval of $(-\infty, \infty)$. In the sequel we shall use the term "periodic" to mean periodic and of period 1.

considered as the sequence of (*symmetric*) *partial sums*

$$s_n(x) = \sum_{k=-n}^{n} \hat{f}(k) e^{2\pi i k x}. \tag{1.2}$$

The reader is undoubtedly familiar with these notions; however, he has probably been introduced to them in a slightly different manner. For example, in classical treatments of Fourier series the term "Fourier transform" is not used. Generally, the sequence of Fourier coefficients

$$c_k = \int_0^1 f(t) e^{-2\pi i k t} dt, \quad k = 0, \pm 1, \pm 2, \pm 3, \ldots,$$

is introduced without emphasizing that one has really defined a function on the integers (the Fourier transform) and the Fourier series of f is usually denoted, simply, by the series $\sum_{k=-\infty}^{\infty} c_k e^{2\pi i k x}$. The above notation and emphasis (as will shortly become apparent) are useful in order to give a unified approach to the theory of Fourier series, Fourier integrals, and their many analogs and extensions. Furthermore, when f is real-valued, the Fourier series of f is often introduced as the series

$$\frac{a_0}{2} + \sum_{k=1}^{\infty} (a_k \cos 2\pi k x + b_k \sin 2\pi k x) \tag{1.1'}$$

considered as the sequence of partial sums

$$s_n(x) = \frac{a_0}{2} + \sum_{k=1}^{n} (a_k \cos 2\pi k x + b_k \sin 2\pi k x), \tag{1.2'}$$

where

$$a_k = 2 \int_0^1 f(t) \cos 2\pi k t \, dt \quad \text{and} \quad b_k = 2 \int_0^1 f(t) \sin 2\pi k t \, dt,$$

$k = 0, 1, 2, 3, \ldots$. It is easy to verify that the two expressions (1.2) and (1.2') are equal.

In most advanced calculus courses it is shown that the Fourier series (1.1) (or (1.1')) converges to $f(x)$ provided f is sufficiently well-behaved at the point x. For example, as we shall show in Sec. 3, this occurs whenever f is differentiable at the point x. This is a solution of a very special case of the central problem in the classical study of Fourier series: to determine whether, and in what sense, the series (1.1) represents the function f. Perhaps one of the best ways of penetrating into the subject of harmonic analysis is by studying this problem.

If we pose this problem in the most obvious way by asking if the series (1.1) converges to $f(x)$ for all x or for almost all x (a much more reasonable question, since altering f on a set of measure zero does not alter the Fourier series), we immediately encounter some serious difficulties. In fact, Kolmogoroff [10, p. 310] has shown that there exists a periodic function, integrable on $(0, 1)$, whose Fourier series diverges everywhere. In general, one must impose fairly strong conditions on

f in order to obtain the convergence of its Fourier series. Perhaps the most important unsolved problem in the classical theory is the following: does there exist a continuous periodic function whose Fourier series diverges on a set of positive measure?

On the other hand, if we consider only functions in $L^2(0,1)$, that is, periodic functions f such that

$$\|f\|_2 = \left(\int_0^1 |f(x)|^2 \, dx \right)^{1/2} < \infty,$$

we obtain a complete and elegant solution to the central problem announced above restricted to this space and its norm. More precisely, we shall show that in this case the partial sums (1.2) converge to f in the L^2-norm; that is,

$$\lim_{n \to \infty} \|f - s_n\|_2 = \lim_{n \to \infty} \left(\int_0^1 |s_n(x) - f(x)|^2 \, dx \right)^{1/2} = 0.$$

This is an immediate consequence of the fact that, with respect to the inner product $(f,g) = \int_0^1 f(x)\overline{g(x)}\,dx$, $L^2(0,1)$ is "essentially" a Hilbert space* and the exponential functions $e_k, k = 0, \pm 1, \pm 2, \pm 3, \ldots$, where $e_k(x) = e^{2\pi i k x}$, form an *orthonormal basis*; that is,

$$(e_k, e_j) = \int_0^1 e^{2\pi i k x} e^{-2\pi i j x} \, dx = \delta_{kj}, \tag{1.3}$$

where δ_{kj} is the "Kronecker δ" ($\delta_{kj} = 0$ when $k \neq j$ and $\delta_{kk} = 1$), and for each $f \in L^2(0,1)$

$$\lim_{n \to \infty} \left\| \sum_{k=-n}^{n} c_k e_k - f \right\|_2^2$$

$$= \lim_{n \to \infty} \left(\sum_{k=-n}^{n} c_k e_k - f, \sum_{k=-n}^{n} c_k e_k - f \right) = 0, \tag{1.4}$$

when $c_k = (f, e_k)$ is the kth Fourier coefficient of $f, k = 0, \pm 1, \pm 2, \ldots$. While the orthogonality relations (1.3) are obvious, the convergence (1.4) will require some proof. A very simple argument, however, shows that the partial sums $s_n = \sum_{k=-n}^{n} c_k e_k$ $= \sum_{k=-n}^{n} \hat{f}(k) e_k$ converge in the L^2-norm to some function in L^2. By the Riesz-Fisher theorem, which asserts that all the L^p spaces, $1 \leq p \leq \infty$, are complete with respect to the L^p-norm, this result will hold provided the sequence $\{s_n\}$ is Cauchy in L^2: $\lim_{n,m \to \infty} \|s_n - s_m\|_2 = 0$. But if, say $m \leq n$, the orthogonality relations (1.3) imply that

$$\|s_n - s_m\|_2^2 = (s_n, s_n) - 2(s_m, s_n) + (s_m, s_m) = \sum_{n \geq |k| > m} |c_k|^2 = \sum_{n \geq |k| > m} |\hat{f}(k)|^2.$$

---

* More precisely, it is the collection of equivalence classes we obtain by identifying functions in $L^2(0,1)$ that are equal almost everywhere that forms a Hilbert space.

That this last sum tends to zero as m and n tend to ∞ follows from *Bessel's inequality* for functions f in $L^2(0,1)$,

$$\sum_{k=-\infty}^{\infty} |\hat{f}(k)|^2 \leqslant \|f\|_2^2 = \int_0^1 |f(x)|^2 dx, \tag{1.5}$$

which is an easy consequence of the orthogonality relations (1.3) and the definition of the Fourier coefficients $c_k = \hat{f}(k)$: since, for any g in $L^2(0,1), (g,g) \geqslant 0$, we have

$$0 \leqslant \left(\sum_{k=-n}^{n} c_k e_k - f, \sum_{k=-n}^{n} c_k e_k - f \right)$$

$$= \left(\sum_{k=-n}^{n} c_k e_k, \sum_{k=-n}^{n} c_k e_k \right)$$

$$- \left(\sum_{k=-n}^{n} c_k e_k, f \right) - \left(f, \sum_{k=-n}^{n} c_k e_k \right) + (f,f)$$

$$= \sum_{k=-n}^{n} |c_k|^2 - \sum_{k=-n}^{n} |c_k|^2 - \sum_{k=-n}^{n} |c_k|^2 + \|f\|_2^2.$$

That is,

$$\sum_{k=-n}^{n} |c_k|^2 \leqslant \|f\|_2^2,$$

and (1.5) follows by letting $n \to \infty$. These arguments give a flavor of the elegance and simplicity of the L^2-theory.

Another satisfactory solution of the problem of representation of functions by their Fourier series is to consider, instead of convergence, some methods of summability of Fourier series at individual points. The two best known types of summability (and the only ones we shall consider) are *Cesàro* and *Abel summability*. The former (often also referred to as the *method of summability by the first arithmetic means* or, simply, as $(C, 1)$ *summability*) is defined in the following way: suppose we are given a numerical series $u_0 + u_1 + u_2 + \cdots$ with partial sums $s_0, s_1, s_2, \ldots$. We then form the $(C, 1)$ *means* (or *first arithmetic means*)

$$\sigma_n = \frac{s_0 + s_1 + \cdots + s_n}{n+1} = \sum_{\nu=0}^{n} \left(1 - \frac{\nu}{n+1} \right) u_\nu$$

and say that the series is $(C, 1)$ summable to l if $\lim\limits_{n \to \infty} \sigma_n = l$. The *Abel means* of the series are defined for each r, $0 \leqslant r < 1$, by setting

$$A(r) = u_0 + u_1 r + u_2 r^2 + \cdots = \sum_{k=0}^{\infty} u_k r^k$$

and we say that the series is Abel summable to l if $\lim\limits_{r \to 1^-} A(r) = l$. It is not hard to show that if $u_0 + u_1 + u_2 + \cdots$ is convergent to the sum l then it must be both $(C, 1)$

and Abel summable to l. On the other hand, there are many series that are summable but not convergent. An illustrative example is the series $1-1+1-1 + \cdots = \sum_{k=0}^{\infty} (-1)^k$, whose $(C,1)$ and Abel means are easily seen to converge to $\frac{1}{2}$. It can also be shown that $(C,1)$ summability implies Abel summability. Thus, many results involving Abel summability follow from corresponding theorems that deal with Cesàro summability. Nevertheless, an independent study of the former is of interest, particularly when we consider Fourier series of functions f. This is true, not only because such series may be Abel summable under weaker conditions on f than are necessary to guarantee their $(C,1)$ summability, but also because Abel summability has special properties, related to the theory of harmonic and analytic functions, that are not enjoyed by Cesàro summability.

We now describe, briefly, how these concepts apply to the study of Fourier series. The two most important results in connection with the problem of representing functions by their Fourier series are the following:

(1.6) *If f is periodic and integrable on $(0,1)$ then the $(C,1)$ means and the Abel means of the Fourier series of f converge to*

$$\tfrac{1}{2}\{f(x_0+0)+f(x_0-0)\}$$

at every point x_0 where the limits $f(x_0\pm 0)$ exist. In particular, they converge at every point of continuity of f.

(1.7) *If f is periodic and integrable on $(0,1)$ then the $(C,1)$ means and the Abel means of the Fourier series of f converge to $f(x)$ for almost every x in $(0,1)$.**

We can obtain more insight into these results by examining more closely the first arithmetic means and the Abel means of the Fourier series of a function f. We first obtain an expression for the partial sums (1.2):

$$s_n(x) = \sum_{k=-n}^{n} \hat{f}(k)e^{2\pi i k x}$$

$$= \sum_{k=-n}^{n} \left(\int_0^1 f(t)e^{-2\pi i k t}\, dt \right) e^{2\pi i k x}$$

$$= \int_0^1 \left(\sum_{k=-n}^{n} e^{2\pi i k(x-t)} \right) f(t)\, dt.$$

By multiplying $D_n(\theta) = \sum_{k=-n}^{n} e^{2\pi i k \theta}$ by $2\sin \pi\theta = i(e^{-i\pi\theta} - e^{i\pi\theta})$ all but the first and last term of the resulting sum cancel and we obtain

$$2D_n(\theta)\sin \pi\theta = i(e^{-(2n+1)\pi i\theta} - e^{(2n+1)\pi i\theta}) = 2\sin(2n+1)\pi\theta;$$

---

* When restricted to Cesàro summability (1.6) is known as the theorem of Fejér and (1.7) as the theorem of Lebesgue. That these results then hold for Abel summability follows from the fact, mentioned above, that $(C,1)$ summability implies Abel summability.

that is,

$$D_n(\theta) = \frac{\sin(2n+1)\pi\theta}{\sin\pi\theta}. \tag{1.8}$$

Hence,

$$s_n(x) = \int_0^1 f(t) D_n(x-t)\, dt$$

$$= \int_0^1 f(t) \frac{\sin(2n+1)\pi(x-t)}{\sin\pi(x-t)}\, dt. \tag{1.9}$$

The expression (1.8) is called the *Dirichlet kernel.* We can now express the $(C,1)$ means in terms of it:

$$\sigma_n(x) = \frac{s_0(x) + s_1(x) + \cdots + s_n(x)}{n+1}$$

$$= \frac{1}{n+1} \int_0^1 f(t) \left(\sum_{k=0}^n D_k(x-t) \right) dt.$$

By multiplying the numerator and denominator of

$$K_n(\theta) = \frac{1}{n+1} \sum_{k=0}^n D_k(\theta) = \frac{1}{n+1} \sum_{k=0}^n \frac{\sin(2k+1)\pi\theta}{\sin\pi\theta}$$

by $\sin\pi\theta$ and replacing the products of sines in the numerator by differences of cosines we obtain

$$K_n(\theta) = \frac{1}{n+1} \sum_{k=0}^n \frac{\cos 2k\pi\theta - \cos 2(k+1)\pi\theta}{2\sin^2\pi\theta}$$

$$= \frac{1}{n+1} \frac{1 - \cos 2(n+1)\pi\theta}{2\sin^2\pi\theta}$$

$$= \frac{1}{n+1} \left[\frac{\sin(n+1)\pi\theta}{\sin\pi\theta} \right]^2. \tag{1.10}$$

Consequently,

$$\sigma_n(x) = \int_0^1 f(t) K_n(x-t)\, dt$$

$$= \frac{1}{n+1} \int_0^1 f(t) \left\{ \frac{\sin(n+1)\pi(x-t)}{\sin\pi(x-t)} \right\}^2 dt. \tag{1.11}$$

$K_n(\theta)$ is called the *Fejér kernel.*

The result (1.6) follows easily from three basic properties of this kernel:

(A) $\int_0^1 K_n(\theta)\, d\theta = 1$;

(B) $K_n(\theta) \geqslant 0$;

(C) *for each $\delta > 0$,* $\displaystyle\max_{\delta \leqslant \theta \leqslant 1-\delta} K_n(\theta) \to 0$ *as $n \to \infty$.*

Property (B) is obvious. Property (A) is a consequence of the corresponding property for the Dirichlet kernel (which is immediate:

$$\int_0^1 D_n(\theta)\, d\theta = \sum_{k=-n}^n \int_0^1 e^{2\pi i k\theta}\, d\theta = 1)$$

and the representation $K_n(\theta) = \sum_{k=0}^n D_k(\theta)/(n+1)$. Finally, (C) follows from the inequality (see (1.10))

$$\max_{\delta \le \theta \le 1-\delta} K_n(\theta) \le \frac{1}{n+1}\sin^2 \pi\delta.$$

Now, to obtain (1.6) we argue as follows: suppose x_0 is a point at which the limits $f(x_0 \pm 0)$ exist and let $a = \frac{1}{2}\{f(x_0+0) + f(x_0-0)\}$. Then, using the periodicity of the functions involved, the change of variables $t = x - s$, and property (A),

$$\sigma_n(x_0) - a = \int_{-1/2}^{1/2} f(s) K_n(x_0-s)\, ds - a \cdot 1$$

$$= \int_{-1/2}^{1/2} f(x_0-t) K_n(t)\, dt - a\int_{-1/2}^{1/2} K_n(t)\, dt$$

$$= 2\int_0^\delta \left\{ \frac{f(x_0-t)+f(x_0+t)}{2} - a \right\} K_n(t)\, dt$$

$$+ \int_{\delta \le |t| \le 1/2} \{f(x_0-t) - a\} K_n(t)\, dt.$$

Hence, if $\delta > 0$ is so chosen that $|f(x_0-t) + f(x_0+t) - 2a| \le \epsilon$ if $|t| \le \delta$, we have, by (B) and (A),

$$|\sigma_n(x_0) - a| \le \epsilon \int_0^\delta K_n(t)\, dt$$

$$+ \left\{ \max_{\delta \le |t| \le 1/2} K_n(t) \right\} \int_{\delta \le |t| \le 1/2} |f(x_0-t) - a|\, dt$$

$$\le \epsilon \int_{-1/2}^{1/2} K_n(t)\, dt$$

$$+ \left\{ \max_{\delta \le |t| \le 1/2} K_n(t) \right\} \int_{-1/2}^{1/2} |f(x_0-t) - a|\, dt$$

$$= \epsilon \cdot 1 + \left\{ \max_{\delta \le |t| \le 1/2} K_n(t) \right\} \int_{-1/2}^{1/2} |f(x_0-t) - a|\, dt;$$

but, by (C), the last term tends to 0 as $n \to \infty$. Since $\epsilon > 0$ is arbitrary we can conclude that $\lim_{n \to \infty} |\sigma_n(x_0) - a| = 0$ and (1.6) is proved.

The theorem of Lebesgue, result (1.7) for the $(C,1)$ means, is somewhat deeper and we postpone its proof until later. Since $(C,1)$ summability implies Abel summability, as remarked above, both the results (1.6) and (1.7) follow once we establish them for Cesàro means. We commented before, however, that an independent study of Abel summability is of interest since it has special properties not enjoyed by $(C, 1)$ summability. This is easily made clear by examining the Abel means of Fourier series more closely; we do this by showing how the study of Fourier series is intimately connected with analytic function theory.

That this connection should exist is not surprising once we make the observation, *when f is real-valued*, that the series (1.1) is the real part of the power series

$$\hat{f}(0) + \sum_{k=1}^{\infty} 2\hat{f}(k)z^k \qquad (1.12)$$

restricted to the unit circle $z = e^{2\pi ix}$. We note that this series defines an analytic function in the interior of the unit circle since the coefficients $\hat{f}(k)$ are uniformly bounded (in fact,

$$|\hat{f}(k)| \leqslant \int_0^1 |f(t)| \, dt = \|f\|_1).$$

Thus, the real part of (1.12) is a harmonic function when $r = |z| < 1$. But this real part is nothing more than the Abel mean of the Fourier series (1.1):

$$A(r,x) = A_f(r,x) = \hat{f}(0) + \sum_{k=1}^{\infty} r^k \hat{f}(k)(e^{2\pi ikx} + e^{-2\pi ikx})$$

$$= \sum_{k=-\infty}^{\infty} r^{|k|}\hat{f}(k)e^{2\pi ikx}.$$

The imaginary part of (1.10), when $z = e^{2\pi ix}$, is (formally),

$$-i \sum_{k=-\infty}^{\infty} (\operatorname{sgn} k)\hat{f}(k)e^{2\pi ikx}, \qquad (1.13)$$

where, for any nonzero complex number z, $\operatorname{sgn} z = z/|z|$ and $\operatorname{sgn} 0 = 0$. This series is called the series *conjugate* to the Fourier series (1.1). Though it is not, in general, a Fourier series, this conjugate series is closely connected (see Sec. 4) to a (not necessarily integrable) function, the *conjugate function*, $\tilde{f}$.

As in the case of the $(C, 1)$ means, the Abel means $A(r,x)$ have an integral representation; that is, a representation similar to (1.11). We have, for $0 \leqslant r < 1$,

$$A(r,x) = \sum_{k=-\infty}^{\infty} r^{|k|}\hat{f}(k)e^{2\pi ikx}$$

$$= \sum_{k=-\infty}^{\infty} r^{|k|}\left(\int_0^1 f(t)e^{-2\pi ikt} \, dt\right)e^{2\pi ikx}$$

$$= \int_0^1 \left(\sum_{k=-\infty}^{\infty} r^{|k|}e^{2\pi ik(x-t)}\right)f(t) \, dt,$$

the change in the order of integration and summation being justifiable by the uniform convergence of the series

$$P(r,\theta) = \sum_{k=-\infty}^{\infty} r^{|k|}e^{2\pi i\theta}$$

for $0 \leqslant r < 1$. But, setting $z = re^{2\pi i\theta}$, $P(r,\theta)$ is simply the real part of

$$1 + \sum_{k=1}^{\infty} 2r^k e^{2\pi i k\theta} = 1 + 2\sum_{k=1}^{\infty} z^k = \frac{1+z}{1-z}.$$

Consequently,

$$P(r,\theta) = \frac{1-r^2}{1-2r\cos 2\pi\theta + r^2} \tag{1.14}$$

and we obtain the desired integral representation for the Abel means

$$A(r,x) = \int_0^1 P(r,x-t)f(t)\,dt = \int_0^1 \frac{1-r^2}{1-2r\cos 2\pi(x-t)+r^2} f(t)\,dt. \tag{1.15}$$

$P(r,\theta)$ is called the *Poisson kernel* and the integral (1.15) is called the *Poisson integral* of f. The reader can easily verify that this kernel satisfies the three properties, completely analogous to those of the Fejér kernel:

(A′) $\displaystyle\int_0^1 P(r,\theta)\,d\theta = 1$;

(B′) $P(r,\theta) \geqslant 0$;

(C′) *for each* $\delta > 0$, $\displaystyle\max_{\delta \leqslant \theta \leqslant 1-\delta} P(r,\theta) \to 0$ *as* $r \to 1$.

From this we see that to the proof of (1.6) given above in the case of the Cesàro means there corresponds a practically identical proof of this result for the Abel means.

Let us observe that the imaginary part of $\dfrac{1+z}{1-z}$ has the form

$$Q(r,\theta) = \frac{2r\sin 2\pi\theta}{1-2r\cos 2\pi\theta + r^2}$$

and one readily obtains the Abel mean of the conjugate Fourier series (1.13) by the integral

$$\tilde{A}(r,x) = \int_0^1 Q(r,x-t)f(t)\,dt$$

$$= \int_0^1 \frac{2r\sin 2\pi(x-t)}{1-2r\cos 2\pi(x-t)+r^2} f(t)\,dt. \tag{1.16}$$

This integral is called the *conjugate Poisson integral of* f and $Q(r,\theta)$ is known as the *conjugate Poisson kernel*.

In this discussion we assumed that f was real-valued. It is clear, however, that the Poisson integral formula (1.15) for the Abel means of the Fourier series of f holds in case f is complex-valued as well. To see this one need only apply it to the real and imaginary parts of f.

Before passing to other aspects of harmonic analysis let us examine more closely the integrals (1.9), (1.11), (1.15), and (1.16) that gave us the partial sums, the

$(C, 1)$ means, the Abel means of the Fourier series of an integrable periodic function f, and the Abel means of the conjugate Fourier series of f. All these integrals have the form

$$(g*f)(x) = \int_0^1 g(x-t) f(t) \, dt, \tag{1.17}$$

where g is a periodic integrable function. In fact, in all these cases g is much better than merely integrable; for example, it is a bounded function and, consequently, it is obvious that, for each x, the integrand in (1.17) is integrable and $(g*f)(x)$ is well defined. We shall see, however, that the latter is well defined for almost all x when g is integrable. We therefore obtain a function, $g*f$ (defined almost everywhere), by forming the integral (1.17) whenever g and f belong to $L^1(0, 1)$ and are periodic. This operation, that assigns to each such pair (g, f) the function $g*f$, is called *convolution* and plays an important role in the theory of Fourier series. The most important elementary properties of convolution are the following:

(i) *If f and g are periodic and in $L^1(0, 1)$ so is $f*g$ and*

$$\|f*g\|_1 = \int_0^1 |(f*g)(x)| \, dx$$

$$\leqslant \left(\int_0^1 |f(t)| \, dt \right) \left(\int_0^1 |g(t)| \, dt \right)$$

$$= \|f\|_1 \|g\|_1;$$

(ii) $f*g = g*f$;
(iii) $(f*g)*h = f*(g*h)$ *whenever f, g, and h are periodic and in $L^1(0, 1)$*;
(iv) *For f, g, and h as in* (iii) *and any two complex numbers a and b*

$$f*(ag + bh) = a(f*g) + b(f*h).$$

That $(f*g)(x)$ is well defined for almost all x, as well as property (i), is an easy consequence of Fubini's theorem: since

$$|(f*g)(x)| \leqslant \int_0^1 |f(x-t)| \, |g(t)| \, dt$$

we have, using the periodicity of f,

$$\int_0^1 |(f*g)(x)| \, dx \leqslant \int_0^1 \left(\int_0^1 |f(x-t)| \, |g(t)| \, dt \right) dx$$

$$= \int_0^1 |g(t)| \left(\int_0^1 |f(x-t)| \, dx \right) dt$$

$$= \int_0^1 |g(t)| \, \|f\|_1 \, dt = \|f\|_1 \|g\|_1.^*$$

---

* We are using the following version of Fubini's theorem: If $h \geqslant 0$ is a measurable function in the square $\{0 \leqslant x \leqslant 1, 0 \leqslant t \leqslant 1\}$ and the iterated integral $\int_0^1 (\int_0^1 h(x, t) \, dt) \, dx$ is finite, then h is integrable, $\int_0^1 h(x, t) \, dt$ is finite for almost every x and $\int_0^1 (\int_0^1 h(x, t) \, dt) \, dx = \int_0^1 (\int_0^1 h(x, t) \, dx) \, dt$. We have tacitly assumed, when $h(x, t) = |f(x-t) g(t)|$, that h is measurable. We leave the proof of this fact to the reader.

The remaining three properties follow from simple transformations of integrals and we omit their proofs.

The relation between Fourier transformation and convolution is very simple and elegant:

Suppose f and g are periodic and in $L^1(0, 1)$, then for all integers k

$$(f*g)^\wedge(k) = \hat{f}(k)\hat{g}(k).^*$$ (1.18)

In order to see this we use Fubini's theorem and the periodicity of the functions involved:

$$(f*g)^\wedge(k) = \int_0^1 \left(\int_0^1 f(x-t)g(t)\,dt \right) e^{-2\pi ikx}\,dx$$

$$= \int_0^1 g(t)e^{-2\pi ikt} \left(\int_0^1 f(x-t)e^{-2\pi ik(x-t)}\,dx \right) dt$$

$$= \int_0^1 g(t)e^{-2\pi ikt}\hat{f}(k)\,dt$$

$$= \hat{f}(k)\hat{g}(k).$$

It is this result, that the Fourier transform of the convolution of two functions is, simply, the product of their Fourier transforms, that makes convolution play such an important role in the study of Fourier series. This, as we shall see, becomes clear very early in the development of harmonic analysis.

2. HARMONIC ANALYSIS ON THE INTEGERS AND ON THE REAL LINE

Up to this point we have considered only functions that were periodic of period 1. It is often useful to think of such functions as defined on the *additive group of real numbers modulo 1*† or on the perimeter of the unit circle $\{z \text{ complex}; z = e^{2\pi i\theta}\}$ of the complex plane. Consequently, the theory of Fourier series is often referred to as the harmonic analysis associated with this circle, or the reals modulo 1. Toward the end of this monograph we shall describe how harmonic analysis can be associated to a wide variety of domains. In this section we shall consider two of these, the integers and the entire real line. The harmonic analysis corresponding to these domains is intimately connected with the theory of Fourier series.

In the case of the real line we obtain the theory of Fourier integrals, a topic that is as important and as well known as the theory of Fourier series. The harmonic analysis associated with functions defined on the integers, however, is not generally studied per se. The main reason is that its elementary aspects (but by no means its

---

* In general, we shall let $(\ )^\wedge$ denote the Fourier transform of the expression in the parentheses.

† If we say that two real numbers are equivalent when their difference is an integer we obtain a partition of the reals into equivalence classes. Let $[x], [y], [z], \dots$, denote the equivalence classes containing the real numbers $x, y, z, \dots$. Then the additive group of real numbers modulo 1 consists of these equivalence classes together with the operation defined by $[x]+[y]=[x+y]$.

deeper ones) consist of results that are essentially on the surface. But precisely this property, this elementary nature of the subject, makes its study very worthwhile for the non specialist as it provides a great deal of motivation for the theories of Fourier series and integrals. Furthermore, some remarks about this topic are necessary in order to understand better the general picture of our subject. Accordingly, we shall not consider this part of harmonic analysis in any detail, but will treat it briefly and use it to motivate the introduction of the inverse Fourier transform, which will enable us to consider the problem of representation of functions by their Fourier series from a more general point of view. Also, we shall use it to motivate our introductory remarks concerning Fourier integrals. We strongly urge the reader, however, to find the analogs, for functions defined on the integers, of results we shall present in the theories of Fourier series and integrals.

In the last section we started out with a periodic function belonging to $L^1(0,1)$ and obtained, by means of a certain integral, a function defined on the integers, the Fourier transform. We then asked if it were possible to obtain the original function from the latter by means of a certain series, the Fourier series (1.1). This indicates a duality between the interval $(0,1)$ and the integers and it is not unreasonable to expect that, by considering originally a function defined on the integers, we can introduce, in analogy to the Fourier transform, a periodic function by means of an appropriate series. Furthermore, we should be able to recapture the original function from this periodic function and a suitable integral. It is only natural, in view of these remarks, to hope that this can be done by interchanging the roles played by the interval $(0,1)$ and the integers. More explicitly, let us examine the result of systematically replacing, in the definitions made at the beginning of the last section,

$$\sum_{k=-\infty}^{\infty} \quad \text{by} \quad \int_0^1, \qquad \int_0^1 \quad \text{by} \quad \sum_{k=-\infty}^{\infty},$$

the continuous variable $x \in (0,1)$ by the integral variable k and k by x.

Let us consider, then, the integers as a measure space in which each point has measure 1 and an integrable function, f, defined on this measure space; that is, f satisfies

$$\sum_{k=-\infty}^{\infty} |f(k)| < \infty. \tag{2.1}$$

The space of such integrable functions is usually denoted by l^1. For f in l^1 we introduce the periodic function $\hat{f}$ whose value at x is

$$\hat{f}(x) = \sum_{j=-\infty}^{\infty} f(j)e^{-2\pi ijx}. \tag{2.2}$$

We shall call $\hat{f}$ the *Fourier transform* of f in this case as well. Because of the convergence (2.1) the series (2.2) converges uniformly; consequently, $\hat{f}$ is a continuous function. Corresponding to the Fourier series (1.1) we have the integral

$$\int_0^1 \hat{f}(x)e^{2\pi ikx}\,dx. \tag{2.3}$$

But the uniform convergence of (2.2), allowing us to integrate term-by-term, and the orthogonality relations (1.3) immediately imply that

$$\int_0^1 \hat{f}(x)e^{2\pi ikx}\,dx = f(k). \tag{2.4}$$

We therefore see that in the present case we do not encounter any of the difficulties described in the first section when we try to express the original function in terms of its Fourier transform. This illustrates the simplicity of the elementary aspects of the harmonic analysis associated with the integers.

In particular, we see that the mapping that assigns to $f \in l^1$ its Fourier transform is one-to-one and, thus, it has an inverse. This inverse, in view of (2.4), has an obvious extension to all of $L^1(0,1)$; namely, the operator, called the *inverse Fourier transform*, that takes a function g in $L^1(0,1)$ into the function $\check{g}$ whose value at $k = 0, \pm 1, \pm 2, \ldots$, is

$$\check{g}(k) = \int_0^1 g(x)e^{2\pi ikx}\,dx. \tag{2.5}$$

We can, therefore, rewrite (2.4) in the following way:

$$(\hat{f})^{\check{}} = f, \tag{2.6}$$

whenever f belongs to l^1.

These considerations lead us to a useful and more general restatement of the problem we studied in the last section concerning the representation of functions by their Fourier series. Suppose we again interchange the roles played by the interval $(0,1)$ and the integers; it is then natural to try to define the inverse Fourier transform of a function, g, whose domain is the integers, by the expression

$$\check{g}(x) = \sum_{k=-\infty}^{\infty} g(k)e^{2\pi ikx}. \tag{2.7}$$

When g is in l^1 the series (2.7) is convergent and we obtain a well defined mapping, $g \to \check{g}$, from l^1 into the class of continuous periodic functions. This mapping, however, is insufficient for our purposes. For example, in view of (2.6), we should expect that whenever g is the Fourier transform of an integrable function f it then follows that $\check{g} = (\hat{f})^{\check{}} = f$. But, because of Kolmogoroff's example of an integrable function with an everywhere divergent Fourier series, this equality cannot be valid if we let $\check{g}$ be defined as the function which, at each x (or at almost every x), satisfies (2.7) in the usual sense (that is, the sequence of partial sums $s_n(x)$ $= \sum_{k=-n}^{n} g(k)e^{2\pi ikx}$ converges). On the other hand, using (1.7), we do obtain an almost everywhere defined $\check{g}(x) = (\hat{f})^{\check{}}(x) = f(x)$ if we interpret (2.7) to mean that the $(C,1)$ or Abel means of the series on the right converge to $\check{g}(x)$. Similarly, if g is the Fourier transform of an f belonging to $L^2(0,1)$, (2.7) gives us a well-defined function $\check{g} = f$ if we interpret the series on the right to be convergent in the L^2-norm. In each of these cases we obtain a mapping which is an inverse to the Fourier transform mapping when the latter is restricted to some important domain of functions (such as $L^2(0,1)$ or $L^1(0,1)$). Thus, a general formulation of the problem we discussed in the last section is the following: given a class C of periodic functions for which the Fourier transform is defined, does there exist a mapping, $g \rightarrow \check{g}$, defined on a class of functions, whose common domain is the integers, such that $(\hat{f})^{\vee}$ is defined for all f in C and $(\hat{f})^{\vee} = f$? We shall refer to the problem stated in this form as the *Fourier inversion problem*.

Let us now turn to the harmonic analysis related to the real line, the theory of Fourier integrals. Suppose that f is integrable over $(-\infty, \infty)$, then its *Fourier transform* is defined for all real x by

$$\hat{f}(x) = \int_{-\infty}^{\infty} f(t) e^{-2\pi i x t} dt.$$

The integral on the right is usually also called the Fourier integral. Since the Fourier transform of a function defined on the entire real line is again such a function it should not surprise the reader, in view of our discussion in this section, that a good heuristic approach for obtaining the basic notions and results in the theory of Fourier integrals is to let the real line play the roles that the interval (0, 1) and the integers played in the theory of Fourier series. Let us examine the Fourier inversion problem from this point of view.

Motivated by the expressions (2.5), (2.6), and (2.7) we would expect that the inverse Fourier transform of a function, g, defined on the real line should be given by the formula

$$\check{g}(x) = \int_{-\infty}^{\infty} g(t) e^{2\pi i x t} dt, \qquad (2.8)$$

and that for each f in $L^1(-\infty, \infty)$ we would then have $(\hat{f})^{\vee} = f$. Just as in the case of Fourier series, however, we immediately encounter the problem of giving relation (2.8) a suitable interpretation. Though f has several nice properties (the reader can easily check that it is uniformly continuous and bounded; in fact,

$$\|\hat{f}\|_{\infty} = \sup_{-\infty < x < \infty} |\hat{f}(x)| \leqslant \int_{-\infty}^{\infty} |f(t)| dt = \|f\|_1 \qquad (2.9)$$

whenever f is in $L^1(-\infty, \infty))$ it is not always true that it is integrable. For example, if f is the characteristic function, $X_{(a,b)}$, of the finite interval (a, b) then

$$\hat{X}_{(a,b)}(x) = \int_a^b e^{-2\pi i x t} dt = \frac{e^{-\pi i x a} - e^{-2\pi i x b}}{2\pi i x}, \qquad (2.10)$$

when $x \neq 0$ and $\hat{X}_{(a,b)}(0) = b - a$. Here, as in the last section, we obtain a satisfactory solution of the Fourier inversion problem if we consider $(C, 1)$ and Abel summability. We see this easily if we let ourselves be guided by the above mentioned heuristic principle of substituting the real line for the interval (0, 1) and the integers. Let us first examine briefly the case of Cesàro summability.

If u is integrable in the intervals $[-R, R]$, for all $R > 0$, the *Cesàro means*, or $(C, 1)$ *means*, of $\int_{-\infty}^{\infty} u(t) dt$ are defined by the integrals

$$\sigma_R = \int_{-R}^{R} \left(1 - \frac{|t|}{R}\right) u(t) dt.$$

We say that $\int_{-\infty}^{\infty} u(t)dt$ is $(C, 1)$ summable to l if $\lim_{R\to\infty} \sigma_R = l$. It is easy to see that if $u \in L^1(-\infty, \infty)$ and its integral is l then $\sigma_R \to l$ as $R \to \infty$.

Let us now consider the Cesàro means of the integral in (2.8) when g is the Fourier transform of an integrable function f. We have

$$\sigma_R(x) = \int_{-R}^{R}\left(1 - \frac{|t|}{R}\right)e^{2\pi ixt}\hat{f}(t)dt$$

$$= \int_{-R}^{R}\left(1 - \frac{|t|}{R}\right)e^{2\pi ixt}\left\{\int_{-\infty}^{\infty} f(y)e^{-2\pi ity}dy\right\}dt$$

$$= \int_{-\infty}^{\infty} f(y)\left\{\int_{-R}^{R}\left(1 - \frac{|t|}{R}\right)e^{2\pi it(x-y)}dt\right\}dy.$$

It is not hard to obtain a simple expression for the inner integral. Using the fact that the sine function is odd and integrating by parts we obtain:

$$K_R(\theta) = \int_{-R}^{R}\left(1 - \frac{|t|}{R}\right)e^{2\pi i\theta t}dt = 2\int_{0}^{R}\left(1 - \frac{t}{R}\right)\cos(2\pi\theta t)dt$$

$$= 2\int_{0}^{R}\frac{1}{R}\frac{\sin(2\pi\theta t)}{2\pi\theta}dt = \frac{1}{2\pi^2 R}\cdot\frac{1 - \cos(2\pi R\theta)}{\theta^2}. \qquad (2.11)$$

Consequently,

$$\sigma_R(x) = \int_{-\infty}^{\infty} f(y)K_R(x-y)dy$$

$$= \frac{1}{2\pi^2 R}\int_{-\infty}^{\infty} f(y)\frac{1 - \cos(2\pi R(x-y))}{(x-y)^2}dy. \qquad (2.12)$$

$K_R(\theta)$ is called the *Fejér kernel* and it satisfies the following three basic properties:

(a) $\int_{-\infty}^{\infty} K_R(\theta)d\theta = 1$;

(b) $K_R(\theta) \geqslant 0$;

(c) *for each* $\delta > 0$, $\int_{|\theta|\geqslant\delta} K_R(\theta)d\theta \to 0$ *as* $R \to \infty$.

The second property is obvious. The first property follows easily from the well-known result: $\lim_{N\to\infty}\int_{0}^{N}(\sin t/t)dt = \pi/2$. To see this we change variables and integrate by parts:

$$\int_{-\infty}^{\infty} K_R(\theta)d\theta = 2\int_{-\infty}^{\infty}\frac{1 - \cos 2\pi R\theta}{R(2\pi\theta)^2}d\theta = \frac{1}{\pi}\int_{-\infty}^{\infty}\frac{1 - \cos s}{s^2}ds$$

$$= \frac{2}{\pi}\lim_{N\to\infty}\int_{0}^{N}\frac{1 - \cos s}{s^2}ds = \frac{2}{\pi}\lim_{N\to\infty}\int_{0}^{N}\frac{\sin t}{t}dt = 1.$$

In order to prove (c) let us first observe that $K_R(\theta) \leqslant 1/R\theta^2$ (thus,

$$\max_{|\theta|\geqslant\delta} K_R(\theta) \leqslant \frac{1}{R}\max_{|\theta|\geqslant\delta}\frac{1}{\theta^2} \to 0 \qquad \text{as } R\to\infty). \qquad (2.13)$$

Consequently,

$$\int_{|\theta|>\delta} K_R(\theta)d\theta \leqslant \frac{1}{R}\int_{|\theta|>\delta}\frac{d\theta}{\theta^2}=\frac{2}{R\delta}\to 0 \qquad \text{as } R\to\infty.$$

If we replace (c) by (2.13) we have three properties that are completely analogous to the properties (A), (B), and (C) of the Fejér kernel obtained in the periodic case. Precisely the same argument that is used in establishing the theorem of Fejér (see (1.6)) will then give us the corresponding result for Fourier integrals. We introduce property (c), however, to show how $(C, 1)$ summability can be used in yet another way in order to obtain a solution of the Fourier inversion problem. More precisely, we shall prove the following result:

(2.14) *If f is integrable, then the $(C, 1)$ means*

$$\sigma_R(x)=\int_{-R}^{R}\left(1-\frac{|t|}{R}\right)e^{2\pi ixt}\hat{f}(t)dt=\int_{-\infty}^{\infty}f(t)K_R(x-t)dt$$

of the integral defining the inverse Fourier transform of $\hat{f}$ converge to f in the L^1 norm. That is,

$$\lim_{R\to\infty}\|f-\sigma_R\|_1=\lim_{R\to\infty}\int_{-\infty}^{\infty}|\sigma_R(x)-f(x)|dx=0.$$

It is convenient, at this point, to introduce the L^p *modulus of continuity* of a function f in $L^p(-\infty, \infty)$:

$$\omega_p(\delta)=\max_{0<t\leqslant\delta}\left\{\int_{-\infty}^{\infty}|f(x+t)-f(x)|^pdx\right\}^{1/p}.$$

It is an elementary fact that $\omega_p(\delta)\to 0$ as $\delta\to 0$ when $1\leqslant p<\infty$. For, given $\delta>0$, we can write $f=f_1+f_2$, where f_1 is continuous, vanishes outside a finite interval, and $\|f_2\|_p<\epsilon/3$. Thus, by Minkowski's inequality,

$$\left\{\int_{-\infty}^{\infty}|f(x+t)-f(x)|^pdx\right\}^{1/p}\leqslant\left\{\int_{-\infty}^{\infty}|f_1(x+t)-f_1(x)|^pdx\right\}^{1/p}$$

$$+\left\{\int_{-\infty}^{\infty}|f_2(x+t)|^pdx\right\}^{1/p}+\left\{\int_{-\infty}^{\infty}|f_2(x)|^pdx\right\}^{1/p}.$$

Each of the last two terms is less than $\epsilon/3$; since f_1 is uniformly continuous and vanishes outside a finite interval, the last term is also smaller than $\epsilon/3$ provided t is close enough to 0. Thus, $\omega_p(\delta)<\epsilon/3+\epsilon/3+\epsilon/3=\epsilon$ if δ is close enough to 0.

We now prove (2.14). Using the change of variables $t=x-s$ and property (a),

$$\sigma_R(x)-f(x)=\int_{-\infty}^{\infty}f(x)K_R(x-s)ds-f(x)\cdot 1$$

$$=\int_{-\infty}^{\infty}[f(x-t)-f(x)]K_R(t)dt.$$

Thus,

$$
\begin{aligned}
\int_{-\infty}^{\infty} |\sigma_R(x) - f(x)| \, dx &= \int_{-\infty}^{\infty} |\int_{-\infty}^{\infty} [f(x-t) - f(x)] K_R(t) \, dt| \, dx \\
&\leqslant \int_{-\infty}^{\infty} \left\{ \int_{-\infty}^{\infty} |f(x-t) - f(x)| K_R(t) \, dt \right\} dx \\
&= \int_{-\infty}^{\infty} \left\{ \int_{-\infty}^{\infty} |f(x-t) - f(x)| \, dx \right\} K_R(t) \, dt \\
&= \int_{|t|<\delta} \left\{ \int_{-\infty}^{\infty} |f(x-t) - f(x)| \, dx \right\} K_R(t) \, dt \\
&\quad + \int_{|t|>\delta} \left\{ \int_{-\infty}^{\infty} |f(x-t) - f(x)| \, dx \right\} K_R(t) \, dt.
\end{aligned}
$$

The first of these last two terms is clearly dominated by

$$
\omega_1(\delta) \int_{-\infty}^{\infty} K_R(t) \, dt = \omega_1 \delta,
$$

which tends to 0 as δ tends to 0. The second term is majorized by

$$
\int_{|t|>\delta} \left\{ \int_{-\infty}^{\infty} |f(x-t)| \, dx + \int_{-\infty}^{\infty} |f(x)| \, dx \right\} K_R(t) \, dt = 2\|f\|_1 \int_{|t|>\delta} K_R(t) \, dt.
$$

Thus, given $\epsilon > 0$, let us first choose $\delta > 0$ so that $\omega_1(\delta) < \epsilon/2$; then, with this δ fixed, property (c) can be used to find $R_0 > 0$ so that

$$
\int_{|t|>\delta} K_R(t) \, dt < \frac{\epsilon}{4\|f\|_1}
$$

when $R \geqslant R_0$. This shows that

$$
\int_{-\infty}^{\infty} |\sigma_R(x) - f(x)| \, dx < \frac{\epsilon}{2} + \frac{2\|f\|_1}{4\|f\|_1} \epsilon = \epsilon,
$$

provided $R \geqslant R_0$, and (2.14) is proved.

We recall that in the case of Fourier series the Abel means behaved very much like the $(C, 1)$ means, yet an independent study of them was of interest, particularly when we examined the relation between the theory of Fourier series and the theory of harmonic and analytic functions of a complex variable. This is equally true for Fourier integrals; consequently it is worthwhile, at this point, to devote a few words to Abel summability and its relation to the Fourier inversion problem.

Guided by our heuristic principle of replacing the integers by the real line and sums by integrals, we would expect the Abel means of the integral $\int_{-\infty}^{\infty} u(t) \, dt$ to be defined by the expression $\int_{-\infty}^{\infty} r^{|t|} u(t) \, dt$ with $0 \leqslant r < 1$. For technical reasons, which will become apparent shortly, it is convenient to put $r = e^{-2\pi y}$, $0 < y < \infty$; thus the Abel means of $\int_{-\infty}^{\infty} u(t) \, dt$ have the form

$$
A(y) = \int_{-\infty}^{\infty} e^{-2\pi y|t|} u(t) \, dt, \quad y > 0,
$$

and we say that our integral is Abel summable to l if $\lim\limits_{y\to 0+} A(y)=l$.

Let us now examine the Abel means of the integral (2.8) when g is the Fourier transform of an integrable function f. We then have

$$f(x,y)=\int_{-\infty}^{\infty} e^{-2\pi y|t|}e^{2\pi ixt}\hat{f}(t)\,dt$$

$$=\int_{-\infty}^{\infty} e^{-2\pi y|t|}e^{2\pi ixt}\left\{\int_{-\infty}^{\infty} f(s)e^{-2\pi its}ds\right\}dt$$

$$=\int_{-\infty}^{\infty} f(s)\left\{\int_{-\infty}^{\infty} e^{-2\pi y|t|}e^{2\pi it(x-s)}dt\right\}ds.$$

As in the case of the $(C,1)$ means we can easily obtain a simple expression for the inner integral:

$$\int_{-\infty}^{\infty} e^{-2\pi y|t|}e^{2\pi itx}dt=\int_0^{\infty} e^{2\pi t(ix-y)}dt+\int_{-\infty}^0 e^{2\pi t(ix+y)}dt$$

$$=\frac{1}{2\pi(y-ix)}+\frac{1}{2\pi(y+ix)}=\frac{1}{\pi}\frac{y}{x^2+y^2}.$$

Hence,

$$f(x,y)=\frac{1}{\pi}\int_{-\infty}^{\infty} f(t)\frac{y}{(x-t)^2+y^2}dt$$

$$=\int_{-\infty}^{\infty} f(t)P(x-t,y)dt \qquad (2.15)$$

where

$$P(x,y)=\frac{1}{\pi}\frac{y}{x^2+y^2}, \qquad (2.16)$$

for $y>0$ and $-\infty<x<\infty$. $P(x,y)$ is called the *Poisson kernel* and the integral (2.15) is called the *Poisson integral of f*.

It is clear that result (2.14) still holds if we replace the Cesàro means of the integral $\int_{-\infty}^{\infty}\hat{f}(t)e^{2\pi ixt}dt$ by the Abel means, provided we can show that the Poisson kernel satisfies

(a') $\displaystyle\int_{-\infty}^{\infty} P(x,y)dx=1$;

(b') $P(x,y)\geqslant 0$;

(c') *for each* $\delta>0$ $\displaystyle\int_{|x|\geqslant\delta} P(x,y)dy\to 0$ *as* $y\to 0$.

But the last two of these properties are obvious. Property (a') is also easy to establish: if we let $s=x/y$ then

$$\int_{-\infty}^{\infty} P(x,y)dx=\frac{1}{\pi}\int_{-\infty}^{\infty}\frac{ds}{1+s^2}$$

$$=\lim_{N\to\infty}\frac{1}{\pi}\left[\tan^{-1}N-\tan^{-1}(-N)\right]=1.$$

We note that the Poisson kernel is a harmonic function in the upper half-plane $\{z = x + iy; y > 0\}$. This can be seen either by computing its Laplacian directly or by observing that it is the real part of the analytic function

$$\frac{i}{\pi}\frac{1}{z} = \frac{1}{\pi}\frac{y}{x^2+y^2} + i\frac{1}{\pi}\frac{x}{x^2+y^2}.$$

The imaginary part,

$$Q(x,y) = \frac{1}{\pi}\frac{x}{x^2+y^2},$$

is called the *conjugate Poisson kernel.*

Now suppose f belongs to $L^1(-\infty, \infty)$ and is real-valued. Let us form the integral

$$F(z) = F(x + iy) = \frac{i}{\pi}\int_{-\infty}^{\infty} f(t)\frac{1}{(x-t)+iy}dt, \qquad (2.17)$$

where $y > 0$ and $-\infty < x < \infty$. It is easy to see that F is an analytic function in the upper half-plane* and that its real part is given by the Poisson integral of f. Thus the latter defines a harmonic function in the upper half-plane. The imaginary part,

$$\tilde{f}(x,y) = \frac{1}{\pi}\int_{-\infty}^{\infty} f(t)\frac{x-t}{(x-t)^2+y^2}dt$$

$$= \int_{-\infty}^{\infty} f(t)Q(x-t,y)dt, \qquad (2.18)$$

is called the *conjugate Poisson integral* of f. We see from these observations that there must be a strong connection between Poisson integrals and the theory of harmonic and analytic functions of a complex variable.

By now we have given a good deal of evidence that the theory of Fourier integrals is not only intimately connected with the theory of Fourier series but is very similar to it. This is indeed the case. We shall see in more detail in the next section, for example, how the L^2-theory of Fourier integrals is as elegant as its analog, the L^2-theory of Fourier series, which we described briefly in the first section. We shall not, however, always discuss a result concerning, say, Fourier series and then also describe the corresponding result in the theory of Fourier integrals. On the contrary, we shall often discuss an aspect of harmonic analysis in one of the two theories, but not in both. The reader should be aware that there exist parallel results in the other theory as well. For example, the operation of convolution of two functions in $L^1(-\infty, \infty)$ plays an equally important role on the real line. Its definition is the obvious one: if f and g are integrable on $(-\infty, \infty)$,

---

* Perhaps the easiest way of proving this fact is to use Morera's theorem: given a simple closed contour C in the upper half-plane we can easily check that

$$\int_C F(z)dz = \frac{i}{\pi}\int_{-\infty}^{\infty} f(t)\left\{\int_C \frac{dz}{z-t}\right\}dt = \frac{i}{\pi}\int_{-\infty}^{\infty} f(t)\cdot 0 dt = 0.$$

This then implies the analyticity of F. We leave the details of this argument to the reader.

then $f*g$ is defined by

$$(f*g)(x) = \int_{-\infty}^{\infty} f(t)g(x-t)\,dt.$$

We leave it to the reader to check that $(f*g)(x)$ is defined for almost all real x and that the properties (i), (ii), (iii), and (iv) announced in the first section hold in this case as well. Moreover, the argument that was used to establish the important relation (1.18) between convolution and Fourier transformation can be used, after some obvious changes, to show that the same relation holds in the case of Fourier integrals.

3. THE L^1 AND L^2 THEORIES

In the last two sections we introduced several concepts but have not studied any of them very deeply. In this section we shall examine in much greater detail the convergence and summability of Fourier series, the Fourier inversion problem, and the L^2-theory. Moreover, we shall describe some of the better-known theorems in the harmonic analysis associated with the real line and the circle.

Let us begin with a closer look at the Fourier coefficients of an integrable and periodic function f. We have seen that when f belongs to $L^2(0, 1)$ the Fourier coefficients $c_k = \hat{f}(k)$ satisfy Bessel's inequality (see (1.5)). Thus, in particular,

$$\sum_{k=-\infty}^{\infty} |c_k|^2 < \infty.$$

It follows, therefore, that $c_k \to 0$ as $|k| \to \infty$. But this result is true even when f is in $L^1(0, 1)$. For, suppose $\epsilon > 0$, we can then write $f = g + h$ where g is in $L^2(0, 1)$ and $\|h\|_1 < \epsilon/2$. Since $|\hat{h}(k)| \leqslant \|h\|_1$ for $k = 0, \pm 1, \pm 2, \ldots$, we have $|\hat{f}(k)| \leqslant |\hat{g}(k)| + \epsilon/2$. Now, using the result just established for $L^2(0, 1)$, we can find an $N > 0$ such that $|\hat{g}(k)| < \epsilon/2$ if $|k| \geqslant N$. Thus, $|\hat{f}(k)| < \epsilon/2 + \epsilon/2 = \epsilon$ if $|k| \geqslant N$. We have proved the following result:

(3.1) THE RIEMANN-LEBESGUE THEOREM: *If f is an integrable and periodic function then* $\lim_{|k| \to \infty} \hat{f}(k) = 0$.

An immediate application of the Riemann-Lebesgue theorem is the following convergence test for Fourier series.

(3.2) DINI TEST: *If a periodic and integrable function, f, satisfies the condition*

$$\int_{-1/2}^{1/2} \left| \frac{f(x-t) - f(x)}{\tan \pi t} \right| dt < \infty \tag{3.3}$$

at a point x, then the partial sums $s_n(x) = \sum_{k=-n}^{n} \hat{f}(k)e^{2\pi i k x}$ *converge to $f(x)$ as $n \to \infty$.*

To see this we let g be the function whose value at $t \in (-\frac{1}{2}, \frac{1}{2})$ is $[f(x-t) - f(x)]/\tan \pi t$, then the integrability of f and condition (3.3) imply that g is integrable. Using (1.9) and the fact that

$$\int_{-1/2}^{1/2} D_n(t)dt = 1$$

we have, for $n \geqslant 1$,

$$S_n(x) - f(x) = \int_{-1/2}^{1/2} f(x-t) \frac{\sin(2n+1)\pi t}{\sin \pi t} dt - f(x) \int_{-1/2}^{1/2} D_n(t) dt$$

$$= \int_{-1/2}^{1/2} \{f(x-t) - f(x)\} \frac{\sin(2n+1)\pi t}{\sin \pi t} dt$$

$$= \int_{-1/2}^{1/2} \{f(x-t) - f(x)\} \left\{ \frac{e^{2\pi int} - e^{-2\pi int}}{2i \tan \pi t} + \frac{e^{2\pi int} + e^{-2\pi int}}{2} \right\} dt$$

$$= \frac{\hat{g}(-n) - \hat{g}(n)}{2i} + \frac{e^{-2\pi inx}\hat{f}(-n) + e^{2\pi inx}\hat{f}(n)}{2}.$$

But it follows from the Riemann-Lebesgue theorem that this last expression tends to 0 as $n \to \infty$. This proves (3.2).

The Dini test is probably the most useful of the various convergence criteria in the literature. One of its consequences is the fact that the *Fourier series of an integrable and periodic function converges to the value of the function at each point of differentiability*. We see this by first noting that, since $\lim\limits_{t\to 0} \dfrac{\tan \pi t}{\pi t} = 1$, condition (3.3) is equivalent to the condition

$$\int_{-1/2}^{1/2} \left| \frac{f(x-t) - f(x)}{t} \right| dt < \infty. \tag{3.3'}$$

But it is obvious that if f is differentiable at x then (3.3') must hold.

Before passing to the topic of summability of Fourier series we state, without proof, what is probably the best-known convergence test in the theory of Fourier series:

(3.4) THE DIRICHLET-JORDAN TEST: *Suppose a periodic function f is of bounded variation over* (0, 1). *Then*

(a) *the partial sums $s_n(x)$ converge to $\frac{1}{2}\{f(x+0)+f(x-0)\}$ at each real number x. In particular, they converge to $f(x)$ at each point of continuity of f;*

(b) *if f is continuous on a closed interval, then $s_n(x)$ converges uniformly on this interval.*

We now pass to a more detailed study of summability of Fourier series. Let us observe that in the proof of the theorem of Fejér (result (1.6) restricted to Cesàro summability) we really have shown that the convergence of $\sigma_n(x)$ to $f(x)$ is uniform in any interval where f is uniformly continuous. From this we easily obtain the following classical result:

(3.5) WEIERSTRASS APPROXIMATION THEOREM: *Suppose f is a continuous periodic function and $\epsilon > 0$. Then there exists a trigonometric polynomial T, that is, a finite* *linear combination of the exponentials* $e^{2\pi i n x}$, $n = 0$, ± 1, $\pm 2, \ldots$, *such that*

$$|f(x) - T(x)| < \epsilon \quad \text{for all } x.$$

For we may take $T(x) = \sigma_n(x)$ for n large enough since the Cesàro means converge to f uniformly in this case.

One important consequence of these considerations is that *the system* $\{e^{2\pi i n x}\}$ *is complete*; that is, *if all the Fourier coefficients of an integrable periodic function f vanish then f must be 0 almost everywhere.* We first note that if f is continuous and $\hat{f}(k) = 0$ for all integers k then $\sigma_n(x) \equiv 0$ for all n. But, since $\sigma_n(x) \to f(x)$ at all x, we must have $f(x) \equiv 0$. If f is merely integrable and periodic we form the indefinite integral $F(x) = \int_0^x f(t) dt$. The condition $\hat{f}(0) = 0$ implies that

$$F(x+1) - F(x) = \int_x^{x+1} f(t) dt = \int_0^1 f(t) dt = 0.$$

Thus, F is a continuous periodic function. We claim that the hypothesis $\hat{f}(k) = 0$ for $k = \pm 1, \pm 2, \pm 3, \ldots$, implies that $\hat{F}(k) = 0$ for $k = \pm 1, \pm 2, \pm 3, \ldots$. For, integrating by parts,

$$\hat{F}(k) = \int_0^1 F(t) e^{-2\pi i k t} dt = F(t) \frac{e^{-2\pi i k t}}{-2\pi i k} \Big]_0^1 + \frac{1}{2\pi i k} \int_0^1 f(t) e^{-2\pi i k t} dt$$

$$= 0 + \frac{\hat{f}(k)}{2\pi i k} = 0 + 0 = 0.$$

From this and the orthogonality relations (1.3) we then conclude that the continuous and periodic function G whose value at x is $F(x) - \hat{f}(0)$ must satisfy $\hat{G}(k) = 0$ for all integers k. But we have shown that this implies that $G(x) \equiv 0$. Since, by Lebesgue's theorem on the differentiation of the integral $F'(x) = f(x)$ for almost every x, it follows that $0 = G'(x) = F'(x) = f(x)$ almost everywhere. This proves the completeness of the system $\{e^{2\pi i n x}\}$.

We now show how to obtain the theorem of Lebesgue (see result (1.7)) that asserts that the $(C,1)$ means of the Fourier series of an integrable and periodic function converge almost everywhere to the values of the function. In order to do this we will have to introduce the *Lebesgue set* of such a function f. We have just used the well-known fact that $F'(x) = f(x)$ for almost every x when $F(x) = \int_0^x f(t) dt$. We can rewrite this fact in the following way:

$$\lim_{h \to 0} \frac{1}{h} \int_0^h \{f(x+t) - f(x)\} \, dt = 0$$

for almost every x. It turns out that a stronger result is true:

$$\lim_{h \to 0} \frac{1}{h} \int_0^h |f(x+t) - f(x)| \, dt = 0 \tag{3.6}$$

for almost every x. It is not hard to show this: For a fixed rational number r let E_r

be the set of all x such that

$$\lim_{h\to 0}\frac{1}{h}\int_0^h |f(x+t)-r|dt=|f(x)-r|$$

fails to hold. Applying Lebesgue's theorem on the differentiation of the integral to $g(t)=|f(x+t)-r|$ we conclude that E_r has measure 0. Let $E=\cup E_r$, the union being taken over all rational numbers r. Then E also has measure 0. We claim that if x does not belong to E then (3.6) holds. For let $\epsilon>0$. Choose a rational number r_0 such that $|f(x)-r_0|<\epsilon/2$. Then

$$\frac{1}{h}\int_0^h |f(x+t)-f(x)|dt \leqslant \frac{1}{h}\int_0^h |f(x+t)-r_0|dt$$

$$+\frac{1}{h}\int_0^h |f(x)-r_0|dt.$$

But the first term of this sum is less than $\epsilon/2$ if h is close to 0 while

$$\frac{1}{h}\int_0^h |f(x)-r_0|dt < \frac{1}{h}\int_0^h \frac{\epsilon}{2}dt=\frac{\epsilon}{2}.$$

Thus,

$$\frac{1}{h}\int_0^h |f(x+t)-f(x)|dt<\epsilon$$

if h is small.

The set of all x such that (3.6) holds is called the *Lebesgue set* of f. We shall show that the $(C, 1)$ means of the Fourier series of f converge to $f(x)$ whenever x is a member of the Lebesgue set.

We shall need the following two estimates on the Fejér kernel:

(a) $K_n(t)\leqslant n+1$;

(b) $K_n(t)\leqslant \dfrac{A}{(n+1)t^2},|t|\leqslant \dfrac{1}{2}$, where A is an absolute constant.

The first one follows from the obvious estimate on the Dirichlet kernel $|D_k(t)|\leqslant 2k+1$: for

$$K_n(t)=\frac{1}{n+1}\sum_{k=0}^n D_k(t)\leqslant \frac{1}{n+1}\sum_{k=0}^n (2k+1)$$

$$=\frac{(n+1)^2}{n+1}=n+1.$$

The second one is a consequence of formula (1.10) and the well-known fact $\lim_{t\to 0}\dfrac{\sin t}{t}=1$.

Now suppose x belongs to the Lebesgue set of f. As in the proof of (1.6), we have, using property (A) of the Fejér kernel,

$$\sigma_n(x)-f(x)=\int_{-1/2}^{1/2}\{f(x-t)-f(x)\}K_n(t)\,dt.$$

Thus, using the estimates (a) and (b),

$$|\sigma_n(x) - f(x)| \leqslant \int_{-1/2}^{1/2} |f(x-t) - f(x)| K_n(t) \, dt$$

$$\leqslant (n+1) \int_{|t| \leqslant 1/(n+1)} |f(x-t) - f(x)| \, dt$$

$$+ \frac{A}{n+1} \int_{1/(n+1) \leqslant |t| \leqslant 1/2} \frac{|f(x-t) - f(x)|}{t^2} \, dt.$$

Given $\epsilon > 0$ let $\delta > 0$ be such that $\dfrac{1}{h} \displaystyle\int_{|t| \leqslant h} |f(x-t) - f(x)| \, dt < \epsilon$ if $h \leqslant \delta$. Then the first term in the above sum is less than ϵ whenever $(n+1)^{-1} \leqslant \delta$. In order to estimate the second term we write the integral as the sum of the two integrals $\displaystyle\int_{(n+1)^{-1}}^{1/2}$ and $\displaystyle\int_{-1/2}^{-(n+1)^{-1}}$. We shall show that the first integral tends to 0 as $n \to \infty$; a similar estimate will then show that the same is true for the second.

Let $G(t) = \int_0^t |f(x-s) - f(x)| \, ds$. Then, integrating by parts, we have

$$\frac{A}{n+1} \int_{1/(n+1)}^{1/2} \frac{|f(x-t) - f(x)|}{t^2} \, dt$$

$$\leqslant \frac{A}{4(n+1)} G(1/2) + \frac{2A}{n+1} \int_{1/(n+1)}^{\delta} \frac{G(t)}{t^3} \, dt + \frac{2A}{n+1} \int_{\delta}^{1/2} \frac{G(t)}{t^3} \, dt.$$

The first and third terms tend to 0 as $n \to \infty$. Since $(1/t) G(t) < \epsilon$ for $|t| \leqslant \delta$ the second term is dominated by

$$\frac{2A\epsilon}{n+1} \int_{1/(n+1)}^{\delta} \frac{dt}{t^2} < 2A\epsilon.$$

Thus, $|\sigma_n(x) - f(x)|$ can be made as small as we wish by choosing n large enough. This proves the theorem of Lebesgue.

An application of this theorem is that we can reverse the inequality in Bessel's inequality (see (1.5)). For, if $f \in L^2(0, 1) \subset L^1(0,1)$ and $c_k = \hat{f}(k)$, $k = 0, \pm 1, \pm 2, \ldots$, then the $(C,1)$ means of f have the form

$$\sigma_n(x) = \sum_{k=-n}^{n} \left(1 - \frac{|k|}{n+1}\right) c_k e^{2\pi i k x}.$$

Using the orthogonality relations (1.3) we have

$$\int_0^1 |\sigma_n(x)|^2 dx = \sum_{k=-n}^{n} \left(1 - \frac{|k|}{n+1}\right)^2 |c_k|^2 \leqslant \sum_{k=-\infty}^{\infty} |c_k|^2.$$

Since $\sigma_n(x) \to f(x)$ almost everywhere, Fatou's lemma implies that

$$\int_0^1 |f(x)|^2 dx \leqslant \lim_{n \to \infty} \int_0^1 |\sigma_n(x)|^2 dx.$$

Consequently,

$$\int_0^1 |f(x)|^2 dx \leqslant \sum_{k=-\infty}^{\infty} |c_k|^2.$$

Together with Bessel's inequality this gives us the following relation, known as *Parseval's formula*:

$$\int_0^1 |f(x)|^2 dx = \sum_{k=-\infty}^{\infty} |\hat{f}(k)|^2. \tag{3.7}$$

We can also show easily that the partial sums of the Fourier series of a function f in $L^2(0, 1)$ converge to f in the L^2-norm. We have already seen that they do converge to a function g in $L^2(0, 1)$ (see the argument preceding (1.5)). This implies that g and f have the same Fourier coefficients $\{c_k\} = \{\hat{f}(k)\}$; for

$$\int_0^1 g(t)e^{-2\pi i k t} dt = \int_0^1 [g(t) - s_n(t)]e^{-2\pi i k t} dt + \int_0^1 s_n(t)e^{-2\pi i k t} dt.$$

The first term of this sum is dominated, in absolute value, by $\|g - s_n\|_2$ (use Schwarz's inequality) and, thus, tends to 0 as $n \to \infty$. The second term equals c_k as long as $n \geqslant k$. From this we conclude that the Fourier coefficients of the function $f - g$ are all 0. But, since the system $\{e^{2\pi i n x}\}$ is complete, this implies $f(x) - g(x) = 0$ almost everywhere.

Let us observe that if we had started with a square summable sequence $\{c_k\}$ (that is, $\sum_{k=-\infty}^{\infty} |c_k|^2 < \infty$), then, by the orthogonality relations (1.3), the partial sums $s_n(x)$ of $\sum_{k=-\infty}^{\infty} c_k e^{2\pi i k x}$ converge in the L^2-norm to a function g. The argument just used shows that $\hat{g}(k) = c_k$ for $k = 0, \pm 1, \pm 2, \ldots$.

We collect these facts together in the following statement:

(3.8) *Suppose f belongs to $L^2(0, 1)$, then its Fourier series converges to f in the L^2-norm; that is,*

$$\|f - s_n\|_2 = \left(\int_0^1 |f(x) - s_n(x)|^2 dx \right)^{1/2}$$

$$= \left(\int_0^1 |f(x) - \sum_{k=-n}^{n} \hat{f}(k)e^{2\pi i k x}|^2 dx \right)^{1/2}$$

tends to 0 as n tends to ∞. Furthermore,

$$\|f\|_2 = \left(\int_0^1 |f(x)|^2 dx \right)^{1/2} = \left(\sum_{k=-\infty}^{\infty} |\hat{f}(k)|^2 \right)^{1/2} = \|\hat{f}\|_2.$$

If a sequence $\{c_k\}$ satisfies $\sum_{k=-\infty}^{\infty} |c_k|^2 < \infty$, then there exists a function f in $L^2(0, 1)$ such that $c_k = \hat{f}(k)$ for all integers k.

Except for not having proved (1.7) in the case of Abel summability we have now established all the results announced in the first section in connection with the Fourier inversion problem. We leave it to the reader to show that essentially the argument used above for Cesàro summability, using the estimates

(a') $P(r,t) \leqslant \dfrac{1}{1-r}$ and

(b') $P(r,t) \leqslant \dfrac{A(1-r)}{t^2}, |t| \leqslant \dfrac{1}{2}, 0 \leqslant r < 1$, where A is an absolute constant,

gives us (1.7) in full.

As we stated toward the end of the last section, we shall not essentially repeat all this material by giving the corresponding results in the theory of Fourier integrals. The reader should have no trouble, for example, in stating and proving the analogs of results (1.6) and (1.7). Nevertheless, some discussion of what happens when we carry over the above material to the case of the real line is in order.

First of all, we cannot adapt the argument we gave to establish the Riemann-Lebesgue theorem to the case of functions in $L^1(-\infty, \infty)$. For one thing, we have not even defined the Fourier transform for functions in $L^2(-\infty, \infty)$; moreover, as we shall see shortly after we define it, it is *not* true in general that $\hat{f}(x) \to 0$ as $|x| \to \infty$ when $f \in L^2(-\infty, \infty)$. We shall show, however, that the Riemann-Lebesgue theorem does extend to the case of the real line, and the simple argument we shall give can be adapted to prove (3.1) as well. We shall prove

(3.9) *If $f \in L^1(-\infty, \infty)$, then $\hat{f}(x) \to 0$ as $|x| \to \infty$.*

Since

$$-\hat{f}(x) = \int_{-\infty}^{\infty} (-1)e^{-2\pi i x t} f(t)\, dt$$

$$= \int_{-\infty}^{\infty} e^{-2\pi i x[t - (1/2x)]} f(t)\, dt$$

$$= \int_{-\infty}^{\infty} e^{-2\pi i x t} f\left(t + \frac{1}{2x}\right) dt,$$

we have

$$|\hat{f}(x)| = \left| \frac{1}{2} \int_{-\infty}^{\infty} \left\{ f(t) - f\left(t + \frac{1}{2x}\right) \right\} e^{-2\pi i x t} dt \right|$$

$$\leqslant \frac{1}{2} \omega_1\left(\frac{1}{2x}\right) \to 0 \qquad as \ x \to \infty.$$

Let us now state the result that corresponds to (3.8):

(3.10) The Plancherel Theorem: *If f belongs to $L^2(-\infty, \infty)$, then there exists a function $\hat{f}$, also in $L^2(-\infty, \infty)$, such that*

$$\int_{-\infty}^{\infty} |\hat{f}(x) - \int_{-N}^{N} e^{-2\pi i x t} f(t)\, dt|^2\, dx \to 0$$

as $N\to\infty$. The function $\hat{f}$ is called the Fourier transform of f and it agrees a. e. with the previously defined Fourier transform whenever $f\in L^1(-\infty, \infty)\cap L^2(-\infty, \infty)$. Furthermore, Parseval's formula holds

$$\|\hat{f}\|_2 = \|f\|_2.$$

Fourier inversion is possible in the L^2-norm:

$$\int_{-\infty}^{\infty}|f(t)-\int_{-N}^{N}e^{2\pi ixt}\hat{f}(x)dx|^2dt\to 0$$

as $N\to\infty$. Finally, each f in $L^2(-\infty, \infty)$ has the form $f=\hat{g}$ for an (almost everywhere) unique g in $L^2(-\infty, \infty)$.

To prove (3.10) let us choose an f in $L^1(-\infty, \infty)\cap L^2(-\infty, \infty)$ and form the convolution $h=f*g$ where $g(t)=\overline{f(-t)}$. It follows from the remarks made at the very end of Sec. 2 that h, being the convolution of two functions in $L^1(-\infty, \infty)$, is integrable and, also, $\hat{h}(x)=\hat{f}(x)\cdot\overline{\hat{f}}(x)=|\hat{f}(x)|^2\geqslant 0$ (by the real-line analog of 1.18)). Moreover, we claim that 0 is a point of the Lebesgue set of h; in fact, it is a point of continuity of h. For

$$|h(\delta)-h(0)|=|\int_{-\infty}^{\infty}\{g(\delta-t)-g(-t)\}f(t)dt|$$

$$\leqslant\left[\int_{-\infty}^{\infty}|g(\delta-t)-g(-t)|^2dt\right]^{1/2}\|f\|_2.$$

But $[\int_{-\infty}^{\infty}|g(\delta-t)-g(-t)|^2dt]^{1/2}$ is dominated by the L^2 modulus of continuity evaluated at δ, $\omega_2(\delta)$, of the function whose value at t is $g(-t)$. Since the latter belongs to $L^2(-\infty, \infty)$, we can conclude that $\lim_{\delta\to 0}|h(\delta)-h(0)|=0$. Consequently, the $(C, 1)$ means of the integral $\int_{-\infty}^{\infty}\hat{h}(x)e^{2\pi ixt}dx$, defining $(\hat{h})^{\vee}(x)$, converge to $h(t)$ when $t=0$. That is,

$$\int_{-R}^{R}\left(1-\frac{|x|}{R}\right)\hat{h}(x)dx\to h(0).$$

But $\hat{h}(x)\geqslant 0$ and the integrand in this integral increases monotonically to $\hat{h}(x)$. Thus, by the Lebesgue monotone convergence theorem $\hat{h}$ is integrable and

$$\int_{-\infty}^{\infty}\hat{h}(x)dx=h(0).$$

Since $h(0)=\int_{-\infty}^{\infty}f(t)\overline{f(t)}dt=\|f\|_2^2$ and $\hat{h}=|\hat{f}|^2$ this shows

$$\int_{-\infty}^{\infty}|\hat{f}(x)|^2dx=\int_{-\infty}^{\infty}|f(x)|^2dx. \tag{3.11}$$

Thus, Parseval's formula holds when $f\in L^1(-\infty, \infty)\cap L^2(-\infty, \infty)$.

In particular, (3.11) tells us that the mapping $f\to\hat{f}$ is bounded, in the L^2-norm, as a linear operator on the dense subset $L^1\cap L^2$ of the Hilbert space L^2 into L^2. It is well known that in such a case there exists a unique, bounded extension of the operator on all of the Hilbert space. Using the same notation for this extension we then can conclude that (3.11) holds for all $f \in L^2(-\infty, \infty)$.

If we let χ_N be the characteristic function of the interval $[-N, N]$ we set $f_N = \chi_N f$, for $f \in L^2(-\infty, \infty)$. Then $f_N \in L^1(-\infty, \infty) \cap L^2(-\infty, \infty)$ and $\|f - f_N\|_2 \to 0$ as $N \to \infty$. Because of the boundedness of the operator we have just defined we then must have $\|\hat{f} - \hat{f}_N\|_2 \to 0$. This proves the first part of (3.10).

The Fourier inversion part of Plancherel's theorem follows easily from the relation

$$\int_{-\infty}^{\infty} f(x)\hat{g}(x)dx = \int_{-\infty}^{\infty} \hat{f}(x) g(x)dx \qquad (3.12)$$

whenever $f, g \in L^2(-\infty, \infty)$. The proof of (3.12) for functions in $L^1 \cap L^2$ is straightforward; thus, we first establish (3.12) for f_N and g_N and obtain the general result by letting $N \to \infty$. It is an immediate consequence of (3.12) that

$$\bar{f} = \left(\bar{\hat{f}} \right)^\wedge \qquad (3.13)$$

for all $f \in L^2(-\infty, \infty)$. For

$$\left\| \bar{f} - \left(\bar{\hat{f}} \right)^\wedge \right\|_2^2 = \left\{ \int f\bar{f} - \int f\left(\bar{\hat{f}}\right)^\wedge \right\} - \overline{\left\{ \int f\left(\bar{\hat{f}}\right)^\wedge - \int \left(\bar{\hat{f}}\right)^\wedge \overline{\left(\bar{\hat{f}}\right)^\wedge} \right\}}.$$

But both expressions in the brackets are 0; applying (3.12) to the first one we obtain $\int f\bar{f} - \int \bar{\hat{f}}\hat{f}$, which is 0 by Parseval's formula. A similar argument shows the second expression is 0 also. Now, because of (3.13) we have that $\bar{f}$ is the limit in the L^2-norm, as $N \to \infty$, of the functions given by the integrals

$$\int_{-N}^{N} e^{-2\pi i x t} \bar{\hat{f}}(t) \, dt = \overline{\int_{-N}^{N} e^{2\pi i x t} \hat{f}(t) \, dt}.$$

By taking complex conjugates we have the Fourier inversion result announced in (3.10).

The last statement follows from this inversion applied to $g = f$, where $\check{f}$ is the limit in L^2, as $N \to \infty$, of

$$\int_{-N}^{N} e^{2\pi i x t} f(t) \, dt = \overline{\left(\bar{f}_N\right)^\wedge(x)}.$$

The material presented up to this point belongs to the foundation of the theories of Fourier series and integrals. It is desirable also to describe some of the directions in which harmonic analysis has been developed. This subject, however, is so rich with results, covering such a wide field of mathematics, that it is impossible to present something that approximates a survey of the highlights in the space that we have available. For this reason we shall, from time to time, select certain topics and present them, mostly without proofs. The bases for our selections are the possibility of extending these topics to other parts of harmonic analysis, the simplicity of the concepts involved, and their applicability to other branches of mathematics. We conclude this section with what is perhaps the one result that best fits this description, a celebrated theorem of Wiener. As we shall see in Sec. 5, this theorem extends to the harmonic analysis associated with any locally compact abelian group, no new concepts are needed to understand it, and from it one can prove rather easily the prime number theorem [4, p. 303]!

Suppose f belongs to $L^1(-\infty, \infty)$, then the collection of all finite linear combinations of translates of f will be denoted by T_f. That is, g belongs to T_f if and only if it has the form

$$g(x) = \Sigma a_k f(x + t_k) \tag{3.14}$$

for some finite set of real numbers t_k and complex numbers a_k. The theorem of Wiener asserts the following:

(3.15) *Suppose f belongs to $L^1(-\infty, \infty)$ and that $\hat{f}(x)$ is never 0, then the closure, in the L^1 topology, of T_f is all of $L^1(-\infty, \infty)$. In other words, any function in $L^1(-\infty, \infty)$ can be approximated arbitrarily closely in the L^1-norm by functions of the form* (3.14).

It is easy to find functions f whose Fourier transforms never vanish. For example, the proof of (2.16), with $y = 1$, consisted, simply, of showing that when $f(t) = e^{-2\pi|t|}$ then $\hat{f}(x) = \dfrac{1}{\pi} \dfrac{1}{1 + x^2}$.

That the condition $\hat{f}(x) \neq 0$ for all real x is necessary is clear. For, if $\hat{f}(x_0) = 0$ for some x_0 and $g \in T_f$ then g has the form (3.14) and, thus,

$$\hat{g}(x) = \Sigma a_k e^{2\pi i t_k} \hat{f}(x).$$

Therefore $\hat{g}(x_0) = 0$. Now suppose h is in the L^1 closure of T_f; then there exists a sequence $\{g_n\}$ in T_f such that $\|g_n - h\|_1 \to 0$ as $n \to \infty$. Thus, by (2.9),

$$|\hat{g}_n(x_0) - \hat{h}(x_0)| \leqslant \|\hat{g}_n - \hat{h}\|_\infty \leqslant \|g_n - h\|_1 \to 0 \qquad \text{as} \quad n \to \infty.$$

Since $g_n(x_0) = 0$ for all n we must have $\hat{h}(x_0) = 0$. We have shown that the Fourier transforms of all h in the closure of T_f vanish at x_0. Since there are integrable functions whose Fourier transforms never vanish, this closure cannot comprise all of $L^1(-\infty, \infty)$.

We shall not give a proof of (3.15). We would like to point out, however, that this proof uses strongly the fact that whenever $f \in L^1(-\infty, \infty)$, then the closure of T_f is a (closed) *ideal* in $L^1(-\infty, \infty)$. By the term "ideal" we mean a linear subspace, I, of $L^1(-\infty, \infty)$ such that $g * h \in I$ whenever $g \in I$ and $h \in L^1(-\infty, \infty)$.

An important class of ideals in $L^1(-\infty, \infty)$ is the collection of closed *maximal ideals* (an ideal M is said to be maximal if it is not contained in any proper ideal in $L^1(-\infty, \infty)$ other than M itself). This class has a very elegant characterization:

(3.16) *M is a closed maximal ideal if and only if there exists a real number x such that M consists of all $f \in L^1(-\infty, \infty)$ such that $\hat{f}(x) = 0$.*

Thus, we have a one-to-one correspondence between the real numbers and the closed maximal ideals in $L^1(-\infty, \infty)$. We shall denote the closed maximal ideal corresponding to x by $M(x)$. It is not hard to see that the following is a generalization of Wiener's theorem:

(3.15') *Every proper closed ideal in $L^1(-\infty, \infty)$ is contained in a closed maximal ideal.*

For if $\hat{f}(x)$ is never 0, then f cannot belong to $M(x)$ for any x. Thus, the closed ideal obtained by taking the closure of T_f cannot be included in any closed maximal ideal. Consequently, this ideal is not proper; that is, it must coincide with $L^1(-\infty, \infty)$.

These considerations lead us to the formulation of a well-known problem in harmonic analysis, the problem of *spectral synthesis*. If the closure of T_f is a proper ideal, I_f, then, (3.15'), it is contained in a certain class of ideals $M(x)$. It is easy to check that the intersection of all closed maximal ideals containing I_f is a closed ideal. The problem of spectral synthesis is to determine for which $f \in L^1(-\infty, \infty)$ it is true that I_f equals this intersection. It has been discovered only recently (in 1959) that there are $f \in L^1(-\infty, \infty)$ for which I_f is not equal to the intersection of all maximal ideals containing it.

It is often useful to rephrase this problem in the following way: *For which f in $L^1(-\infty, \infty)$ is it true that, if $\hat{g}(x) = 0$ whenever $\hat{f}(x) = 0$, then g is in the closure of T_f?*

4. SOME OPERATORS THAT ARISE
IN HARMONIC ANALYSIS

Suppose $F(z) = a_0 + a_1 z + a_2 z^2 + \cdots + a_n z^n + \cdots$ is an analytic function in the interior of the unit circle. Suppose, further, that F is bounded in this domain; say, $|F(z)| \leqslant B < \infty$ for $|z| < 1$. Let us write $z = re^{2\pi i\theta}$, $0 \leqslant r < 1$, $0 \leqslant \theta < 1$. Then, using the orthogonality relations (1.3),

$$\sum_{k=0}^{\infty} |a_k|^2 r^{2k} = \int_0^1 \left(\sum_{k=0}^{\infty} a_k r^k e^{2\pi ik\theta} \right) \left(\sum_{k=0}^{\infty} \bar{a}_k r^k e^{-2\pi ik\theta} \right) d\theta$$

$$= \int_0^1 |F(re^{2\pi i\theta})|^2 d\theta \leqslant B^2 \qquad \text{for} \quad 0 \leqslant r < 1.$$

Letting $r \to 1$ we therefore obtain $\sum\limits_{k=0}^{\infty} |a_k|^2 < \infty$. By (3.8) we thus can conclude that there exists an f belonging to $L^2(0, 1)$ such that $\hat{f}(k) = a_k$, $k = 0, 1, 2, \ldots$, and $\hat{f}(k) = 0$ for all negative integers k. This shows that

$$F(re^{2\pi i\theta}) = \sum_{k=0}^{\infty} \hat{f}(k) r^k e^{2\pi ik\theta} = \sum_{k=-\infty}^{\infty} \hat{f}(k) r^k e^{2\pi ik\theta}, \qquad 0 \leqslant r < 1,$$

are the Abel means of the Fourier series of f. By (1.7), therefore, $\lim\limits_{r \to 1} F(re^{2\pi i\theta}) = f(\theta)$ for almost every θ. In particular, we have proved

(4.1) FATOU'S THEOREM: *If F is a bounded analytic function in the interior of the unit circle, then the radial limits* $\lim\limits_{r\to 1} F(re^{2\pi i\theta})$ *exist for almost every* θ *in* [0, 1].

We shall use this theorem to define an important operator, the *conjugate function mapping*, acting on integrable and periodic functions. Suppose f is such a function. It follows from our discussion concerning the Poisson kernel and the conjugate Poisson kernel that the function G defined by

$$G(z) = \int_0^1 \frac{1 + re^{2\pi i(\theta - t)}}{1 - re^{2\pi i(\theta - t)}} f(t)\, dt$$

$$= \int_0^1 P(r, \theta - t) f(t)\, dt + i \int_0^1 Q(r, \theta - t) f(t)\, dt, \qquad (4.2)$$

$z = re^{2\pi i\theta}$, is analytic in the interior of the unit circle. We already know that the first expression in the last sum has radial limits, as $r \to 1$, for almost all θ. The following theorem asserts that this is also true for the second term.

(4.3) *Suppose* $f \in L^1(0, 1)$; *then the limits*, $\tilde{f}(\theta)$, *as* $r \to 1$, *of*

$$\tilde{A}(r, \theta) = \int_0^1 Q(r, \theta - t) f(t)\, dt$$

$$= \int_0^1 \frac{2r \sin 2\pi(\theta - t)}{1 - 2r \cos 2\pi(\theta - t) + r^2} f(t)\, dt$$

exist for almost all θ. *The function* $\tilde{f}$ *is called the conjugate function of* f*.

By decomposing f into its real and imaginary parts and considering separately the positive and negative parts of each of these, we see that it suffices to prove (4.3) for $f \geqslant 0$. Thus, letting $A(r, \theta)$ be the Poisson integral and $\tilde{A}(r, \theta)$ the conjugate Poisson integral of f, we obtain an analytic function for $|z| < 1$, $z = re^{2\pi i\theta}$, and its values lie in the right half-plane (by property (B′) of the Poisson kernel). Thus,

$$F(z) = e^{-A(r, \theta) - i\tilde{A}(r, \theta)}$$

---

* If we let $r \to 1$ we obtain, formally,

$$\tilde{f}(\theta) = \int_0^1 \frac{2 \sin 2\pi(\theta - t)}{2(1 - \cos 2\pi(\theta - t))} f(t)\, dt = \int_0^1 \frac{f(t)}{4 \tan \pi(\theta - t)} dt.$$

This last integral, however, is not defined even when f is an extremely well-behaved function (for example, if f is constant and nonzero in a neighborhood of θ the integral fails to exist). One can show, however, that if we take the principle value integral

$$\lim_{\substack{\epsilon \to 0+ \\ 0 < t < 1}} \int_{\epsilon < |\theta - t|} \frac{f(t)}{4 \tan \pi(\theta - t)} dt \qquad (*)$$

we do obtain a value for almost all θ. In fact, an argument not unlike that used to prove (1.7) shows that the existence of these limits, as $\epsilon = 1 - r \to 0$, is equivalent almost everywhere to the existence of the limits of (4.3). One may take (*) as the definition of the conjugate function, therefore, and avoid the use of analytic function theory, However, the real-variable proof of the existence of $\tilde{f}$ is by no means easy.

is a bounded ($|F(z)| \leqslant 1$) analytic function in the interior of the unit circle. By Fatou's theorem (4.1) the radial limits of F exist almost everywhere. Since the radial limits of $A(r, \theta)$ also exist almost everywhere and are finite (they equal $f(\theta)$), the limits of F must be nonzero almost everywhere. But this implies the existence of $\lim_{r \to 1} \tilde{A}(r, \theta)$ for almost all θ, and (4.3) is proved.

The conjugate function mapping is obviously linear. If $f \in L^2(0, 1)$, then, using the fact that $\tilde{A}(r, \theta)$, $0 \leqslant r < 1$, $0 \leqslant \theta < 1$, are the Abel means of the conjugate Fourier series of f and, also, the result (3.8), we can show very easily that

$$\|\tilde{f}\|_2^2 = \sum_{|k| \geqslant 1} |\tilde{f}(k)|^2 \leqslant \sum_{k=-\infty}^{\infty} |\hat{f}(k)|^2 = \|f\|_2^2. \tag{4.4}$$

Thus the mapping $f \to \tilde{f}$ is a bounded linear transformation when restricted to the space $L^2(0, 1)$. One can show, however, that there are functions in $L^1(0,1)$ for which the conjugate function is not integrable. In particular, it follows that this mapping is not bounded as an operator from $L^1(0, 1)$ into $L^1(0, 1)$. We do have the following theorem, however.

(4.5) THEOREM OF M. RIESZ: *If $f \in L^p(0, 1)$, $1 < p < \infty$, then $\tilde{f} \in L^p(0, 1)$ and*

$$\|\tilde{f}\|_p \leqslant A_p \|f\|_p,$$

where A_p depends only on p.

For exactly the same reasons that we gave in the proof of (4.3) it suffices to consider the case $f \geqslant 0$. Furthermore, we claim that it is sufficient to show that

$$\int_0^1 |\tilde{A}(r, \theta)|^p d\theta \leqslant c_p \int_0^1 |A(r, \theta)|^p d\theta = c_p \int_0^1 [A(r, \theta)]^p d\theta \tag{4.6}$$

for $0 \leqslant r < 1$, where c_p depends only on p. For, an argument very similar to that used to prove (2.14), shows that the Poisson integrals $A(r, \theta)$ converge to $f(\theta)$ in the $L^p(0, 1)$ norms. In particular,

$$\lim_{r \to 1} \int_0^1 [A(r, \theta)]^p d\theta = \|f\|_p^p.$$

Since $\tilde{A}(r, \theta) \to \tilde{f}(\theta)$ almost everywhere as $r \to 1$, an application of Fatou's lemma then gives us the inequality $\|\tilde{f}\|_p^p \leqslant c_p \|f\|_p^p$, from which the theorem follows.

To show (4.6) we argue in the following manner. Let

$$F(z) = A(r, \theta) + i\tilde{A}(r, \theta) \qquad \text{for} \quad x + iy = z = re^{2\pi i\theta}, 0 \leqslant r < 1,$$

and let $\Delta = \partial^2/\partial x^2 + \partial^2/\partial y^2$ denote the Laplacian operator. Treating A, $\tilde{A}$, and F as functions of x and y we have, by a simple calculation which uses the Cauchy-Riemann equations (recall that F is analytic),

$$\Delta A^p = p(p-1)A^{p-2}|F'|^2 \qquad \text{and} \qquad \Delta |F|^p = p^2|F|^{p-2}|F'|^2.$$

Let us assume, first, that $1 < p \leqslant 2$. Then, since $|F| \geqslant A$,

$$\Delta |F|^p \leqslant q \Delta A^p,$$

where $(1/q = 1 - (1/p))$. We claim that this inequality and Green's formula imply

$$\int_0^1 |F(re^{2\pi i\theta})|^p d\theta \leqslant q \int_0^1 [A(r,\theta)]^p d\theta$$

for $0 \leqslant r < 1$, which certainly implies (4.6). The form of Green's formula we need is the following. Suppose u is a continuous function defined in the unit circle which has continuous first and second derivatives, S is the circle $\{(x,y); x^2 + y^2 \leqslant r^2 < 1\}$ and C its circumference. Then

$$\int_c \frac{\partial u}{\partial r} ds = \int \int_S \Delta u \, dx \, dy,$$

where $\partial/\partial r$ denotes differentiation in the direction of the radius vector and $ds = r \, d\theta$. Applying this formula to $u = A^p$ and $u = |F|^p$ we obtain, because of the inequality $\Delta |F|^p \leqslant q \Delta A^p$,

$$\int_0^1 \left(\frac{\partial}{\partial r} |F(re^{2\pi i\theta})|^p \right) r \, d\theta \leqslant q \int_0^1 \left(\frac{\partial}{\partial r} [A(r,\theta)]^p \right) r \, d\theta.$$

Thus, because of the smoothness of the functions involved,

$$\frac{d}{dr} \int_0^1 |F(re^{2\pi i\theta})|^p d\theta \leqslant \frac{d}{dr} q \int_0^1 A[(r,\theta)]^p d\theta.$$

Since $F(0) = A(0,\theta)$ we obtain the desired inequality by integrating with respect to r.

It remains for us to show that the theorem holds for $p \geqslant 2$. But it is an easy exercise, using the fact that L^p and L^q are dual when $1/p + 1/q = 1$, to show that whenever a bounded operator acting on L^p is given by a convolution, then it is defined on L^q and is a bounded operator on this space as well. The mapping $f \to \tilde{A}(r,\theta)$ is such an operator and it satisfies

$$\int_0^1 |\tilde{A}(r,\theta)|^p d\theta \leqslant B_p \int_0^1 |A(r,\theta)|^p d\theta \leqslant B_p \|f\|_p^{p*}$$

for $1 < p \leqslant 2$. Thus, it satisfies this inequality for the indices conjugate to p; that is, for p replaced by $q = p/(p-1)$:

$$\int_0^1 |\tilde{A}(r,\theta)|^q d\theta \leqslant C_q \|f\|_q^q$$

whenever $f \in L^q(0,1)$, $q \geqslant 2$. But, by Fatou's lemma, this implies

$$\int_0^1 |\tilde{f}(\theta)|^q d\theta = \int_0^1 \lim_{r \to 1} |\tilde{A}(r,\theta)|^q d\theta \leqslant C_q \|f\|_q^q$$

and (4.5) is proved.

---

* One can give several direct proofs of the inequality

$$\int_0^1 |A(r,\theta)|^p d\theta \leqslant \int_0^1 |f(\theta)|^p d\theta, p \geqslant 1.$$

Since it is an immediate consequence of Young's inequality (4.8) (for $g(\theta) = P(r,\theta)$ defines a function in L^1 and $A(r,\theta) = (g*f)(\theta)$) we will not prove it here.

This development gives us a glimpse of the role that "complex methods" (that is, the use of the theory of analytic functions of a complex variable) play in the theory of Fourier series.

Let us examine some more operators that arise naturally in harmonic analysis. For example, let us study the Fourier transform mapping acting on functions defined on the entire real line. Inequality (2.9) tells us that it is a bounded transformation defined on $L^1(-\infty, \infty)$ with values in $L^\infty(-\infty, \infty)$. The Plancherel theorem (3.10) tells us that it is a bounded transformation from $L^2(-\infty, \infty)$ into itself. A natural question, then, is whether it can be defined on other classes L^p and, if so, whether we obtain a bounded transformation with values in some classes L^q. But any function in L^p, $1<p<2$, can be written as a sum of a function in L^1 and one in L^2: put $f=f_1+f_2$, where $f_2(x)=f(x)$ when $|f(x)|\leqslant 1$ and $f_2(x)=0$ otherwise; then $f_1\in L^1$ and $f_2\in L^2$. Thus, we can write $\hat{f}=\hat{f}_1+\hat{f}_2$, where $\hat{f}_1$ is defined as the Fourier transform of a function in L^1 while $\hat{f}_2$ is defined by (3.10). The fact that these two definitions agree when a function belongs to $L^1\cap L^2$ implies that $\hat{f}$ is well defined. The following theorem tells us that the Fourier transformation defined on $L^p(-\infty, \infty)$, $1<p<2$, is bounded as a mapping into $L^q(-\infty, \infty)$, where q is the conjugate index to p.

(4.7) THE HAUSDORFF-YOUNG THEOREM: *If $f\in L^p(-\infty, \infty)$, $1\leqslant p\leqslant 2$, then $\hat{f}\in L^q(-\infty, \infty)$, where $1/p+1/q=1$, and*

$$\|\hat{f}\|_q \leqslant \|f\|_p.$$

We shall not prove (4.7) immediately. Instead, we shall give examples of some other inequalities that occur in harmonic analysis and then state some general results from which all these inequalities, including (4.5) and (4.7), follow as relatively easy consequences.

(4.8) YOUNG'S THEOREM: *Suppose $\dfrac{1}{r}=\dfrac{1}{p}+\dfrac{1}{q}-1$, where $\dfrac{1}{p}+\dfrac{1}{q}\geqslant 1$. If $f\in L^p(-\infty, \infty)$ and $g\in L^q(-\infty, \infty)$ then $f*g$ belongs to $L^r(-\infty, \infty)$ and*

$$\|f*g\|_r \leqslant \|f\|_p\|g\|_q.$$

The same result holds for periodic functions if we replace the interval $(-\infty, \infty)$ by the interval $(0, 1)$.

The operator on functions defined on $(-\infty, \infty)$ that corresponds to the conjugate function operator satisfies the same inequality (4.5). Using (2.18) and arguments that are completely analogous to those we gave at the beginning of this section, we see that this operator, called the *Hilbert transform*, can be defined by letting

$$\tilde{f}(x)=\lim_{y\to 0+}\tilde{f}(x,y)=\lim_{y\to 0+}\frac{1}{\pi}\int_{-\infty}^{\infty}f(t)\frac{x-t}{(x-t)^2+y^2}dt$$

correspond to $f\in L^p(-\infty, \infty)$, $1\leqslant p$, and that the following result holds:

(4.9) *If* $f \in L^p(-\infty, \infty)$, $1 < p < \infty$, *then its Hilbert transform* $\tilde{f}$ *also belongs to* $L^p(-\infty, \infty)$ *and*

$$\|\tilde{f}\|_p \leqslant A_p \|f\|_p,$$

where A_p *depends only on* p.

All the operators we have encountered up to this point are linear. There are several important transformations in harmonic analysis, however, that are not linear. Perhaps the best-known example of such a transformation is the *Hardy-Littlewood maximal* function. This operator is defined in the following way: if $f \in L^p(-\infty, \infty)$, $1 \leqslant p \leqslant \infty$, then its maximal function is the function whose value at $x \in (-\infty, \infty)$ is

$$f^*(x) = \sup_{h \neq 0} \frac{1}{h} \int_x^{x+h} |f(t)| \, dt.$$

Lebesgue's theorem on the differentiation of the integral guarantees that $f^*(x) < \infty$ for almost every x. It can be shown that

(4.10) *If* $f \in L^p(-\infty, \infty)$, $1 < p \leqslant \infty$, *then* $f^* \in L^p(-\infty, \infty)$ *and*

$$\|f^*\|_p \leqslant A_p \|f\|_p,$$

where A_p *depends only on* p.

The usefulness of the maximal function lies in the fact that it majorizes several important operators. Thus, it is clear why a theorem like (4.10) is desirable, as it immediately implies the boundedness of these operators.

Although the mapping $f \to f^*$ is not linear, it does satisfy the inequality $(f+g)^* \leqslant f^* + g^*$. This property is generally referred to as *sublinearity*. More generally, we say that an operator T mapping functions into functions is *sublinear* if, whenever Tf and Tg are defined, so is $T(f+g)$ and

$$|T(f+g)| \leqslant |Tf| + |Tg|.$$

In all these instances special cases of the inequalities involved are fairly easy to establish. For the conjugate function mapping the case $p = 2$ was seen to be an easy consequence of Theorem (3.8) (see (4.4)). A similar argument, using the Plancherel theorem, shows that the same is true for the Hilbert transform. We have pointed out that the cases $p = 1$, $q = \infty$ and $p = 2 = q$ of the Hausdorff-Young theorem had already been obtained by us in the previous sections. The inequality $\|f*g\|_1 \leqslant \|f\|_1 \|g\|_1$, which was the first result (property (i)) we established after introducing the operation of convolution, is the special case $r = p = q = 1$ of Young's theorem. Another special case of this theorem that is immediate is obtained when p and q are conjugate indices, $1/p + 1/q = 1$, and, thus, $r = \infty$; for this is simply a consequence of Hölder's inequality. Finally, it is clear that (4.10) holds when $p = \infty$.

It was M. Riesz who first discovered (in 1927) a general principle that asserted, in part, that in a wide variety of inequalities of the type we are discussing, special cases, such as those described in the previous paragraph, imply the general case. In order to state his theorem, known as the *M. Riesz convexity theorem*, we need to establish some notation. Suppose (M, μ) and (N, ν) are two measure spaces, where M and N are the point sets and μ and ν the measures. An operator T mapping measurable functions on M into measurable functions on N is said to be of *type* (p, q) if it is defined on $L^p(M)$ and there exists a constant A, independent of $f \in L^p(M)$, such that

$$\|Tf\|_q = \left(\int_N |Tf|^q d\nu \right)^{1/q} \leqslant A \left(\int_M |f|^p d\mu \right)^{1/p} = A\|f\|_p. \tag{4.11}$$

The least A for which (4.11) holds is called the *bound*, or *norm*, of T. The general principle can then be stated in the following way.

(4.12) THE M. RIESZ CONVEXITY THEOREM: *Suppose a linear operator T is of types (p_0, q_0) and (p_1, q_1), with bounds A_0 and A_1, respectively. Then it is of type (p_t, q_t), with bound $A_t \leqslant A_0^{1-t} A_1^t$, for $0 \leqslant t \leqslant 1$, where*

$$\frac{1}{p_t} = \frac{1-t}{p_0} + \frac{t}{p_1} \quad \text{and} \quad \frac{1}{q_t} = \frac{1-t}{q_0} + \frac{t}{q_1}.$$

The Hausdorff-Young theorem is an immediate consequence of this result. Since $T: f \to \hat{f}$ is of types $(1, \infty)$ and $(2, 2)$, it must be of type $(\frac{2}{2-t}, \frac{2}{t})$ for $0 \leqslant t \leqslant 1$. But if $p = \frac{2}{2-t}$, then the conjugate index is $q = \frac{p}{p-1} = \frac{2}{t}$. Since the "end-point" (i.e., $t=0$ and $t=1$) bounds are 1, we have

$$\|\hat{f}\|_q \leqslant 1^{(1-t)} 1^t \|f\|_p = \|f\|_p,$$

which is the inequality in (4.7).

Similarly, Young's theorem follows from (4.12). First, let us fix $g \in L^1$ and define $Tf = f * g$. We have seen that T is of type $(1, 1)$, with bound $\|g\|_1$, and of type (∞, ∞), also with bound $\|g\|_1$. Thus, T is of type $(\frac{1}{1-t}, \frac{1}{1-t})$, $0 \leqslant t \leqslant 1$, with a bound less than or equal to $\|g\|_1^{1-t} \|g\|_1^t = \|g\|_1$. Putting $p = \frac{1}{1-t}$ this gives us (4.8) with $r = p$ and $q = 1$. To obtain the general case we fix $f \in L^p$ and define $Tg = f * g$. We have just shown that T is of type $(1, p)$ with bound $\|f\|_p$. Letting $g \in L^q$, where q is conjugate to p, we also have T of type (q, ∞) with bound $\|f\|_p$. Thus, T is of type (p_t, q_t), with bound no greater than $\|f\|_p$, where $p_t = \frac{p}{p-t}$ and $q_t = \frac{p}{1-t}$, $0 \leqslant t \leqslant 1$. That is,

$$\|f * g\|_{q_t} \leqslant \|f\|_p \|g\|_{p_t}.$$

Since it follows immediately that $\frac{1}{q_t} = \frac{1}{p} + \frac{1}{p_t} - 1$, and, as t ranges between 0 and 1, $\frac{1}{p} + \frac{1}{p_t}$ ranges from $\frac{1}{p} + 1$ to 1, this is precisely the inequality of (4.8).

Unfortunately none of the other inequalities we stated can be derived from the special cases discussed above and the M. Riesz convexity theorem. For example, the conjugate function mapping, as we have seen, is easily seen to be of type (2, 2). Were we able to show that it is of type (1, 1), it would then follow that it is of type (p, p), $1 < p < 2$, and this, in turn, would imply the result for $p > 2$ (as we saw at the end of the proof of (4.5)). But we have already stated that this operator is not a bounded transformation on $L^1(0, 1)$. Nevertheless, there is a substitute result, due to Kolmogoroff, and an extension of the M. Riesz convexity theorem, due to Marcinkiewicz, that does allow us to obtain Theorem (4.5) much in the same way we obtained (4.7) and (4.8). Furthermore, this method is applicable to Theorems (4.9) and (4.10) as well.

The substitute result of Kolmogoroff is a condition that is weaker than type (1, 1). We shall consider this condition in a more general setting. First, however, we need to introduce the concept of the distribution function of a measurable function. Let g be a measurable function defined on the measure space (N, ν) and, for $y > 0$, $E_y = \{x \in N; \ |g(x)| > y\}$. Then the *distribution function of* g is the nonincreasing function $\lambda = \lambda_g$ defined for all $y > 0$ by

$$\lambda(y) = \nu(E_y).$$

It is an easy exercise in measure theory to show that if $g \in L^q(N)$, then

$$\|g\|_q = \left(\int_N |g(x)|^q d\nu \right)^{1/q} = \left(q \int_0^\infty y^{q-1} \lambda(y) \, dy \right)^{1/q}. \tag{4.13}$$

Suppose, now, that T is an operator of type (p, q), with bound A, $1 \le q < \infty$, mapping functions defined on M into functions defined on N. Let $f \in L^p(M)$, $g = Tf$, and λ the distribution function of g. Then

$$y^q \lambda(y) = \int_{E_y} y^q d\nu \le \int_{E_y} |g(x)|^q d\nu \le \int_N |g(x)|^q d\nu$$

$$\le \left(A \left[\int_M |f(t)|^p d\mu \right]^{1/p} \right)^q.$$

That is,

$$\lambda_g(y) = \lambda(y) \le \left(\frac{A}{y} \|f\|_p \right)^q. \tag{4.14}$$

This condition is easily seen to be weaker than boundedness. An operator that satisfies (4.14) for all $f \in L^p(M)$ is said to be of *weak-type* (p, q). If $q = \infty$, it is convenient to identify weak-type with type.

Kolmogoroff showed that the conjugate function mapping is of weak-type (1, 1). It is then immediate that the following theorem can be used to prove (4.5):

(4.15) THE MARCINKIEWICZ INTERPOLATION THEOREM: *Suppose T is a sublinear operator of weak types (p_0, q_0) and (p_1, q_1), where $1 \le p_i \le q_i \le \infty$ for $i = 0, 1$, and $q_0 \ne q_1, p_0 \ne p_1$. Then T is of type (p, q) whenever*

$$\frac{1}{p} = \frac{1-t}{p_0} + \frac{t}{p_1} \quad \text{and} \quad \frac{1}{q} = \frac{1-t}{q_0} + \frac{t}{q_1}, \quad 0 < t < 1.$$

Similarly, the *Hilbert transform* can be shown to be of weak type (1, 1); thus (4.9) is also a consequence of (4.15). The same is true of (4.10). We shall not prove any of these facts. The reader, however, should have no difficulty in checking that the maximal function mapping cannot be of type (1, 1) (take for f the characteristic function of a finite interval; then f^* is not integrable). The proof that it is of weak type (1, 1) is not hard. The corresponding results for the conjugate function and for the Hilbert transform, however, are somewhat more difficult.

The M. Riesz convexity theorem amd the Marcinkiewicz interpolation theorem have many more applications. The examples discussed in this section, however, are sufficient to illustrate the role they play in harmonic analysis.

5. HARMONIC ANALYSIS ON LOCALLY COMPACT ABELIAN GROUPS

We have discussed harmonic analysis associated with three different domains, the circle group (or the group of reals modulo one), the group of integers, and the (additive) group of real numbers. All of these are examples of *locally compact abelian groups*. These are abelian (commutative) groups G, with elements x, y, $z,\ldots$, endowed with a locally compact Hausdorff topology in such a way that the maps $x \to x^{-1}$ and $(x, y) \to xy$ (defined on G and $G \times G$, respectively) are continuous (we are following the usual custom of writing the operation on G as multiplication and not as addition—which was the case in our three examples; this should not be a source of confusion to the reader). In this section we shall indicate how harmonic analysis can be extended to functions defined on such groups.

On each such group G there exists a nontrivial regular measure M that, in analogy with Lebesgue measure, has the property that it is invariant with respect to translation. By this we mean that whenever A is a measurable subset of G then $m(A) = m(Ax)$ for all $x \in G$. This is equivalent to the assertion

$$\int_G f(yx)\,dm(y) = \int_G f(y)\,dm(y) \qquad (5.1)$$

for all $x \in G$ whenever f is an integrable function. It is obvious that any constant multiple of m also has this property. Conversely, it can be shown that any regular measure satisfying this invariance property must be a constant multiple of m. Such measures are known as *Haar measures*.

The operation of convolution of two functions f and g in $L^1(G)$ is defined, as in the classical case, by the integral

$$(f*g)(x) = \int_G f(xy^{-1}) g(y)\,dy.$$

The four properties (i), (ii), (iii), and (iv) (see the end of the first section) hold in this case as well. In particular, $f*g \in L^1(G)$ and $\|f*g\|_1 \leqslant \|f\|_1 \|g\|_1$.

Moreover, we shall now show that it is possible to give a definition of the Fourier transform so that (1.18) also holds; that is, $(f*g)^\hat{} = \hat{f}\hat{g}$ for all f and g in $L^1(G)$. We have seen that the Fourier transform of f is not usually defined on the domain of f. In case of the circle group, for example, Fourier transformation gave us functions defined on the integers. In order to describe the general situation we shall need the concept of a character: By a *character* of a locally compact group G we mean a continuous function, $\hat{x}$, on G such that $|\hat{x}(x)| = 1$ for all x in G and $\hat{x}(xy) = \hat{x}(x)\hat{x}(y)$ for all $x, y \in G$.

The collection of all characters of G is usually denoted by $\hat{G}$. If we define multiplication in $\hat{G}$ by letting $\hat{x}_1\hat{x}_2(x) = \hat{x}_1(x)\hat{x}_2(x)$, for all $x \in G$, whenever $\hat{x}_1$, $\hat{x}_2 \in \hat{G}$, $\hat{G}$ then becomes an abelian group. We introduce a topology on $\hat{G}$ by letting the sets

$$U(\epsilon, C, x_0) = \{\hat{x} \in \hat{G}; |\hat{x}(x) - \hat{x}_0(x)| < \epsilon, x \in C\},$$

where $\hat{x}_0 \in \hat{G}$, $\epsilon > 0$, and C is a compact subset of G, form a basis. With this topology $\hat{G}$ is then also a locally compact abelian group. $\hat{G}$ is usually called the *character group of G* or the *dual group of G*.

For example, when G is the group of real numbers we easily see that if we let a be a real number, then the mapping $\hat{x}: x \to e^{2\pi i a x}$, defined for all real x, is a character. One can show that all characters are of this type. Thus, there is a natural one-to-one correspondence between the group of real numbers and $\hat{G}$. Furthermore, this correspondence is a homeomorphism. Hence, we can identify G with $\hat{G}$ in this case.

If G is the group of reals modulo 1, the mappings $\hat{x}: x \to e^{2\pi i a x}$, $x \in G$, where a is an integer, are characters, and each character has this form. Thus, $\hat{G}$ and the integers are in a one-to-one correspondence that, in this case also, can be shown to be a homeomorphism. Therefore, we can identify $\hat{G}$ with the integers. Similarly the dual group of integers can be identified with the group of reals modulo 1.

In general, if we fix an x in G and consider the mapping $\hat{x} \to \hat{x}(x)$ we obtain a character on $\hat{G}$. It can be shown that every character has this form and that this correspondence between G and $(\hat{G})^{\wedge}$ is a homeomorphism. This result is known as the *Pontrjagin duality theorem* and it is usually stated, simply, by writing the equality $G = (\hat{G})^{\wedge}$. Because of this duality the functional notation $\hat{x}(x)$ is discarded and the symbol

$$\langle x, \hat{x} \rangle$$

is used instead. Thus, $\langle x, \hat{x} \rangle$ may be thought of as the value of the function x at $\hat{x}$, $x(\hat{x})$, as well as the value of $\hat{x}$ at x; these two values are clearly equal.

It is now clear, if we let ourselves be motivated by our three classical examples of locally compact abelian groups, that a natural definition of the Fourier transform for $f \in L^1(G)$, when G is a general locally compact abelian group, is to let it be the function $\hat{f}$ on $\hat{G}$ given by

$$\hat{f}(\hat{x}) = \int_G f(x) \overline{\langle x, \hat{x} \rangle} \, dm(x).$$

Many of the results we presented in the previous section hold in this case as well. For example, $\hat{f}$ is a continuous function on $\hat{G}$; when $\hat{G}$ is not compact the Riemann-Lebesgue theorem holds:

(5.2) *If $\hat{G}$ is not compact, $f \in L^1(G)$, and $\epsilon > 0$, then there exists a compact set $C \subset \hat{G}$ such that $|\hat{f}(\hat{x})| < \epsilon$ if $\hat{x}$ is outside of C.*

The basic relation (1.18) between convolution and Fourier transformation is true in general:

(5.3) *If f and g belong to $L^1(G)$ then $(f * g)\hat{} = \hat{f}\hat{g}$.*

Wiener's theorem (3.15) is still valid:

(5.4) *If $f \in L^1(G)$ and $\hat{f}(\hat{x})$ is never 0, then any $g \in L^1(G)$ can be approximated arbitrarily closely in the L^1-norm by functions of the form*

$$\sum a_k f(xt_k),$$

where the a_k's are a finite collection of complex numbers and the t_k's belong to G.

The Plancherel theorem also has an analog to this general case:

(5.5) *If we restrict the transformation $f \to \hat{f}$ to $L^1(G) \cap L^2(G)$, then the L^2 norms are preserved; that is, $\hat{f} \in L^2(\hat{G})$ and Parseval's formula holds:*

$$\|f\|_2 = \|\hat{f}\|_2.$$

Furthermore, this transformation can be extended to a norm preserving transformation of $L^2(G)$ onto $L^2(G)$.

Harmonic analysis can be generalized still further. For example, locally compact groups that are not abelian are associated with important versions of harmonic analysis (the theory of spherical harmonics is associated with the group of rotations in 3-space). We will not, however, pursue this topic further.

6. A SHORT GUIDE TO THE LITERATURE

So many books and papers have been written in harmonic analysis that no attempt will be made here to give anything like a comprehensive bibliography. Rather, our intention is to give some *very* brief suggestions to the reader who would like to pursue the subject further.

All that has been discussed here concerning Fourier series is contained in A. Zygmund's two-volume *Trigonometric Series* [10]. This scholarly book contains essentially all the important work that has been done on the subject. Anyone seriously interested in classical (or, for that matter, modern) harmonic analysis would do well to become acquainted with it. It is often worthwhile, however, to read a short treatment of a subject when learning it. R. R. Goldberg's *Fourier Transforms* [3] does an excellent job of presenting that part of Fourier integral theory that generalizes to locally compact abelian groups. In this book the reader will find a proof of Wiener's theorem and a more thorough discussion of the problem of spectral synthesis. For more comprehensive treatments of Fourier integral theory we refer the reader to S. Bochner's *Lectures on Fourier Integrals* [2] and E. C. Titchmarsh's *The Theory of the Fourier Integral* [8].

The literature dealing with the more abstract forms of harmonic analysis is also very large. Pontrjagin's classic *Topological Groups* [6] is still highly recommended reading. The same is true of A. Weil's *L'intégration dans les groupes topologiques et ses applications* [9]. Two very readable modern works that treat the subject of harmonic analysis on groups are Rudin's *Fourier Analysis on Groups* [7] and *Abstract Harmonic Analysis* by Hewitt and Ross [5]. We also recommend an excellent survey on this subject by J. Braconnier [1].

References

1. Braconnier, J., L'analyse harmonique dans les groupes abéliens, Monographies de l'Enseignement mathématique, No. 5.

2. Bochner, S., Lectures on Fourier Integrals. Princeton, N. J.: Princeton University Press, 1959.

3. Goldberg, R. R., Fourier Transforms. New York: Cambridge University Press, 1961.

4. Hardy, G. H., Divergent Series. Oxford: Clarendon Press, 1949.

5. Hewitt, E., and K. A. Ross, Abstract Harmonic Analysis. Berlin: Springer, 1963.

6. Pontrjagin, L., Topological Groups. Princeton, N. J.: Princeton University Press, 1946.

7. Rudin, W., Fourier Analysis on Groups. New York: Interscience Publishers, 1962.

8. Titchmarsh, E. C., The Theory of the Fourier Integral. Oxford: Clarendon Press, 1937.

9. Weil, A., L'intégration dans les groupes topologiques et ses applications. Paris: Hermann, 1940.

10. Zygmund, A., Trigonometric Series, 2nd ed. Cambridge: Cambridge University Press, 1959, 2 vols.

17

CAN ONE HEAR THE SHAPE OF A DRUM?

MARK KAC, The Rockefeller University, New York

To George Eugene Unlenbeck on the occasion of his sixty-fifth birthday

> "La Physique ne nous donne pas seulement
> l'occasion de résoudre des problèmes ... , elle nous fait
> pressentir la solution." H. POINCARÉ.

Before I explain the title and introduce the theme of the lecture I should like to state that my presentation will be more in the nature of a leisurely excursion than of an organized tour. It will not be my purpose to reach a specified destination at a scheduled time. Rather I should like to allow myself on many occasions the luxury of stopping and looking around. So much effort is being spent on streamlining mathematics and in rendering it more efficient, that a solitary transgression against the trend could perhaps be forgiven.

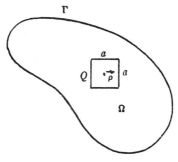

FIG. 1

1. And now to the theme and the title.

It has been known for well over a century that if a membrane Ω, held fixed along its boundary Γ (see Fig. 1), is set in motion its displacement (in the direction perpendicular to its original plane)

$$F(x, y; t) \equiv F(\vec{\rho}; t)$$

obeys the wave equation

$$\frac{\partial^2 F}{\partial t^2} = c^2 \nabla^2 F,$$

where c is a certain constant depending on the physical properties of the membrane and on the tension under which the membrane is held.

I shall choose units to make $c^2 = \frac{1}{2}$.

433

Of special interest (both to the mathematician and to the musician) are solutions of the form

$$F(\vec{\rho}; t) = U(\vec{\rho})e^{i\omega t},$$

for, being harmonic in time, they represent the *pure tones* the membrane is capable of producing. These special solutions are also known as normal modes.

To find the normal modes we substitute $U(\vec{\rho})e^{i\omega t}$ into the wave equation and see that U must satisfy the equation $\frac{1}{2}\nabla^2 U + \omega^2 U = 0$ with the boundary condition $U = 0$ on the boundary Γ of Ω, corresponding to the membrane being held fixed along its boundary.

The meaning of "$U = 0$ on Γ" should be made clear; for sufficiently smooth boundaries it simply means that $U(\vec{\rho}) \to 0$ as $\vec{\rho}$ approaches a point of Γ (from the inside). To show that a membrane is capable of producing a discrete spectrum of pure tones i.e. that there is a discrete sequence of ω's $\omega_1 \leq \omega_2 \leq \omega_3 \leq \cdots$ for which nontrivial solutions of

$$\tfrac{1}{2}\nabla^2 U + \omega^2 U = 0, \qquad U = 0 \text{ on } \Gamma,$$

exist, was one of the great problems of 19th century mathematical physics. Poincaré struggled with it and so did many others.

The solution was finally achieved in the early years of our century by the use of the theory of integral equations.

We now know and I shall ask you to believe me if you do not, that for regions Ω bounded by a smooth curve Γ there is a sequence of numbers $\lambda_1 \leq \lambda_2 \leq \cdots$ called eigenvalues such that to each there corresponds a function $\psi(\vec{\rho})$, called an eigenfunction, such that

$$\tfrac{1}{2}\nabla^2 \psi_n + \lambda_n \psi_n = 0$$

and $\psi_n(\vec{\rho}) \to 0$ as $\vec{\rho} \to$ a point of Γ.

It is customary to normalize the ψ's so that

$$\iint_\Omega \psi_n^2(\vec{\rho})\,d\vec{\rho} = 1.$$

Note that I use $d\vec{\rho}$ to denote the element of integration (in Cartesian coordinates, e.g., $d\vec{\rho} \equiv dxdy$).

2. The focal point of my exposition is the following problem:

Let Ω_1 and Ω_2 be two plane regions bounded by curves Γ_1 and Γ_2 respectively, and consider the eigenvalue problems:

$$\tfrac{1}{2}\nabla^2 U + \lambda U = 0 \text{ in } \Omega_1 \qquad\qquad \tfrac{1}{2}\nabla^2 V + \mu V = 0 \text{ in } \Omega_2$$

with

$$U = 0 \text{ on } \Gamma_1 \qquad\qquad\qquad V = 0 \text{ on } \Gamma_2.$$

Assume that for each n the eigenvalue λ_n for Ω_1 is equal to the eigenvalue μ_n

for Ω_2. Question: Are the regions Ω_1 and Ω_2 congruent in the sense of Euclidean geometry?

I first heard the problem posed this way some ten years ago from Professor Bochner. Much more recently, when I mentioned it to Professor Bers, he said, almost at once: "You mean, if you had perfect pitch could you find the shape of a drum."

You can now see that the "drum" of my title is more like a tambourine (which really is a membrane) and that stripped of picturesque language the problem is whether we can determine Ω if we know all the eigenvalues of the eigenvalue problem

$$\tfrac{1}{2}\nabla^2 U + \lambda U = 0 \text{ in } \Omega,$$

$$U = 0 \text{ on } \Gamma.$$

3. Before I go any further let me say that as far as I know the problem is still unsolved. Personally, I believe that one cannot "hear" the shape of a tambourine but I may well be wrong and I am not prepared to bet large sums either way.

What I propose to do is to see how much about the shape can be inferred from the knowledge of all the eigenvalues, and to impress upon you the multitude of connections between our problem and various parts of mathematics and physics.

It should perhaps be stated at this point that throughout the paper only *asymptotic properties* of large eigenvalues will be used. This may represent, of course, a serious loss of information and it may perhaps be argued that *precise* knowledge of *all* the eigenvalues may be sufficient to determine the shape of the membrane. It should also be pointed out, however, that quite recently John Milnor constructed two noncongruent sixteen dimensional tori whose Laplace-Betrami operators have exactly the same eigenvalues (see his one page note "Eigenvalues of the Laplace operator on certain manifolds" Proc. Nat. Acad. Sc., 51 (1964) 542).

4. The first pertinent result is that one can "hear" the area of Ω. This is an old result with a fascinating history which I shall now relate briefly.

At the end of October 1910 the great Dutch physicist H. A. Lorentz was invited to Göttingen to deliver the Wolfskehl lectures. Wolfskehl, by the way, endowed a prize for proving, or disproving, Fermat's last theorem and stipulated that in case the prize is not awarded the proceeds from the principal be used to invite eminent scientists to lecture at Göttingen.

Lorentz gave five lectures under the overall title "Alte und neue Fragen der Physik"—Old and new problems of physics—and at the end of the fourth lecture he spoke as follows (in free translation from the original German): "In conclusion there is a mathematical problem which perhaps will arouse the interest of mathematicians who are present. It originates in the radiation theory of Jeans.

"In an enclosure with a perfectly reflecting surface there can form standing electromagnetic waves analogous to tones of an organ pipe; we shall confine our attention to very high overtones. Jeans asks for the energy in the frequency interval $d\nu$. To this end he calculates the number of overtones which lie between the frequencies ν and $\nu+d\nu$ and multiplies this number by the energy which belongs to the frequency ν, and which according to a theorem of statistical mechanics is the same for all frequencies.

"It is here that there arises the mathematical problem to prove that the number of sufficiently high overtones which lies between ν and $\nu+d\nu$ is independent of the shape of the enclosure and is simply proportional to its volume. For many simple shapes for which calculations can be carried out, this theorem has been verified in a Leiden dissertation. There is no doubt that it holds in general even for multiply connected regions. Similar theorems for other vibrating structures like membranes, air masses, etc. should also hold."

If one expresses this conjecture of Lorentz in terms of our membrane, it emerges in the form:

$$N(\lambda) = \sum_{\lambda_n < \lambda} 1 \sim \frac{|\Omega|}{2\pi} \lambda.$$

Here $N(\lambda)$ is the number of eigenvalues less than λ, $|\Omega|$ the area of Ω and $\sim$ means that

$$\lim_{\lambda \to \infty} \frac{N(\lambda)}{\lambda} = \frac{|\Omega|}{2\pi}.$$

There is an apocryphal report that Hilbert predicted that the theorem would not be proved in his life time. Well, he was wrong by many, many years. For less than two years later Herman Weyl, who was present at the Lorentz' lecture and whose interest was aroused by the problem, proved the theorem in question, i.e. that as $\lambda \to \infty$

$$N(\lambda) \sim \frac{|\Omega|}{2\pi} \lambda.$$

Weyl used in a masterly way the theory of integral equations, which his teacher Hilbert developed only a few years before, and his proof was a crowning achievement of this beautiful theory. Many subsequent developments in the theory of differential and integral equations (especially the work of Courant and his school) can be traced directly to Weyl's memoir on the conjecture of Lorentz.

5. Let me now consider briefly a different physical problem which too is closely related to the problem of the distribution of eigenvalues of the Laplacian.

It can be taken as a basic postulate of classical statistical mechanics that if a system of M particles confined to a volume Ω is in equilibrium with a thermostat of temperature T the probability of finding specified particles at $\vec{r}_1, \vec{r}_2, \cdots, \vec{r}_M$ (within volume elements $\vec{dr}_1, \vec{dr}_2, \cdots, \vec{dr}_M$) is

$$\frac{\exp\left[-\dfrac{1}{kT} V(\vec{r}_1, \cdots, \vec{r}_M)\right] d\vec{r}_1 \cdots d\vec{r}_M}{\displaystyle\int_{\Omega} \cdots \int_{\Omega} \exp\left[-\dfrac{1}{kT} V(\vec{r}_1 \cdots \vec{r}_M)\right] d\vec{r}_1 \cdots d\vec{r}_M},$$

where $V(\vec{r}_1, \cdots, \vec{r}_M)$ is the interaction potential of the particles and $k = R/N$ with R the "gas constant" and N the Avogadro number.

For identical particles each of mass m obeying the so called Boltzmann statistics the corresponding assumption in quantum statistical mechanics seems much more complicated. One starts with the Schrödinger equation

$$\frac{\hbar^2}{2m} \nabla^2 \psi - V(\vec{r}_1, \cdots, \vec{r}_M)\psi = -E\psi \quad \left(\hbar = \frac{h}{2\pi}, \text{ where } h \text{ is the Planck constant}\right)$$

with the boundary condition $\lim \psi(\vec{r}_1. \cdots, \vec{r}_M) = 0$, whenever at least one $\vec{r}_k$ approaches the boundary of Ω. (This boundary condition has the effect of confining the particles to Ω.) Let $E_1 \leqq E_2 \leqq E_3 \leqq \cdots$ be the eigenvalues and $\psi_1, \psi_2, \cdots$ the corresponding normalized eigenfunctions. Then the basic postulate is that the probability of finding specified particles at $\vec{r}_1, \vec{r}_2, \cdots, \vec{r}_M$ (within $d\vec{r}_1, \cdots, d\vec{r}_M$) is

$$\frac{\displaystyle\sum_{s=1}^{\infty} e^{-E_s/kT} \psi_s^2(\vec{r}_1, \cdots, \vec{r}_M) d\vec{r}_1 \cdots d\vec{r}_M}{\displaystyle\sum_{s=1}^{\infty} e^{-E_s/kT}}.$$

There are actually no known particles obeying the Boltzmann statistics. But don't let this worry you—for our purposes this regrettable fact is immaterial.

Now, let us specialize our discussion to the case of an *ideal* gas which, by *definition*, means that $V(\vec{r}_1, \cdots, \vec{r}_M) \equiv 0$.

Classically, the probability of finding specified particles at $\vec{r}_1, \cdots, \vec{r}_M$ is clearly

$$\frac{d\vec{r}_1 \cdots d\vec{r}_M}{|\Omega|^M},$$

where $|\Omega|$ is now the volume of Ω.

Quantum mechanically the answer is not nearly so explicit. The Schrödinger equation for an ideal gas is

$$\frac{\hbar^2}{2m} \nabla^2 \psi = -E\psi$$

and the equation is obviously separable.

If I now consider the three-dimensional (rather than the $3M$-dimensional) eigenvalue problem

$$\tfrac{1}{2} \nabla^2 \psi(\vec{r}) = -\lambda\psi(\vec{r}), \qquad \vec{r} \in \Omega,$$

$$\psi(\vec{r}) \to 0 \quad \text{as} \quad \vec{r} \to \text{the boundary of } \Omega,$$

it is clear that the E_s as well as the $\psi_s(\vec{r}_1, \cdots, \vec{r}_M)$ are easily expressible in terms of the λ's and corresponding $\psi(r)$'s.

The formula for the probability of finding specified particles at $\vec{r}_1, \cdots, \vec{r}_M$ turns out to be

$$\prod_{k=1}^{M} \frac{\displaystyle\sum_{n=1}^{\infty} \exp\left[-\frac{\lambda_n\hbar^2}{mkT}\right]\psi_n^2(r_k)}{\displaystyle\sum_{n=1}^{\infty} \exp\left[-\frac{\lambda_n\hbar^2}{mkT}\right]} \, d\vec{r}_k.$$

Now, as $\hbar \to 0$ (or as $T \to \infty$) the quantum mechanical result should go over into the classical one and this immediately leads to the conjecture that as

$$\tau \to 0 \qquad \left[\tau = \frac{\hbar^2}{mkT}\right],$$

$$\sum_{n=1}^{\infty} e^{-\lambda_n\tau}\psi_n^2(\vec{r}) \sim \frac{1}{|\Omega|} \sum_{n=1}^{\infty} e^{-\lambda_n\tau}.$$

If instead of a realistic three-dimensional container Ω I consider a two-dimensional one, the result would still be the same

$$\sum_{n=1}^{\infty} e^{-\lambda_n\tau}\psi_n^2(\vec{r}) \sim \frac{1}{|\Omega|} \sum_{n=1}^{\infty} e^{-\lambda_n\tau}, \qquad \tau \to 0,$$

except that now $|\Omega|$ is the area of Ω rather than the volume.

Clearly the result is expected to hold only for $\vec{r}$ in the interior of Ω.

If we believe Weyl's result that (in the two-dimensional case)

$$N(\lambda) \sim \frac{|\Omega|}{2\pi}\lambda, \qquad \lambda \to \infty,$$

it follows immediately by an Abelian theorem that

$$\frac{1}{|\Omega|} \sum_{n=1}^{\infty} e^{-\lambda_n\tau} \sim \frac{1}{2\pi\tau}, \qquad \tau \to 0,$$

and hence that

$$\sum_{n=1}^{\infty} e^{-\lambda_n\tau}\psi_n^2(\vec{r}) \sim \frac{1}{2\pi\tau} = \frac{1}{2\pi}\int_0^{\infty} e^{-\lambda\tau}d\lambda.$$

Setting $A(\lambda) = \sum_{\lambda_n < \lambda} \psi_n^2(\vec{r})$, we can record the last result as

$$\int_0^\infty e^{-\lambda \tau} \, dA(\lambda) \sim \frac{1}{2\pi} \int_0^\infty e^{-\lambda \tau} \, d\lambda, \qquad \tau \to 0.$$

Since $A(\lambda)$ is nondecreasing we can apply the Hardy-Littlewood-Karamata Tauberian theorem and conclude what everyone would be tempted to conclude, namely that

$$A(\lambda) = \sum_{\lambda_n < \lambda} \psi_n^2(\vec{r}) \sim \frac{\lambda}{2\pi}, \qquad \lambda \to \infty,$$

for every $\vec{r}$ in the interior of Ω.

Though this asymptotic formula is thus nearly "obvious" on "physical grounds," it was not until 1934 that Carleman succeeded in supplying a rigorous proof.

In concluding this section it may be worthwhile to say a word about the "strategy" of our approach.

We are primarily interested, of course, in asymptotic properties of λ_n for large n. This can be approached by the device of studying the Dirichlet series

$$\sum_{n=1}^\infty e^{-\lambda_n t}$$

for small t. This in turn is most conveniently approached through the series

$$\sum_{n=1}^\infty e^{-\lambda_n t} \psi_n^2(\vec{\rho}) = \int_0^\infty e^{-\lambda t} \, dA(\lambda)$$

and thus we are led to the Abelian-Tauberian interplay described above.

6. It would seem that the physical intuition ought not only provide the mathematician with interesting and challenging conjectures, but also show him the way toward a proof and toward possible generalizations.

The context of the theory of black body radiation or that of quantum statistical mechanics, however, is too far removed from elementary intuition and too full of daring and complex physical extrapolations to be of much use even in seeking the kind of understanding that makes a mathematician comfortable, let alone in pointing toward a rigorous proof.

Fortunately, in a much more elementary context the problem of the distribution of eigenvalues of the Laplacian becomes quite tractable. Proofs emerge as natural extensions of physical intuition and interesting generalizations come within reach.

7. The physical context in question is that of *diffusion theory*, another branch of nineteenth century mathematical physics.

Imagine "stuff," initially concentrated at $\vec{\rho}(\equiv(x_0, y_0))$, diffusing through a plane region Ω bounded by Γ. Imagine furthermore that the stuff gets absorbed ("eaten") at the boundary.

The concentration $P_\Omega(\vec{\rho}\,|\,\vec{r};\,t)$ of matter at $\vec{r}(\equiv(x,\,y))$ at time t obeys the differential equation of diffusion

(a)
$$\frac{\partial P_\Omega}{\partial t} = \frac{1}{2}\,\nabla^2 P_\Omega,$$

the boundary condition

(b) $P_\Omega(\vec{\rho}\,|\,\vec{r};\,t) \to 0$ as $\vec{r}$ approaches a boundary point,

and the initial condition

(c) $P_\Omega(\vec{\rho}\,|\,\vec{r};\,t) \to \delta(\vec{r} - \vec{\rho})$ as $t \to 0$;

here $\delta(\vec{r}-\vec{\rho})$ is the Dirac "delta function," with "value" ∞ if $\vec{r} = \vec{\rho}$ and 0 if $\vec{r} \neq \vec{\rho}$.

The boundary condition (b) expresses the fact that the boundary is absorbing and the initial condition (c) the fact that initially all the "stuff" was concentrated at $\vec{\rho}$.

I have again chosen units so as to make the diffusion constant equal to $\frac{1}{2}$.

As is well known the concentration $P_\Omega(\vec{\rho}\,|\,\vec{r};\,t)$ can be expressed in terms of the eigenvalues λ_n and normalized eigenfunctions $\psi_n(\vec{r})$ of the problem

$$\tfrac{1}{2}\nabla^2\psi + \lambda\psi = 0 \text{ in } \Omega,$$

$$\psi = 0 \text{ on } \Gamma.$$

In fact, $P_\Omega(\vec{\rho}\,|\,\vec{r};\,t) = \sum_{n=1}^{\infty} e^{-\lambda_n t}\psi_n(\vec{\rho})\psi_n(\vec{r})$.

Now, for small t, it appears intuitively clear that particles of the diffusing stuff will not have had enough time to have felt the influence of the boundary Ω. As particles begin to diffuse they may not be aware, so to speak, of the disaster that awaits them when they reach the boundary.

We may thus expect that in some approximate sense

$$P_\Omega(\vec{\rho}\,|\,\vec{r};\,t) \sim P_0(\vec{\rho}\,|\,\vec{r};\,t), \text{ as } t \to 0,$$

where $P_0(\vec{\rho}\,|\,\vec{r};\,t)$ still satisfies the same diffusion equation

(a′)
$$\frac{\partial P_0}{\partial t} = \tfrac{1}{2}\,\nabla^2 P_0$$

and the same initial condition

(c′) $P_0(\vec{\rho}\,|\,\vec{r};\,t) = \delta(\vec{r} - \vec{\rho}), \qquad t \to 0,$

but is otherwise unrestricted.

Actually there is a slight additional restriction without which the solution is not unique (a remarkable fact discovered some years ago by D. V. Widder). The restriction is that $P_0 \geqq 0$ (or more generally that P_0 be bounded from below).

A similar restriction for P_Ω is not needed since for diffusion in a *bounded* region it follows automatically.

An explicit formula for P_0 is, of course, well known. It is

$$P_0(\vec{\rho} \mid \vec{r}; t) = \frac{1}{2\pi t} \exp\left[-\frac{\|\vec{r} - \vec{\rho}\|^2}{2t}\right],$$

where $\|\vec{r} - \vec{\rho}\|$ denotes the Euclidean distance between $\vec{\rho}$ and $\vec{r}$.

I can now state a little more precisely the principle of "not feeling the boundary" explained a moment ago.

The statement is that as $t \to 0$

$$P_\Omega(\vec{\rho} \mid \vec{r}; t) = \sum_{n=1}^\infty e^{-\lambda_n t} \psi_n(\vec{\rho})\psi_n(\vec{r}) \sim \frac{1}{2\pi t} \exp\left[-\frac{\|\vec{r} - \vec{\rho}\|^2}{2t}\right] = P_0(\vec{\rho} \mid \vec{r}; t),$$

where $\sim$ stands here for "is approximately equal to." This is a bit vague but let it go at that for the moment.

If we can trust this formula even for $\vec{\rho} = \vec{r}$ we get

$$\sum_{n=1}^\infty e^{-\lambda_n t} \psi_n^2(\vec{r}) \sim \frac{1}{2\pi t}$$

and if we display still more optimism we can integrate the above and, making use of the normalization condition

$$\int_\Omega \psi_n^2(\vec{r})\, d\vec{r} = 1,$$

obtain

$$\sum_{n=1}^\infty e^{-\lambda_n t} \sim \frac{|\Omega|}{2\pi t}.$$

We recognize immediately the formulas discussed a while back in connection with the quantum-statistical-mechanical treatment of the ideal gas. If we apply the Hardy-Littlewood-Karamata theorem, alluded to before, we obtain as corollary the theorems of Carleman and Weyl.

To do this, however, we must be allowed to interpret $\sim$ as meaning "asymptotic to."

8. Now, a little mathematical soul-searching. Aren't we as far from a rigorous treatment as we were before? True, diffusion is more familiar than black body radiation or quantum statistics. But familiarity gives comfort, at best, and comfort may still be (and often is) miles away from the rigor demanded by mathematics.

Let us see then what we can do about tightening the loose talk.

First let me dispose of a few minor items which may cause you worry.

When I write $\psi = 0$ on Γ or $P(\vec{\rho} \mid \vec{r}; t) \to 0$ as $\vec{r}$ approaches a boundary point of Ω there is always a question of interpretation.

Let me assume that Γ is sufficiently regular so that no ambiguity arises i.e.

$$P(\vec{\rho}\,|\,\vec{r};t) \to 0 \quad \text{as} \quad \vec{r} \to \text{a boundary point of } \Omega,$$

means exactly what it says, while $\psi = 0$ on Γ means

$$\psi \to 0 \quad \text{as} \quad \vec{r} \to \text{a boundary point of } \Omega.$$

Likewise, $P(\vec{\rho}\,|\,\vec{r};t) \to \delta(\vec{r}-\vec{\rho})$ as $t \to 0$, has the obvious interpretation, i.e.

$$\lim_{t \to 0} \iint_A P(\vec{\rho}\,|\,\vec{r};t)\,d\vec{r} = 1$$

for every open set A containing $\vec{\rho}$.

Now, to more pertinent items. If the mathematical theory of diffusion corresponds in any way to physical reality we should have the inequality

$$P_{\Omega}(\vec{\rho}\,|\,\vec{r};t) \leq P_0(\vec{\rho}\,|\,\vec{r};t) = \frac{\exp\left[-\dfrac{\|\vec{\rho}-\vec{r}\|^2}{2t}\right]}{2\pi t}.$$

For surely less stuff will be found at $\vec{r}$ at time t if there is a possibility of matter being destroyed (on the boundary Γ of Ω) than if there were no possibility of such destruction.

Now let Q be a square with center at $\vec{\rho}$ totally contained in Ω. Let its boundary act as an absorbing barrier and denote by $P_Q(\vec{\rho}\,|\,\vec{r};t)$, $\vec{r} \in Q$, the corresponding concentration at $\vec{r}$ at time t.

In other words, P_Q satisfies the differential equation

(a'')
$$\frac{\partial P_Q}{\partial t} = \frac{1}{2}\,\nabla^2 P_Q$$

and the initial condition

(c'')
$$P_Q(\vec{\rho}\,|\,\vec{r};t) \to \delta(\vec{r}-\vec{\rho}) \quad \text{as} \quad t \to 0.$$

It also satisfies the boundary condition

(b'')
$$P_Q(\vec{\rho}\,|\,\vec{r};t) \to 0 \quad \text{as} \quad \vec{r} \to \text{a boundary point of } Q.$$

Again it appears obvious that

$$P_Q(\vec{\rho}\,|\,\vec{r};t) \leq P_{\Omega}(\vec{\rho}\,|\,\vec{r};t), \qquad \vec{r} \in Q,$$

for the diffusing stuff which reaches the boundary of Q is lost as far as P_Q is concerned but *need not* be lost as a contribution to P_{Ω}.

Q has been chosen so simply because $P_Q(\vec{\rho}\,|\,\vec{r};t)$ is known explicitly, and, in particular

$$P_Q(\vec{\rho} \,|\, \vec{\rho}; t) = \frac{4}{a^2} \sum_{\substack{m,n \\ \text{odd integers}}} \exp\left[-\frac{(m^2 + n^2)\pi^2}{2a^2} t\right],$$

where a is the side of the square.

The combined inequalities

$$P_Q(\vec{\rho} \,|\, \vec{r}; t) \leqq P_\Omega(\vec{\rho} \,|\, \vec{r}; t) \leqq \frac{\exp\left[-\dfrac{\|\vec{r} - \vec{\rho}\|^2}{2t}\right]}{2\pi t}$$

hold for all $\vec{r} \in Q$ and in particular for $\vec{r} = \vec{\rho}$. In this case we get

$$\frac{4}{a^2} \sum_{\substack{m,n \\ \text{odd integers}}} \exp\left[-\frac{(m^2 + n^2)\pi^2}{2a^2} t\right] \leqq \sum_{n=1}^{\infty} e^{-\lambda_n t} \psi_n^2(\vec{\rho}) \leqq \frac{1}{2\pi t}$$

and it is a simple matter to prove that as $t \to 0$ we have *asymptotically*

$$\frac{4}{a^2} \sum_{\substack{m,n \\ \text{odd integers}}} \exp\left[-\frac{(m^2 + n^2)\pi^2}{2a^2} t\right] \sim \frac{1}{2\pi t}.$$

Thus asymptotically for $t \to 0$ $\sum_{n=1}^{\infty} e^{-\lambda_n t} \psi_n^2(\vec{\rho}) \sim 1/2\pi t$ and Carleman's theorem follows.

It is only a little harder to prove Weyl's theorem.

If one integrates over Q the inequality

$$\frac{4}{a^2} \sum_{\substack{m,n \\ \text{odd}}} \exp\left[-\frac{(m^2 + n^2)\pi^2}{2a^2} t\right] \leqq \sum_{n=1}^{\infty} e^{-\lambda_n t} \psi_n^2(\vec{\rho})$$

one obtains

$$4 \sum_{\substack{m,n \\ \text{odd}}} \exp\left[-\frac{(m^2 + n^2)\pi^2}{2a^2} t\right] \leqq \sum_{n=1}^{\infty} e^{-\lambda_n t} \iint_Q \psi_n^2(\vec{\rho}) \, d\vec{\rho}.$$

We now cover Ω with a net of squares of side a, as shown in Fig. 2, and keep only those contained in Ω. Let $N(a)$ be the number of these squares and let $\Omega(a)$ be the union of all these squares. We have

$$\sum_{n=1}^{\infty} e^{-\lambda_n t} = \sum_{n=1}^{\infty} e^{-\lambda_n t} \iint_\Omega \psi^2(\vec{\rho}) \, d\vec{\rho} \geqq \sum_{n=1}^{\infty} e^{-\lambda_n t} \iint_{\Omega(a)} \psi_n^2(\vec{\rho}) \, d\vec{\rho}$$

$$\geqq 4N(a) \sum_{\substack{m,n \\ \text{odd}}} \exp\left[-\frac{(m^2 + n^2)\pi^2}{2a^2} t\right]$$

and, integrating the inequality $P_\Omega(\vec{\rho}\,|\,\vec{\rho};\ t) \leqq 1/2\pi t$ over Ω we get $\sum_{n=1}^{\infty} e^{-\lambda_n t}$ $\leqq |\Omega|/2\pi t$.

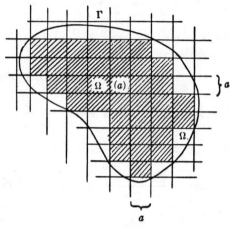

Noting that $N(a)a^2 = |\Omega(a)|$ we record the fruits of our latest labor in the form of the inequality

$$|\Omega(a)|\,\frac{4}{a^2} \sum_{\substack{m,n \\ \text{odd}}} \exp\left[-\frac{(m^2+n^2)\pi^2}{2a^2}\,t\right] \leq \sum_{n=1}^{\infty} e^{-\lambda_n t} \leq \frac{|\Omega|}{2\pi t}\,.$$

From the fact (already noted above) that

$$\lim_{t\to 0} 2\pi t\,\frac{4}{a^2} \sum_{\substack{m,n \\ \text{odd}}} \exp\left[-\frac{(m^2+n^2)\pi^2}{2a^2}\,t\right] = 1$$

we conclude easily that

$$|\Omega(a)| \leq \liminf_{t\to 0} 2\pi t \sum_{n=1}^{\infty} e^{-\lambda_n t} \leq \limsup_{t\to 0} 2\pi t \sum_{n=1}^{\infty} e^{-\lambda_n t} \leq |\Omega|\,;$$

and since, by choosing a sufficiently small, we can make $|\Omega(a)|$ arbitrarily close to $|\Omega|$, we must have $\lim_{t\to 0} 2\pi t \sum_{n=1}^{\infty} e^{-\lambda_n t} = |\Omega|$ or, in other words,

$$\sum_{n=1}^{\infty} e^{-\lambda_n t} \sim \frac{|\Omega|}{2\pi t}\,, \qquad t\to 0.$$

9. Are we now through with rigor? Not quite. For while the inequalities

$$P_\Omega(\vec{\rho}\,|\,\vec{r};\ t) \leqq \frac{\exp\left[-\dfrac{\|\vec{r}-\vec{\rho}\|^2}{2t}\right]}{2\pi t}$$

$$P_\Omega(\vec{\rho} \mid \vec{r}; t) \geqq P_Q(\vec{\rho} \mid \vec{r}; t), \qquad \vec{r} \in Q,$$

are utterly obvious on intuitive grounds they must be proved. Let me indicate a way of doing it which is probably by far not the simplest. I am choosing it to exhibit yet another physical context.

It has been known since the early days of this century, through the work of Einstein and Smoluchowski, that diffusion is but a macroscopic manifestation of microscopic Brownian motion.

Under suitable physical assumptions $P_\Omega(\vec{\rho} \mid \vec{r}; t)$ can be interpreted as the probability density of finding a free Brownian particle at $\vec{r}$ at time t if it started on its erratic journey at $t = 0$ from $\vec{\rho}$ and if it gets absorbed when it comes to the boundary of Ω.

If a large number N of independent free Brownian particles are started from $\vec{\rho}$ then

$$N \iint_A P(\vec{\rho} \mid \vec{r}; t) \, d\vec{r}$$

is the average number of these particles which are found in A at time t. Since the statistical percentage error is of the order $1/\sqrt{N}$ continuous diffusion theory is an excellent approximation when N is large.

A significant deepening of this point of view was achieved in the early twenties by Norbert Wiener. Instead of viewing the problem as a problem in *statistics* of *particles* he viewed it as a problem in *statistics* of *paths*. Without entering into details let me review briefly what is involved here.

Consider the set of all continuous curves $\vec{r}(\tau)$, $0 \leqq \tau < \infty$, starting from some arbitrarily chosen origin O. Let $\Omega_1, \Omega_2, \cdots, \Omega_n$ be open sets and $t_1 < t_2 < \cdots < t_n$ ordered instants of time. The Einstein-Smoluchowski theory required that (with suitable units)

Prob. $\{ \vec{\rho} + \vec{r}(t_1) \in \Omega_1, \vec{\rho} + \vec{r}(t_2) \in \Omega_2, \cdots, \vec{\rho} + \vec{r}(t_n) \in \Omega_n \}$

$$= \int_{\Omega_1} \cdots \int_{\Omega_n} P_0(\vec{\rho} \mid \vec{r_1}; t_1) P_0(\vec{r_1} \mid \vec{r_2}; t_2 - t_1) \cdots P_0(\vec{r_{n-1}} \mid \vec{r_n}; t_n - t_{n-1}) \, d\vec{r_1} \cdots d\vec{r_n}$$

where, as before,

$$P_0(\vec{\rho} \mid \vec{r}; t) = \frac{1}{2\pi t} \exp \left[-\frac{\|\vec{r} - \vec{\rho}\|^2}{2t} \right].$$

Wiener has shown that it is possible to construct a completely additive measure on the space of all continuous curves $\vec{r}(\tau)$ emanating from the origin such that the set of curves $\vec{\rho} + \vec{r}(\tau)$ which at times $t_1 < t_2 < \cdots < t_n$ find themselves in open sets $\Omega_1, \Omega_2, \cdots, \Omega_n$ respectively, has measure given by the Einstein-Smoluchowski formula above.

The set of curves such that $\vec{\rho} + \vec{r}(\tau) \in \Omega$, $0 \leqq \tau \leqq t$, and $\vec{\rho} + \vec{r}(t) \in A$ (A—an open set) turns out to be measurable and it can be shown, if Ω has sufficiently smooth boundaries, that this measure is equal to

$$\int_A P_\Omega(\vec{\rho} \mid \vec{r}; t)\, d\vec{r}.$$

This is not a trivial statement and it should come as no surprise that it trivially implies the inequalities we needed a while back to make precise the principle of not feeling the boundary.

In fact, as the reader no doubt sees, the inequalities in question are simply a consequence of the fact that if sets $\mathcal{A}$, $\mathcal{B}$, $\mathcal{C}$ are such that

$$\mathcal{A} \subset \mathcal{B} \subset \mathcal{C}$$

then meas. $\mathcal{A} \leq$ meas. $\mathcal{B} \leq$ meas. $\mathcal{C}$.

One final remark before we go on. The set of curves for which

$$\vec{\rho} + \vec{r}(\tau) \in \Omega, \qquad 0 \leq \tau \leq t \quad \text{and} \quad \vec{\rho} + \vec{r}(t) \in A$$

is measurable even if the boundary of Ω is quite wild. The measure can still be written as $\int_A P_\Omega(\vec{\rho} \mid \vec{r}; t)d\vec{r}$ and it can be shown that in the interior of Ω, $P_\Omega(\vec{\rho} \mid \vec{r}; t)$ satisfies the diffusion equation $\partial P_\Omega / \partial t = \frac{1}{2} \nabla^2 P_\Omega$ as well as the initial condition

$$\lim_{t \to 0} \int_A P_\Omega(\vec{\rho}; \vec{r}; t)\, d\vec{r} = 1,$$

for all open sets A such that $\vec{\rho} \in A$.

It is, however, no longer clear how to interpret the boundary condition that

$$P_\Omega(\vec{\rho} \mid \vec{r}; t) \to 0 \quad \text{when} \quad \vec{r} \to \Gamma.$$

This difficulty forces the classical theory of diffusion to consider reasonably smooth boundaries. The probabilistic interpretation of $P_\Omega(\vec{\rho} \mid \vec{r}; t)$ provides a natural definition of a *generalized solution* of the boundary value problem under consideration.

10. We are now sure that we can hear the area of a drum and it may seem that we spent a lot of effort to achieve so little.

Let me now show you that the approach we used can be extended to yield more, but to avoid certain purely geometrical complications I shall restrict myself to convex drums.

We have achieved our first success by introducing the principle of not feeling the boundary. But if $\vec{\rho}$ is close to the boundary Γ of Ω then the diffusing particles starting from $\vec{\rho}$ will, to some extent, begin to be influenced by Γ.

Let $\vec{q}$ be the point on Γ closest to $\vec{\rho}$ and let $l(\vec{\rho})$ be the straight line perpendicular to the line joining $\vec{\rho}$ and $\vec{q}$. (See Fig. 3.) Then a diffusing particle starting from ρ will see for a short time the boundary Γ as the straight line $l(\vec{\rho})$.

One may say, using again somewhat picturesque language, that, for small t, the particle has not had time to feel the curvature of the boundary.

If this principle is valid I should be allowed to approximate (for small t)

$$P_{\Omega}(\vec{\rho}\,|\,\vec{r};\,t) \quad \text{by} \quad P_{l(\vec{\rho})}(\vec{\rho}\,|\,\vec{r};\,t),$$

where $P_{l(\vec{\rho})}(\vec{\rho}\,|\,\vec{r};\,t)$ satisfies again the diffusion equation

$$\frac{\partial P}{\partial t} = \frac{1}{2}\,\nabla^2 P$$

with the initial condition $P \to \delta(\vec{\rho}-\vec{r})$ as $t \to 0$, but with the boundary condition

$$P_{l(\vec{\rho})}(\vec{\rho}\,|\,\vec{r};\,t) \to 0 \quad \text{as } \vec{r} \text{ approaches a point on } l(\vec{\rho}).$$

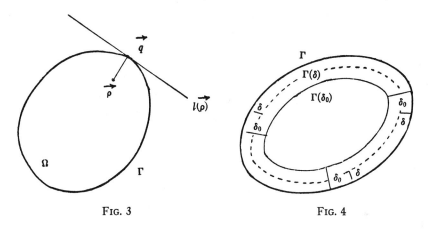

FIG. 3   FIG. 4

Carrying this optimism as far as possible we would expect that to a good approximation

$$\int_{\Omega} P_{\Omega}(\vec{\rho}\,|\,\vec{\rho};\,t)\,d\vec{\rho} \sim \int_{\Omega} P_{l(\vec{\rho})}(\vec{\rho}\,|\,\vec{\rho};\,t)\,d\vec{\rho}.$$

It is well known that

$$P_{l(\vec{\rho})}(\vec{\rho}\,|\,\vec{\rho};\,t) = \frac{1 - e^{-2\delta^2/t}}{2\pi t},$$

where $\delta = \|\vec{q}-\vec{\rho}\|$ = minimal distance from $\vec{\rho}$ to Γ. Thus (hopefully!)

$$\iint_{\Omega} P_{\Omega}(\vec{\rho}\,|\,\vec{\rho};\,t)\,d\vec{\rho} = \sum_{n=1}^{\infty} e^{-\lambda_n t} \sim \frac{|\Omega|}{2\pi t} - \frac{1}{2\pi t}\int_{\Omega} e^{-2\delta^2/t}\,d\vec{\rho}.$$

Here $|\Omega|/2\pi t$ is our old friend from before and it remains to calculate asymptotically (as $t \to 0$) the integral $\int_{\Omega} e^{-2\delta/t}\,d\vec{\rho}$. To do this consider the curve $\Gamma(\delta)$ of points in Ω whose "distance" from Γ is δ. (See Fig. 4.)

For small enough δ, $\Gamma(\delta)$ is well defined (and even convex) and the major contribution to our integral comes from small δ.

If $L(\delta)$ denotes the length of $\Gamma(\delta)$ we have

$$\int_\Omega e^{-2\delta^2/t} \, d\vec{\rho} = \int_0^{\delta_0} e^{-2\delta^2/t} L(\delta) \, d\delta + \text{something less than } \left| \Omega \right| e^{-2\delta_0^2/t}$$

and hence, neglecting an exponentially small term (as well as terms of order t)

$$\int_\Omega e^{-2\delta^2/t} \, d\vec{\rho} \sim \sqrt{t} \int_0^{\delta_0/\sqrt{t}} e^{-2x^2} L(x\sqrt{t}) \, dx \sim \sqrt{t} L \int_0^\infty e^{-2x^2} \, dx = \frac{L}{4} \sqrt{2\pi t},$$

where $L = L(0)$ is the length of Γ.

We are finally led to the formula

$$\sum_{n=1}^\infty e^{-\lambda_n t} \sim \frac{\left| \Omega \right|}{2\pi t} - \frac{L}{4} \frac{1}{\sqrt{2\pi t}}, \quad \text{for } t \to 0,$$

and so we can also "hear" the length of the circumference of the drum!

The last asymptotic formula was proved only a few years ago by the Swedish mathematician Ake Pleijel [2] using an entirely different approach.

It is worth remarking that we can now prove that if all the frequencies of a drum are equal to those of a circular drum then the drum must itself be circular. This follows at once from the classical isoperimetric inequality which states that $L^2 \geqq 4\pi \left| \Omega \right|$, with equality occurring *only* for a *circle*.

By pitch alone one can thus determine whether a drum is circular or not!

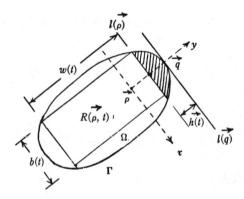

Fig. 5

11. Can the heuristic argument again be made rigorous? Indeed it can. First, we use the inequality

$$P_\Omega(\vec{\rho} \mid \vec{\rho}; t) \leqq P_{l(\rho)}(\vec{\rho} \mid \vec{\rho}; t) = \frac{1}{2\pi t} - \frac{1}{2\pi t} e^{-2\delta^2/t},$$

which is simply a refinement of the one used previously, namely, $P_\Omega(\vec{\rho}\,|\,\vec{\rho};\,t)$ $\leq 1/2\pi t$, and which can be proven the same way.

Next we need a precise lower estimate for $P_\Omega(\vec{\rho}\,|\,\vec{\rho};\,t)$ and this is a little more difficult. We "inscribe" the rectangle $R(\vec{\rho},\,t)$ as shown in Fig. 5, where $h(t)$, the height of the shaded segment, is to be determined a little later.

Let the side of R along the base of the segment be $b(t)$ and the other side be $w(t)$. It should be clear from the picture that the y-axis *bisects* the sides of the rectangle which are parallel to the x-axis.

Now consider $P_R(\vec{\rho}\,|\,\vec{\rho};\,t)$. This notation is perhaps confusing since it suggests that we are dealing with a boundary value problem in which the boundary varies with time. This is not the case. What we have in mind is the following: *fix t, find $P_{R(t)}(\vec{\rho}\,|\,\vec{r};\,\tau)$* which is defined unambiguously, and finally set $\tau = t$. The result is $P_R(\vec{\rho}\,|\,\vec{r};\,t)$. A convenient expression is

$$P_R(\vec{\rho}\,|\,\vec{\rho};\,t) = \frac{1}{2\pi t}\left\{\sum_{-\infty}^{\infty}\left(\exp\left[-\frac{2b^2}{t}n^2\right] - \exp\left[-\frac{2b^2}{t}\left(n+\frac{1}{2}\right)^2\right]\right)\right\}$$
$$\times\left\{\sum_{-\infty}^{\infty}\left(\exp\left[-\frac{2w^2}{t}n^2\right] - \exp\left[-\frac{2w^2}{t}\left(n+\frac{\bar{\delta}}{w}\right)^2\right]\right)\right\}$$

where $\bar{\delta} = \delta - h(t) = \|\vec{q}-\vec{\rho}\| - h(t)$. Now let $h(t) = \epsilon\sqrt{t}$ and, assuming that $l(\vec{\rho})$ is actually *tangent* to the curve (which for a convex curve will happen with at most a denumerable number of exceptional points $\vec{q}$), we have

$$\lim_{t\to 0}\frac{b(t)}{h(t)} = \lim_{t\to 0}\frac{b(t)}{\epsilon\sqrt{t}} = \infty,$$

and consequently

$$\sum_{-\infty}^{\infty}\left(\exp\left[-\frac{2b^2}{t}n^2\right] - \exp\left[-\frac{2b^2}{t}\left(n+\frac{1}{2}\right)^2\right]\right) = 1 + o(1).$$

This is not quite enough, however, and one needs the stronger estimate

$$\sum_{-\infty}^{\infty}\left(\exp\left[-\frac{2b^2}{t}n^2\right] - \exp\left[-\frac{2b^2}{t}\left(n+\frac{1}{2}\right)^2\right]\right) = 1 + o(\sqrt{t}).$$

This will surely be the case, for example, if the curvature exists at $\vec{q}$, for this would imply that $h(t)\sim b^2(t)$ and the $o(\sqrt{t})$ term above would then be an enormous overestimate. Very mild additional regularity conditions at nearly all points $\vec{q}$ would insure $o(\sqrt{t})$. Without entering into a discussion of these conditions let us simply assume the boundary to be such as to guarantee at least $o(\sqrt{t})$.

Since $w(t)$ remains bounded from below as $t\to 0$, we also have

$$\sum_{-\infty}^{\infty}\left(\exp\left[-\frac{2w^2}{t}n^2\right] - \exp\left[-\frac{2w^2}{t}\left(n+\frac{\bar{\delta}}{w}\right)^2\right]\right)$$
$$= 1 - e^{-2\bar{\delta}^2/t} + \text{exponentially small terms.}$$

We are now almost through. We write (cf. Fig. 6)

$$\sum_{n=1}^{\infty} e^{-\lambda_n t} = \int_{\Omega} P_{\Omega}(\vec{\rho}\mid\vec{\rho};t)\,d\vec{\rho} > \int_{\Omega(\epsilon\sqrt{t})} P_R(\vec{\rho}\mid\vec{\rho};t)\,d\vec{\rho}$$

$$= \frac{(1 + o(\sqrt{t}))}{2\pi t}\int_{\Omega(\epsilon\sqrt{t})} (1 - e^{-2\bar{\delta}^2/t} + \text{exponentially small terms})\,d\vec{\rho}.$$

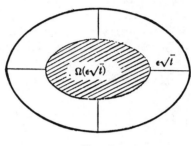

FIG. 6

Except then for exponentially small terms and the factor $1 + o(\sqrt{t})$ in front we have the integral

$$\int_{\Omega(\epsilon\sqrt{t})} (1 - e^{-2\bar{\delta}^2/t})\,d\vec{\rho}$$

which, as before, can be seen to be asymptotically

$$\left|\,\Omega(\epsilon\sqrt{t})\,\right| - \frac{L}{4}\sqrt{2\pi t},$$

where one neglects terms of order t and exponentially small terms. Since asymptotically $\left|\Omega(\epsilon\sqrt{t})\right| \sim \left|\Omega\right| - L\epsilon\sqrt{t}$ one can obtain the inequality

$$\sum_{n=1}^{\infty} e^{-\lambda_n t} > \frac{\left|\,\Omega\,\right|}{2\pi t} - \frac{(L + \epsilon')}{4}\frac{1}{\sqrt{2\pi t}},$$

where ϵ' is related in a simple way to ϵ. Since ϵ' can be made arbitrarily small, the asymptotic formula

$$\sum_{n=1}^{\infty} e^{-\lambda_n t} \sim \frac{\left|\,\Omega\,\right|}{2\pi t} - \frac{L}{4}\frac{1}{\sqrt{2\pi t}}$$

follows.

12. If our overall strategy of attack on the problem is right we should be able to go on and for points very close to a *smooth* boundary replace the boundary locally by suitable circles of curvature.

A result of Pleijel suggests strongly that for a simply connected drum with a smooth boundary (i.e. without corners and with curvature existing at every point) one has

$$\sum_{n=1}^{\infty} e^{-\lambda_n t} \sim \frac{|\Omega|}{2\pi t} - \frac{L}{4} \frac{1}{\sqrt{2\pi t}} + \frac{1}{6} .$$

Unfortunately I am unable to obtain this, for the exasperating reason that I am unable to get a workable expression for $P_\Omega(\vec{\rho}\,|\,\vec{\rho}; t)$ if Ω is a circle.

Rather than yield to despair over this sad state of affairs let me devote the remainder of the lecture to *polygonal drums*, i.e. drums whose boundaries are polygons. This study will show beyond the shadow of a doubt that the constant term in our asymptotic expansion owes its existence to the overall curvature of the boundary.

13. Before I go on I need an expression for $P_{S(\theta_0)}(\vec{\rho}\,|\,\vec{r}; t)$ where $S(\theta_0)$ is an infinite wedge of angle θ_0. In other words $P_{S(\theta_0)}$ is the solution of

$$\frac{\partial P}{\partial t} = \frac{1}{2} \nabla^2 P$$

subject to the usual initial condition $P_{S(\theta_0)}(\vec{\rho}\,|\,\vec{r}; t) \to \delta(\vec{\rho}-\vec{r})$, $t\to 0$, and vanishing as $\vec{r}$ approaches a point on either side of the angle θ_0.

This is a very old, very classical, problem and if $\theta_0 = \pi/m$, with m an integer, it can be solved by the familiar method of images. For m not an integer, Sommerfeld invented a method which, so to speak, extends the method of images to a Riemann surface. A little later, in 1899 to be precise, H. S. Carslaw gave a more elementary approach in which $P_{S(\theta_0)}(\vec{\rho}\,|\,\vec{r}; t)$ is represented by a suitable contour integral. Carslaw transforms the integral into an infinite series of Bessel functions but for our purposes it is best to resist the temptation of Bessel functions and to reduce the integral to a different form. I shall skip the details (though some are quite instructive) and simply reproduce the final result.

Set

$$v(\alpha) = (1/2\pi t) \sum_{\substack{\theta-\alpha-\pi<2k\theta_0 \\ <\theta-\alpha+\pi}} \exp\left[-\frac{r^2 - 2r\rho\cos(\theta-\alpha-2k\theta_0)+\rho^2}{2t}\right]$$

$$- \left(\sin\frac{\pi^2}{\theta_0}\right) \frac{\exp\left[-\dfrac{r^2+\rho^2}{2t}\right]}{4\pi\theta_0 t} \int_{-\infty}^{\infty} \frac{\exp\left[-\dfrac{r\rho}{t}\cosh y\right]}{\cosh\left\{\dfrac{\pi}{\theta_0}y + \dfrac{i\pi}{\theta_0}(\theta-\alpha)\right\} - \cos\dfrac{\pi^2}{\theta_0}}\, dy ,$$

where the summation $\sum$ is extended over k's satisfying the inequality under the summation sign and $\vec{\rho} = (\rho, \alpha)$, $\vec{r} = (r, \theta)$.

Then

$$P_{S(\theta_0)}(\vec{\rho}\,|\,\vec{r};\,t) = v(\alpha) - v(-\alpha).$$

Note that if $\theta_0 = \pi/m$, with m an integer, the complicated integral is out, since the factor in front of it, to wit $\sin \pi^2/\theta_0 = \sin \pi m$, is zero; what remains in the resulting expression for $v(\alpha) - v(-\alpha)$ is a collection of terms easily identifiable with those obtained by the method of images.

Let us now assume that $\pi/2 < \theta_0 < \pi$ and see what $P_S(\vec{\rho}\,|\,\vec{\rho};\,t)$ is in this case. In the expression for $v(\alpha)$ when we set $\theta = \alpha$ the inequality under the $\sum$ sign becomes $-\pi < 2k\theta_0 < \pi$ and only $k=0$ is allowed. In $v(-\alpha)$ the inequality is $2\alpha - \pi < 2k\theta_0 < 2\alpha + \pi$ and what k's to take depends on α.

We see that:

$$0 < \alpha < \theta_0 - \frac{\pi}{2}, \text{ only } k = 0 \text{ is allowed,}$$

$$\frac{\pi}{2} < \alpha < \theta_0, \text{ only } k = 1 \text{ is allowed,}$$

but for $\theta_0 - \pi/2 < \alpha < \pi/2$ both $k=0$ and $k=1$ are allowed. (See Fig. 7.)

Let us now put $\vec{r} = \vec{\rho}$ (so that $\rho = r$) and write down in detail the expressions for $P_{S(\theta_0)}(\vec{\rho}\,|\,\vec{\rho};\,t)$ in the three sectors. For $0 < \alpha < \theta_0 - \pi/2$

$$P_S(\vec{\rho}\,|\,\vec{\rho};\,t) = \frac{1}{2\pi t} - \frac{\exp\left[-\dfrac{r^2}{t}(1 - \cos 2\alpha)\right]}{2\pi t}$$

$$- \left(\sin \frac{\pi^2}{\theta_0}\right) \frac{\exp\left[-\dfrac{r^2}{t}\right]}{4\pi\theta_0 t} \int_{-\infty}^{\infty} \frac{\exp\left[-\dfrac{r^2}{t}\cosh y\right]}{\cosh \dfrac{\pi}{\theta_0}y - \cos \dfrac{\pi^2}{\theta_0}}\, dy$$

$$+ \left(\sin \frac{\pi^2}{\theta_0}\right) \frac{\exp\left[-\dfrac{r^2}{t}\right]}{4\pi\theta_0 t} \int_{-\infty}^{\infty} \frac{\exp\left[-\dfrac{r^2}{t}\cosh y\right]}{\cosh \left\{\dfrac{\pi}{\theta_0}y + 2\pi i \dfrac{\alpha}{\theta_0}\right\} - \cos \dfrac{\pi^2}{\theta_0}}\, dy.$$

For $\pi/2 < \alpha < \theta_0$

$$P_S(\vec{\rho}\,|\,\vec{\rho};\,t) = \frac{1}{2\pi t} - \frac{\exp\left[-\dfrac{r^2}{t}(1 - \cos 2(\theta_0 - \alpha)\right]}{2\pi}$$

$$+ \text{ the same two integrals as above}$$

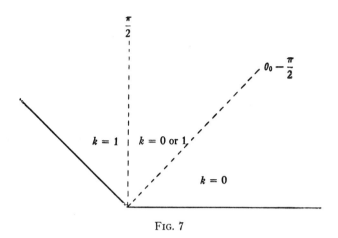

FIG. 7

and finally, for $\theta_0 - \pi/2 < \alpha < \pi/2$

$$P_S(\vec{\rho} \mid \vec{\rho}; t) = \frac{1}{2\pi t} - \frac{\exp\left[-\dfrac{r^2}{t}(1 - \cos 2\alpha)\right]}{2\pi t} - \frac{\exp\left[-\dfrac{r^2}{t}(1 - \cos 2(\theta_0 - \alpha))\right]}{2\pi t}$$

$+$ again the same two integrals.

We should recognize $r^2(1 - \cos 2\alpha)$ (and $r^2(1 - \cos 2(\theta_0 - \alpha))$) as being $2\delta^2$ where δ is the distance from $\vec{\rho}$ to a side of the wedge.

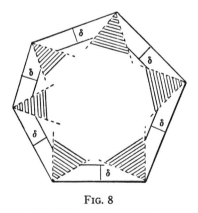

FIG. 8

14. To simplify matters somewhat let me assume that the polygonal drum is convex and that every angle is obtuse.

At each vertex we draw perpendiculars to the sides of the polygon thus obtaining N shaded sectors (where N is the number of sides or vertices of our polygon).

Now let ρ be a point in Ω. Stuff diffusing from $\vec{\rho}$ will either "see" the boundary as a straight line or, if $\vec{\rho}$ is near a vertex, as an infinite wedge.

We may as well say that the boundary will appear to the diffusing particle as the nearest wedge, and that consequently we may replace $P_{\Omega}(\vec{\rho}\,|\,\vec{\rho};\,t)$ by $P_{S(\theta_0)}(\vec{\rho}\,|\,\vec{\rho};\,t)$, where $S(\theta_0)$ is the wedge nearest to $\vec{\rho}$.

Now, each $P_S(\vec{\rho}\,|\,\vec{\rho};\,t)$ has $1/2\pi t$ as a term and after integration over Ω this gives the principal term $|\Omega|/2\pi t$. Next, each $P_S(\vec{\rho}\,|\,\vec{\rho};\,t)$ contains two complicated looking integrals which have to be integrated over the wedge.

Fortunately, the second of these integrates out to 0, while the first yields, upon integration over $S(\theta_0)$,

$$-\frac{1}{8\pi}\left(\sin\frac{\pi^2}{\theta_0}\right)\int_{-\infty}^{\infty}\frac{dy}{(1+\cosh y)\left(\cosh\dfrac{\pi}{\theta_0}y-\cos\dfrac{\pi^2}{\theta_0}\right)}.$$

This is only the contribution of one wedge; to get the total contribution one must sum over all wedges.

Thus the total contribution is

$$-\frac{1}{8\pi}\sum_{\theta_0}\left(\sin\frac{\pi^2}{\theta_0}\right)\int_{-\infty}^{\infty}\frac{dy}{(1+\cosh y)\left(\cosh\dfrac{\pi}{\theta_0}y-\cos\dfrac{\pi^2}{\theta_0}\right)}.$$

Finally, if $\vec{\rho}$ is in the shaded sector of the wedge $S(\theta_0)$ we get, on integrating over the sector,

$$-\int_{\theta_0-\pi/2}^{\pi/2}d\alpha\int_0^{\infty}\left\{\frac{\exp\left[-\dfrac{r^2}{t}(1-\cos 2\alpha)\right]}{2\pi t}\right.$$
$$\left.+\frac{\exp\left[-\dfrac{r^2}{t}(1-\cos 2(\theta_0-\alpha))\right]}{2\pi t}\right\}r\,dr=-\tfrac{1}{2}\frac{1}{2\pi}\cot\left(\theta_0-\frac{\pi}{2}\right),$$

and the total contribution from the shaded sectors is $-\tfrac{1}{2}1/2\pi\sum_{\theta_0}\cot(\theta_0-\pi/2)$.

The remaining contribution is easily seen to be

$$-\frac{1}{2\pi t}\int_0^{\infty}\left(L-2\delta\sum_{\theta_0}\cot\left(\theta_0-\frac{\pi}{2}\right)\right)e^{-2\delta^2/t}\,d\delta$$

$$=-\frac{L}{4}\frac{1}{\sqrt{2\pi t}}+\tfrac{1}{2}\frac{1}{2\pi}\sum_{\theta_0}\cot\left(\theta_0-\frac{\pi}{2}\right).$$

Finally, for a polygonal drum

$$\sum_{n=1}^{\infty} e^{-\lambda_n t} \sim \frac{|\Omega|}{2\pi t} - \frac{L}{4} \frac{1}{\sqrt{2\pi t}}$$

$$-\frac{1}{8\pi} \sum_{\theta_0} \left(\sin \frac{\pi^2}{\theta_0}\right) \int_{-\infty}^{\infty} \frac{dy}{(1 + \cosh y)(\cos \pi/\theta_0\, y - \cos \pi^2/\theta_0)} ,$$

with the understanding that each θ_0 satisfies the inequality $\pi/2 < \theta_0 < \pi$. If the polygon has N sides, and if we let $N \to \infty$ in such a way that each $\theta_0 \to \pi$, then the constant term approaches

$$+\frac{2\pi}{8\pi} \int_{-\infty}^{\infty} \frac{dy}{(1 + \cosh y)^2} = \tfrac{1}{6} .$$

This should strengthen our belief that for simply connected smooth drums the constant is universal and equal to $\tfrac{1}{6}$.

15. What happens for multiply connected drums?

If the drum as well as the holes are polygonal the answer is easily obtained. One only needs $P_{S(\theta_0)}(\vec{\rho} \mid \vec{\rho}; t)$ for θ_0 satisfying the inequality $\pi < \theta_0 < 2\pi$ and this is easily gotten from the general formula quoted above.

Near the holes the diffusing particles will "see" concave wedges but nothing will change in principle.

If we let all polygons approach smooth curves it turns out the constant approaches $(1-r)\tfrac{1}{6}$, where r is the number of holes. It is thus natural to conjecture that for a *smooth* drum with r *smooth* holes

$$\sum_{n=1}^{\infty} e^{-\lambda_n t} \sim \frac{|\Omega|}{2\pi t} - \frac{L}{4} \frac{1}{\sqrt{2\pi t}} + (1 - r)\tfrac{1}{6} ,$$

and that therefore one can "hear" the connectivity of the drum!

One can, of course, speculate on whether in general one can hear the Euler-Poincaré characteristic and raise all sorts of other interesting questions.

As our study of the polygonal drum shows, the structure of the constant term is quite complex since it combines metric and topological features. Whether these can be properly disentangled remains to be seen.

This is an expanded version of a lecture which was filmed under the auspices of the Committee on Educational Media of the Mathematical Association of America.

References

1. M. Kac, On some connections between probability theory and differential and integral equations, Proc. Second Berkeley Symposium on Mathematical Statistics and Probability, 1957, pp. 189–215.

2. A. Pleijel, A study of certain Green's functions with applications in the theory of vibrating membranes, Arkiv för Matematik, 2 (1954) 553–569.

18

SHIING-SHEN CHERN

Professor Chern was born on October 26, 1911, in Kashing, China. He received the Bachelor of Science degree from Nankai University in Tientsin, the Master of Science degree in 1934 from Tsinghua University in Peiping, and the Doctor of Science degree in 1936 from the University of Hamburg, Germany. After serving as Professor of Mathematics from 1937 to 1943 at Tsinghua University in China, he came to the United States in 1943, when he became a member of the Institute for Advanced Study in Princeton. He stayed there until 1946, when he returned to China to join the Institute of Mathematics of the Academia Sinica in Nanking as Professor of Mathematics and its Acting Director.

In 1949 he came back to the United States as Professor of Mathematics at the University of Chicago, where he remained until 1960, except for visiting positions at Harvard University in 1952, at the *Eidgenössische Technische Hochschule* in Zürich in 1953, and at the Massachusetts Institute of Technology in 1957. Since 1960 he has been at the University of California, Berkeley, with the exception of the fall of 1964, when he was once again a member of the Institute for Advanced Study, the fall of 1966, which he spent at the University of California, Los Angeles, and March and April of 1967, when he was a member of the *Institut des Hautes Études Scientifiques* in Paris. He became a naturalized citizen in March of 1961.

Professor Chern held Guggenheim Fellowships in both 1954–55 and 1966–67. He was elected a member of the National Academy of Sciences in 1961 and a Fellow of the American Academy of Arts and Sciences in 1963. He was also a member of *Academia Sinica* since 1948, and a corresponding member, Brazilian Academy of Sciences since 1971. He served as Vice-President of the American Mathematical Society in 1962–64, as Editor of its *Proceedings* in 1955–56, and as Editor of its *Transactions* in 1957–58. He was Editor of the *Illinois Journal of Mathematics* from 1959 to 1965.

Among the numerous honors which have been bestowed upon Professor Chern are the awards of the LL.D. degree by the Chinese University of Hong Kong, the Doctor of Science by the University of Chicago, both in 1969, and D.Sc. (hon.), University of Hamburg, 1971. He received the National Medal of Science in 1975.

Professor Chern's many substantial contributions to geometry and related fields are contained in his numerous papers—already 108 in number—which have appeared in many scientific publications and in several different languages throughout the world.

Professor Chern was the Editor of *MAA Studies in Mathematics*, Volume 4, "Studies in Global Geometry and Analysis," which contains the article for which he has received the Chauvenet Prize.

In accepting, Professor Chern indicated that he felt greatly pleased and honored by having been voted recipient of the 1970 Chauvenet Prize.

CURVES AND SURFACES IN EUCLIDEAN SPACE

SHIING-SHEN CHERN, University of California, Berkeley

INTRODUCTION

This article contains a treatment of some of the most elementary theorems in differential geometry in the large. They are the seeds for further developments and the subject should have a promising future. We shall consider the simplest cases, where the geometrical ideas are most clear.

1. THEOREM OF TURNING TANGENTS

Let E be the euclidean plane, which is oriented so that there is a prescribed sense of rotation. We define a smooth curve by expressing its position vector $X = (x_1, x_2)$ as a function of its arc length s. We suppose the function $X(s)$—that is, the functions $x_1(s), x_2(s)$—to be twice continuously differentiable and the vector $X'(s)$ to be nowhere 0. The latter allows the definition of the unit tangent vector $e_1(s)$, which is the unit vector in the direction of $X'(s)$ and, since E is oriented, the unit normal vector $e_2(s)$, so that the rotation from e_1 to e_2 is positive. The vectors $X(s), e_1(s), e_2(s)$ are related by the so-called Frenet formulas

$$\frac{dX}{ds} = e_1, \quad \frac{de_1}{ds} = ke_2, \quad \frac{de_2}{ds} = -ke_1. \tag{1}$$

The function $k(s)$ is called the *curvature*. It is defined together with its sign and changes its sign if the orientation of the curve or of the plane is reversed.

The curve C is called *closed*, if $X(s)$ is periodic of period L, L being the length of C. It is called *simple* if $X(s_1) \neq X(s_2)$, when $0 < s_1 - s_2 < L$. It is said to be *convex* if it lies in one side of every tangent line.

Let C be an oriented closed curve of length L, with the position vector $X(s)$ as a function of the arc length s. Let O be a fixed point in the plane, which we take as the origin of our coordinate system. Denote by Γ the unit circle about O. We define the tangential mapping $T: C \rightarrow \Gamma$ as the one which maps a point P of C to the endpoint of the unit vector through O parallel to the tangent vector to C at P. Obviously T is a continuous mapping. It is intuitively clear that when a point goes around C once its image point goes around Γ a number of times. This number will be called the rotation index of C. The theorem of turning tangents asserts that if C is simple, the rotation index is ± 1. We begin by giving a rigorous definition of the rotation index.

We choose a fixed vector through O, say Ox, and denote by $\tau(s)$ the angle which Ox makes with the vector $e_1(s)$. We assume that $0 \leq \tau(s) < 2\pi$, so that $\tau(s)$ is uniquely determined. This function $\tau(s)$ is, however, not continuous, for in every neighborhood of s_0 at which $\tau(s_0) = 0$ there may be values of $\tau(s)$ differing from 2π by an arbitrarily small quantity. There exists nevertheless a continuous function $\tilde{\tau}(s)$ closely related to $\tau(s)$, as given by the following lemma.

LEMMA. *There exists a continuous function $\tilde{\tau}(s)$ such that $\tilde{\tau}(s) \equiv \tau(s)$, mod 2π.*

Proof. To prove the lemma, we observe that the mapping T, being continuous, is uniformly continuous. Therefore, there exists a number $\delta > 0$, such that, for $|s_1 - s_2| < \delta$, $T(s_1)$ and $T(s_2)$ lie in the same open half-plane. From our conditions on $\tilde{\tau}(s)$, it follows that, if $\tilde{\tau}(s_1)$ is known, $\tilde{\tau}(s_2)$ is completely determined. We divide the interval $0 \leqq s \leqq L$ by the points $s_0 \, (=0) < s_1 < \cdots < s_m (=L)$ such that $|s_i - s_{i-1}| < \delta, i = 1, \ldots, m$. To define $\tilde{\tau}(s)$, we assign to $\tilde{\tau}(s_0)$ the value $\tau(s_0)$. Then it is determined in the subinterval $s_0 \leqq s \leqq s_1$, in particular at s_1, which determines it in the second subinterval, etc. The function $\tilde{\tau}(s)$ so defined clearly satisfies the conditions of the lemma.

The difference $\tilde{\tau}(L) - \tilde{\tau}(0)$ is an integral multiple of 2π, say, $= \gamma 2\pi$. We assert that the integer γ is independent of the choice of the function $\tilde{\tau}(s)$. In fact, let $\tilde{\tau}'(s)$ be a function satisfying the same conditions. Then we have

$$\tilde{\tau}'(s) - \tilde{\tau}(s) = n(s) \cdot 2\pi,$$

where $n(s)$ is an integer. Since $n(s)$ is continuous in s, it must be a constant. It follows that

$$\tilde{\tau}'(L) - \tilde{\tau}'(0) = \tilde{\tau}(L) - \tilde{\tau}(0),$$

which proves the independence of γ from the choice of $\tilde{\tau}(s)$. We define γ to be the rotation index of C. The *theorem of turning tangents* follows.

THEOREM. *The rotation index of a simple closed curve is ± 1.*

Proof. To prove this theorem, we consider the mapping Σ which carries an ordered pair of points of $C, X(s_1), X(s_2), 0 \leqq s_1 \leqq s_2 \leqq L$, into the endpoint of the unit vector through O parallel to the secant joining $X(s_1)$ to $X(s_2)$. These ordered pairs of points can be represented as a triangle Δ in the (s_1, s_2)-plane defined by $0 \leqq s_1 \leqq s_2 \leqq L$. The mapping Σ of Δ into Γ is continuous. We also observe that its restriction to the side $s_1 = s_2$ is the tangential mapping T.

To a point $p \in \Delta$, let $\tau(p)$ be the angle which Ox makes with $O\Sigma(p)$, such that $0 \leqq \tau(p) < 2\pi$. Again this function need not be continuous. We shall, however, prove that there exists a continuous function $\tilde{\tau}(p), p \in \Delta$, such that $\tilde{\tau}(p) \equiv \tau(p) \bmod 2\pi$.

In fact, let m be an interior point of Δ. We cover Δ by the radii through m. By the arguments used in the proof of the preceding lemma, we can define a function $\tilde{\tau}(p), p \in \Delta$, such that $\tilde{\tau}(p) \equiv \tau(p), \bmod 2\pi$, and such that it is continuous along every radius through m. It remains to prove that it is continuous in Δ. For this purpose, let p_0 be a point of Δ. Since Σ is continuous, it follows from the compactness of the segment mp_0 that there exists a number $\eta = \eta(p_0) > 0$, such that, for $q_0 \in mp_0$, and for any point of $q \in \Delta$ for which the distance $d(q, q_0) < \eta$, the points $\Sigma(q)$ and $\Sigma(q_0)$ are never antipodal. The latter condition is equivalent to the relation

$$\tilde{\tau}(q) - \tilde{\tau}(q_0) \not\equiv 0, \quad \bmod \pi. \tag{2}$$

Now let $\varepsilon > 0, \varepsilon < \pi/2$, be given. We choose a neighborhood U of p_0, such that U is contained in the η-neighborhood of p_0, and such that, for $p \in U$, the angle between $O\Sigma(p_0)$ and $O\Sigma(p)$ is less than ε. This is possible, because the mapping Σ is continuous. The last condition can be expressed in the form

$$\tilde{\tau}(p) - \tilde{\tau}(p_0) = \varepsilon' + 2k(p)\pi, \quad |\varepsilon'| < \varepsilon, \tag{3}$$

where $k(p)$ is an integer. Let q_0 be any point on the segment mp_0. Draw the segment $q_0 q$ parallel to $p_0 p$, with q on mp. The function $\tilde{\tau}(q) - \tilde{\tau}(q_0)$ is continuous in q along mp and is zero when q coincides with m. Since $d(q, q_0)$ is less than η, it follows from Equation (2) that $|\tilde{\tau}(q) - \tilde{\tau}(q_0)| < \pi$. In particular, for $q_0 = p_0, |\tilde{\tau}(p) - \tilde{\tau}(p_0)| < \pi$. Combining this result with Equation (3), we get $k(p) = 0$, which proves that $\tilde{\tau}(p)$ is continuous in Δ. Since $\tilde{\tau}(p) \equiv \tau(p)$, mod 2π, it is easy to see that $\tilde{\tau}(p)$ is differentiable.

Now let $A(0,0)$, $B(0,L)$, and $D(L,L)$ be the vertices of Δ. The rotation index γ of C is defined by the line integral

$$2\pi\gamma = \int_{AD} d\tilde{\tau}.$$

Since $\tilde{\tau}(p)$ is defined in Δ, we have

$$\int_{AD} d\tilde{\tau} = \int_{AB} d\tilde{\tau} + \int_{BD} d\tilde{\tau}.$$

To evaluate the line integrals on the right-hand side, we make use of a suitable coordinate system. We can suppose $X(0)$ to be the "lowest" point of C—that is, the point when the vertical coordinate is a minimum, and we choose $X(0)$ to be the origin O. The tangent vector to C at $X(0)$ is horizontal, and we call it Ox. The curve C then lies in the upper half-plane bounded by Ox, and the line integral $\int_{AB} d\tilde{\tau}$ is equal to the angle rotated by OP as P traverses once along C. Since OP never points downward, this angle is $\varepsilon\pi$, with $\varepsilon = \pm 1$. Similarly, the integral $\int_{BD} d\tilde{\tau}$ is the angle rotated by PO as P goes once along C. Its value is also equal to $\varepsilon\pi$. Hence, the sum of the two integrals is $\varepsilon 2\pi$ and the rotation index of C is ± 1, which completes our proof.

We can also define the rotation index by an integral formula. In fact, using the function $\tilde{\tau}(s)$ in our lemma, we can express the components of the unit tangent and normal vectors as follows:

$$e_1 = (\cos\tilde{\tau}(s), \sin\tilde{\tau}(s)), \quad e_2 = (-\sin\tilde{\tau}(s), \cos\tilde{\tau}(s)).$$

It follows that

$$d\tilde{\tau}(s) = de_1 \cdot e_2 = k\, ds.$$

From this equation, we derive the following formula for the rotation index:

$$2\pi\gamma = \int_C k\, ds. \tag{4}$$

This formula holds for closed curves which are not necessarily simple.

The accompanying figure gives an example of a closed curve with rotation index zero.

SHIING-SHEN CHERN

Fig. 1.

Many interesting theorems in differential geometry are valid for a more general class of curves, the so-called *sectionally smooth curves*. Such a curve is the union of a finite number of smooth arcs $A_0A_1, A_1A_2, \ldots, A_{m-1}A_m$, where the tangents of the two arcs through a common vertex $A_i, i = 1, \ldots, m-1$, may be different. The curve is called *closed*, if $A_0 = A_m$. The simplest example of a closed sectionally smooth curve is a rectilinear polygon.

The notion of rotation index and the theorem of turning tangents can be extended to closed sectionally smooth curves; we summarize, without proof, the result as follows. Let $s_i, i = 1, \ldots, m$, be the arc length measured from A_0 to A_i, so that $s_m = L$ is the length of the curve. The curve supposedly being oriented, the tangential mapping is defined at all points different from A_i. At a vertex A_i there are two unit vectors, tangent respectively to $A_{i-1}A_i$ and A_iA_{i+1}. (We define $A_{m+1} = A_1$.) The corresponding points on Γ we denote by $T(A_i)^-$ and $T(A_i)^+$. Let φ_i be the angle from $T(A_i)^-$ to $T(A_i)^+$, with $0 < \varphi_i < \pi$, briefly the exterior angle from the tangent to $A_{i-1}A_i$ to the tangent to A_iA_{i+1}. For each arc $A_{i-1}A_i$, a continuous function $\tilde{\tau}(s)$ can be defined which is one of the determinations of the angle from Ox to the tangent at $X(s)$. The number γ defined by the equation

$$2\pi\gamma = \sum_{i=1}^m \{\tilde{\tau}(s_i) - \tilde{\tau}(s_{i-1})\} + \sum_{i=1}^m \varphi_i \tag{5}$$

is an integer, which will be called the *rotation index* of the curve. The theorem of turning tangents is again valid.

THEOREM. *If a sectionally smooth curve is simple, the rotation index is equal to* ± 1.

As an application of the theorem of turning tangents, we wish to give the following characterization of a simple closed convex curve.

Remark. A simple closed curve is convex, if and only if it can be so oriented that its curvature is greater than, or equal to, 0.

Let us first remark that the theorem is not true without the assumption that the curve is simple. In fact, the accompanying figure gives a nonconvex curve with $k > 0$.

Proof. To prove the theorem, we let $\tilde{\tau}(s)$ be the function constructed, so that we have $k = d\tilde{\tau}/ds$. The condition $k \geqq 0$ is equivalent to the assertion that $\tilde{\tau}(s)$ is a monotone nondecreasing function. Because C is simple, we can suppose that $\tilde{\tau}(s)$, $0 \leqq s \leqq L$, increases from 0 to 2π. It follows that if the tangents at $X(s_1)$ and

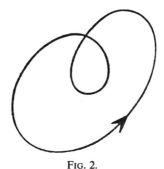

FIG. 2.

$X(s_2)$, $0 \leqq s_1 < s_2 < L$, are parallel in the same sense, the arc of C from $X(s_1)$ to $X(s_2)$ is a straight line segment and these tangents must coincide.

Suppose $\tilde{\tau}(s)$, $0 \leqq s \leqq L$, is monotone nondecreasing and C is not convex. There is a point $A = X(s_0)$ on C such that there are points of C at both sides of the tangent t to C at A. Choose a positive side of t and consider the oriented perpendicular distance from a point $X(s)$ of C to t. This is a continuous function in s and attains a maximum and a minimum at the points M and N of C, respectively. Clearly M and N are not on t and the tangents to C at M and N are parallel to t. Among these two tangents and t itself, there are two tangents parallel in the same sense, which, according to the preceding remark, is impossible.

Next we let C be convex. To prove that $\tilde{\tau}(s)$ is monotone, we suppose $\tilde{\tau}(s_1) = \tilde{\tau}(s_2), s_1 < s_2$. Then the tangents at $X(s_1)$ and $X(s_2)$ are parallel in the same sense. But there exists a tangent parallel to them in the opposite sense. From the convexity of C it follows that two of them coincide.

We are thus led to the consideration of a line t tangent to C at two distinct points, A and B. We claim that the segment AB must be a part of C. In fact, suppose this is not the case and let D be a point of AB not on C. Draw through D a perpendicular u to t in the half-plane which contains C. Then u intersects C in at least two points. Among these points of intersection, let F be the farthest from t and G the nearest, so that $F \neq G$. Then G is an interior point of the triangle ABF. The tangent to C at G must have points of C in both sides, which contradicts the convexity of C.

It follows that, under the hypothesis of the last paragraph, the segment AB is a part of C and that the tangents at A and B are parallel in the same sense. This proves that the segment joining $X(s_1)$ to $X(s_2)$ belongs to C. The latter implies that $\tilde{\tau}(s)$ remains constant in the interval $s_1 \leqq s \leqq s_2$. Hence, the function $\tilde{\tau}(s)$ is monotone, and our theorem is proved.

The first half of the theorem can also be stated as follows.

Remark. A closed curve with $k(s) \geqq 0$ and rotation index equal to 1 is convex.

The theorem of turning tangents was essentially known to Riemann. The above proof was given by H. Hopf, *Compositio Mathematica* 2 (1935), pp. 50–62. For further reading, see:

1. H. Whitney, "On regular closed curves in the plane," *Compositio Mathematica* 4 (1937), pp. 276–84.

2. S. Smale, "Regular curves on a Riemannian manifold," *Transactions of the American Mathematical Society* 87 (1958), pp. 492–511.

3. S. Smale, "A classification of immersions of the two-sphere," *Transactions of the American Mathematical Society* 90 (1959), pp. 281–90.

2. THE FOUR-VERTEX THEOREM

An interesting theorem on closed plane curves is the so-called "four-vertex theorem." By a *vertex* of an oriented closed plane curve we mean a point at which the curvature has a relative extremum. Since the curve forms a compact point set, it has at least two vertices, corresponding respectively to the absolute minimum and maximum of the curvature. Our theorem says that there are at least four.

THEOREM. *A simple closed convex curve has at least four vertices.*

This theorem was first presented by Mukhopadhyaya (1909); the proof we shall give was the work of G. Herglotz. It is also true for nonconvex curves, but the proof is more difficult. The theorem cannot be improved, because an ellipse with unequal axes has exactly four vertices, which are its points of intersection with the axes.

Proof. We suppose that the curve C has only two vertices, M and N, and we shall show that this leads to a contradiction. The line MN does not meet C in any other point, for if it does, the tangent line to C at the middle point must contain the other two points. By the last section, this condition is possible only when the segment MN is a part of C. It would follow that the curvature vanishes at M and N, which is not possible, since they are the points where the curvature takes the absolute maximum and minimum respectively.

We denote by 0 and s_0 the parameters of M and N respectively and take MN to be the x_1-axis. Then we can suppose

$$x_2(s) < 0, \quad 0 < s < s_0,$$
$$x_2(s) > 0, \quad s_0 < s < L,$$

where L is the length of C. Let $(x_1(s), x_2(s))$ be the position vector of a point of C with the parameter s. Then the unit tangent and normal vectors have the components

$$e_1 = (x_1', x_2'), \quad e_2 = (-x_2', x_1'),$$

where primes denote differentiations with respect to s. From the Frenet formulas we get

$$x_1'' = -kx_2', \quad x_2'' = kx_1'. \tag{6}$$

It follows that

$$\int_0^L kx_2' \, ds = -x_1' \Big|_0^L = 0.$$

The integral in the left-hand side can be written as a sum:

$$\int_0^L kx_2' \, ds = \int_0^{s_0} kx_2' \, ds + \int_{s_0}^L kx_2' \, ds.$$

To each summand we apply the second mean value theorem, which is stated as follows. Let $f(x), g(x), a \leqq x \leqq b$, be two functions in x such that $f(x)$ and $g'(x)$ are continuous and $g(x)$ is monotone. Then there exists $\xi, a < \xi < b$, satisfying the equation,

$$\int_a^b f(x) g(x) dx = g(a) \int_a^\xi f(x) dx + g(b) \int_\xi^b f(x) dx.$$

Since $k(s)$ is monotone in each of the intervals $0 \leqq s \leqq s_0, s_0 \leqq s \leqq L$, we get

$$\int_0^{s_0} k x_2' ds = k(0) \int_0^{\xi_1} x_2' ds + k(s_0) \int_{\xi_1}^{s_0} x_2' ds$$

$$= x_2(\xi_1)(k(0) - k(s_0)), \qquad 0 < \xi_1 < s_0$$

$$\int_{s_0}^{L} k x_2' ds = k(s_0) \int_{s_0}^{\xi_2} x_2' ds + k(L) \int_{\xi_2}^{L} x_2' ds$$

$$= x_2(\xi_2)(k(s_0) - k(0)), \qquad s_0 < \xi_2 < L.$$

Since the sum of the left-hand members is zero, these equations give

$$(x_2(\xi_1) - x_2(\xi_2))(k(0) - k(s_0)) = 0,$$

which is a contradiction, because

$$x_2(\xi_1) - x_2(\xi_2) < 0, \quad k(0) - k(s_0) > 0.$$

It follows that there is at least one more vertex on C. Since the relative extrema occur in pairs, there are at least four vertices and the theorem is proved.

At a vertex we have $k' = 0$. Hence, we can also say that on a simple closed convex curve there are at least four points at which $k' = 0$.

The four-vertex theorem is also true for simple closed nonconvex plane curves; see:

1. S. B. Jackson, "Vertices for plane curves," *Bulletin of the American Mathematical Society* 50 (1944), pp. 564–578.

2. L. Vietoris, "Ein einfacher Beweis des Vierscheitelsatzes der ebenen Kurven," *Archiv der Mathematik* 3 (1952), pp. 304–306.

For further reading, see:

1. P. Scherk, "The four-vertex theorem," *Proceedings of the First Canadian Mathematical Congress*. Montreal: 1945, pp. 97–102.

3. ISOPERIMETRIC INEQUALITY
   FOR PLANE CURVES

The theorem can be stated as follows.

THEOREM. *Among all simple closed curves having a given length the circle bounds the largest area. In other words, if L is the length of a simple closed curve C, and A is the area it bounds, then*

$$L^2 - 4\pi A \geqq 0. \tag{7}$$

Moreover, the equality sign holds only when C is a circle.

Many proofs have been given of this theorem, differing in degree of elegance and in the range of curves under consideration—that is, whether differentiability or convexity is supposed. We shall give two proofs, the work of E. Schmidt (1939) and A. Hurwitz (1902), respectively.

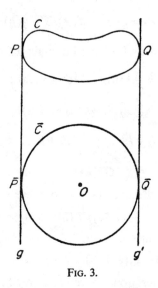

Fig. 3.

Schmidt's Proof. We enclose C between two parallel lines, g and g', such that C lies between g and g' and is tangent to them at the points P and Q, respectively. We let $s = 0$, s_0 being the parameters of P and Q, and construct a circle $\overline{C}$ tangent to g and g' at $\overline{P}$ and $\overline{Q}$, respectively. Denote its radius by r and take its center to be the origin of a coordinate system. Let $X(s) = (x_1(s), x_2(s))$ be the position vector of C, so that $(x_1(0), x_2(0)) = (x_1(L), x_2(L))$. As the position vector of $\overline{C}$ we take $(\bar{x}_1(s), \bar{x}_2(s))$, such that

$$\bar{x}_1(s) = x_1(s),$$

$$\bar{x}_2(s) = -\sqrt{r^2 - x_1^2(s)}\,, \qquad 0 \leqq s \leqq s_0$$

$$= +\sqrt{r^2 - x_1^2(s)}\,, \qquad s_0 \leqq s \leqq L. \tag{8}$$

Denote by $\overline{A}$ the area bounded by $\overline{C}$. Now the area bounded by a closed curve can be expressed by the line integral

$$A = \int_0^L x_1 x_2'\, ds = -\int_0^L x_2 x_1'\, ds = \tfrac{1}{2}\int_0^L (x_1 x_2' - x_2 x_1')\, ds.$$

Applying this to our two curves C and $\overline{C}$, we get

$$A = \int_0^L x_1 x_2'\, ds$$

$$\overline{A} = \pi r^2 = -\int_0^L \bar{x}_2 \bar{x}_1'\, ds = -\int_0^L \bar{x}_2 x_1'\, ds.$$

Adding these two equations, we have

$$A + \pi r^2 = \int_0^L (x_1 x_2' - \bar{x}_2 x_1') \, ds \leq \int_0^L \sqrt{(x_1 x_2' - \bar{x}_2 x_1')^2} \; ds$$

$$\leq \int_0^L \sqrt{(x_1^2 + \bar{x}_2^2)(x_1'^2 + x_2'^2)} \; ds$$

$$= \int_0^L \sqrt{x_1^2 + \bar{x}_2^2} \; ds = Lr. \qquad (9)$$

Since the geometric mean of two positive numbers is less than or equal to their arithmetic mean, it follows that

$$\sqrt{A} \; \sqrt{\pi r^2} \leq \tfrac{1}{2}(A + \pi r^2) \leq \tfrac{1}{2} Lr,$$

which gives, after squaring and cancellation of r^2, the inequality in Equation (7).

Suppose now that the equality sign in Equation (7) holds; then A and πr^2 have the same geometric and arithmetic mean, so that $A = \pi r^2$ and $L = 2\pi r$. The direction of the lines g and g' being arbitrary, this means that C has the same "width" in all directions. Moreover, we must have the equality sign everywhere in Equation (9). It follows, in particular, that

$$(x_1 x_2' - \bar{x}_2 x_1')^2 = (x_1^2 + \bar{x}_2^2)(x_1'^2 + x_2'^2),$$

which gives

$$\frac{x_1}{x_2'} = \frac{-\bar{x}_2}{x_1'} = \frac{\sqrt{x_1^2 + \bar{x}_2^2}}{\sqrt{x_1'^2 + x_2'^2}} = \pm r.$$

From the first equality in Equation (9), the factor of proportionality is seen to be r, that is,

$$x_1 = rx_2', \quad \bar{x}_2 = -rx_1',$$

which remains true when we interchange x_1 and x_2, so that

$$x_2 = rx_1'.$$

Therefore, we have

$$x_1^2 + x_2^2 = r^2,$$

which means that C is a circle.

Hurwitz's proof makes use of the theory of Fourier series. We shall first prove the lemma of Wirtinger.

LEMMA. *Let $f(t)$ be a continuous periodic function of period 2π, possessing a continuous derivative $f'(t)$. If $\int_0^{2\pi} f(t) \, dt = 0$, then*

$$\int_0^{2\pi} f'(t)^2 \, dt \geq \int_0^{2\pi} f(t)^2 \, dt. \qquad (10)$$

Moreover, the equality sign holds if and only if

$$f(t) = a \cos t + b \sin t. \qquad (11)$$

Proof. To prove the lemma, we let the Fourier series expansion of $f(t)$ be

$$f(t) \sim \frac{a_0}{2} + \sum_{n=1}^{\infty} (a_n \cos nt + b_n \sin nt).$$

Since $f'(t)$ is continuous, its Fourier series can be obtained by differentiation term by term, and we have

$$f'(t) \sim \sum_{n=1}^{\infty} (nb_n \cos nt - na_n \sin nt).$$

Since

$$\int_0^{2\pi} f(t)\,dt = \pi a_0,$$

it follows from our hypothesis that $a_0 = 0$. By Parseval's formula, we get

$$\int_0^{2\pi} f(t)^2\,dt = \sum_{n=1}^{\infty} (a_n^2 + b_n^2),$$

$$\int_0^{2\pi} f'(t)^2\,dt = \sum_{n=1}^{\infty} n^2(a_n^2 + b_n^2).$$

Hence,

$$\int_0^{2\pi} f'(t)^2\,dt - \int_0^{2\pi} f(t)^2\,dt = \sum_{n=1}^{\infty} (n^2 - 1)(a_n^2 + b_n^2),$$

which is greater than, or equal to, 0. It is equal to zero, only if $a_n = b_n = 0$ for all $n > 1$. Therefore, $f(t) = a_1 \cos t + b_1 \sin t$, which proves the lemma.

Hurwitz's Proof. In order to prove the inequality in Equation (7), we assume, for simplicity, that $L = 2\pi$, and that

$$\int_0^{2\pi} x_1(s)\,ds = 0.$$

The latter means that the center of gravity lies on the x_1-axis, a condition which can always be achieved by a proper choice of the coordinate system. The length and the area are given by the integrals,

$$2\pi = \int_0^{2\pi} (x_1'^2 + x_2'^2)\,ds, \quad \text{and} \quad A = \int_0^{2\pi} x_1 x_2'\,ds.$$

From these two equations we get

$$2(\pi - A) = \int_0^{2\pi} (x_1'^2 - x_1^2)\,ds + \int_0^{2\pi} (x_1 - x_2')^2\,ds.$$

The first integral is greater than, or equal to, 0 by our lemma and the second integral is clearly greater than, or equal to, 0. Hence, $A \leq \pi$, which is our isoperimetric inequality.

The equality sign holds only when

$$x_1 = a \cos s + b \sin s, \quad x_2' = x_1,$$

which gives

$$x_1 = a \cos s + b \sin s, \quad x_2 = a \sin s - b \cos s + c.$$

Thus, C is a circle.

For further reading, see:

1. E. Schmidt, "Beweis der isoperimetrischen Eigenschaft der Kugel im hyperbolischen und sphärischen Raum jeder Dimensionenzahl," *Math. Zeit.* 49 (1943), pp. 1–109.

4. TOTAL CURVATURE OF A SPACE CURVE

The *total curvature* of a closed space curve C of length L is defined by the integral

$$\mu = \int_0^L |k(s)| \, ds, \tag{12}$$

where $k(s)$ is the curvature. For a space curve, only $|k(s)|$ is defined.

Suppose C is oriented. Through the origin O of our space we draw vectors of length 1 parallel to the tangent vectors of C. Their end-points describe a closed curve Γ on the unit sphere, to be called the *tangent indicatrix* of C. A point of Γ is singular (that is, with either no tangent or a tangent of higher contact) if it is the image of a point of zero curvature of C. Clearly the total curvature of C is equal to the length of Γ.

Fenchel's theorem concerns the total curvature.

THEOREM. *The total curvature of a closed space curve C is greater than, or equal to, 2π. It is equal to 2π if and only if C is a plane convex curve.*

The following proof of this theorem was found independently by B. Segre (*Bolletino della Unione Matematica Italiana* 13 (1934), 279–283), and by H. Rutishauser and H. Samelson (*Comptes Rendus Hebdomadaires des Séances de l'Académie des Sciences* 227 (1948), 755–757). See also W. Fenchel, *Bulletin of the American Mathematical Society* 57 (1951), 44–54. The proof depends on the following lemma:

LEMMA. *Let Γ be a closed rectifiable curve on the unit sphere, with length $L < 2\pi$. There exists a point m on the sphere such that the spherical distance $\overline{mx} \leqq L/4$ for all points x of Γ. If Γ is of length 2π but is not the union of two great semicircular arcs, there exists a point m such that $\overline{mx} < \pi/2$ for all x of Γ.*

We use the notation $\overline{ab}$ to denote the spherical distance of two points, a and b. If $\overline{ab} < \pi$, their *midpoint* m is the point defined by the conditions $\overline{am} = \overline{bm} = \frac{1}{2}\overline{ab}$. Let x be a point such that $\overline{mx} \leqq \frac{1}{2}\pi$. Then $2\overline{mx} \leqq \overline{ax} + \overline{bx}$. In fact, let x' be the symmetry of x relative to m. Then,

$$\overline{x'a} = \overline{xb}, \quad \overline{x'x} = \overline{x'm} + \overline{mx} = 2\overline{mx}.$$

If we use the triangle inequality, it follows that

$$2\overline{mx} = \overline{x'x} \leqq \overline{x'a} + \overline{ax} = \overline{ax} + \overline{bx}, \tag{13}$$

as was to be proved.

Lemma Proof. To prove the first part of the lemma, we take two points, a and b, on Γ which divide the curve into two equal arcs. Then $\overline{ab} < \pi$, and we denote the midpoint by m. Let x be a point of Γ such that $2\overline{mx} < \pi$. Such points exist—for example, the point a. Then we have

$$\overline{ax} \leqq \overset{\frown}{ax}, \quad \overline{bx} \leqq \overset{\frown}{bx},$$

where $\overset{\frown}{ax}$ and $\overset{\frown}{bx}$ are respectively the arc lengths along Γ. From Equation (13), it follows that

$$2\overline{mx} \leqq \overset{\frown}{ax} + \overset{\frown}{bx} = \overset{\frown}{ab} = \frac{L}{2}.$$

Hence, the function $f(x) = \overline{mx}, x \in \Gamma$, is either $\geqq \pi/2$ or $\leqq L/4 < \pi/2$. Since Γ is connected and $f(x)$ is a continuous function in Γ, the range of the function $f(x)$ is connected in the interval $(0, \pi)$. Therefore, we have $f(x) = \overline{mx} \leqq L/4$.

Consider next the case that Γ is of length 2π. If Γ contains a pair of antipodal points, then, being of length 2π, it must be the union of two great semicircular arcs. Suppose that there is a pair of points, a and b, which bisect Γ such that

$$\overline{ax} + \overline{bx} < \pi$$

for all $x \in \Gamma$. Again, let m denote the midpoint of a and b. If $f(x) = \overline{mx} \leqq \frac{1}{2}\pi$, we have, from Equation (13),

$$2\overline{mx} \leqq \overline{ax} + \overline{bx} < \pi,$$

which means that $f(x)$ omits the value $\pi/2$. Since its range is connected and since $f(a) < \pi/2$, we have $f(x) < \pi/2$ for all $x \in \Gamma$. Thus the lemma is true in this case.

It remains to consider the case that Γ contains no pair of antipodal points, and that for any pair of points a and b which bisect Γ, there is a point $x \in \Gamma$ with

$$\overline{ax} + \overline{bx} = \pi.$$

An elementary geometrical argument, which we leave to the reader, will show that this is impossible. Thus, the lemma if proved.

Theorem Proof. To prove Fenchel's theorem, we take a fixed unit vector A and put

$$g(s) = AX(s),$$

where the right-hand side denotes the scalar product of the vectors A and $X(s)$. The function $g(s)$ is continuous on C and hence must have a maximum and a minimum. Since $g'(s)$ exists, we have, at such an extremum s_0,

$$g'(s_0) = AX'(s_0) = 0.$$

Thus A, as a point on the unit sphere, has a distance $\pi/2$ from at least two points of the tangent indicatrix. Since A is arbitrary, the tangent indicatrix is met by every great circle. It follows from the lemma that its length is greater than, or equal to, 2π.

Suppose next that the tangent indicatrix Γ is of length 2π. By our lemma, it must be the union of two great semicircular arcs. It follows that C itself is the union of two plane arcs. Since C has a tangent everywhere, it must be a plane curve. Suppose C be so oriented that its rotation index

$$\frac{1}{2\pi}\int_0^L k\,ds \geqq 0.$$

Then we have

$$0 \leqq \int_0^L \{|k| - k\}\,ds = 2\pi - \int_0^L k\,ds$$

so that the rotation index is either 0 or 1. To a given vector in the plane there is parallel to it a tangent t of C such that C lies to the left of t. Then t is parallel to the vector in the same sense, and at its point of contact we have $k \geqq 0$, implying that $\int_{k>0} k\,ds \geqq 2\pi$. Since $\int_C |k|\,ds = 2\pi$, there is no point with $k < 0$, and $\int k\,ds = 2\pi$. From the remark at the end of Section 1, we conclude that C is convex.

As a corollary we have the following theorem.

COROLLARY. *If* $|k(s)| \leq 1/R$ *for a closed space curve* C, C *has a length* $L \geqq 2\pi R$.

We have

$$L = \int_0^L ds \geqq \int_0^L R|k|\,ds = R\int_0^L |k|\,ds \geqq 2\pi R.$$

Fenchel's theorem holds also for sectionally smooth curves. As the total curvature of such a curve we define

$$\mu = \int_0^L |k|\,ds + \sum_i a_i \tag{14}$$

where the a_i are the angles at the vertices. In other words, in this case the tangent indicatrix consists of a number of arcs each corresponding to a smooth arc of C; we join successive vertices by the shortest great circular arc on the unit sphere. The length of the curve so obtained is the total curvature of C. It can be proved that for a closed sectionally smooth curve we have also $\mu \geqq 2\pi$.

We wish to give another proof of Fenchel's theorem and a related theorem of Fary-Milnor on the total curvature of a knot.* The basis is Crofton's theorem on the measure of great circles which cut an arc on the unit sphere. Every oriented great circle determines uniquely a "pole," the endpoint of the unit vector normal to

---

* I. Fary (*Bulletin de la Société Mathématique de France*, 77 (1949), pp. 128–138), and J. Milnor (*Annals of Mathematics*, 52 (1950), pp. 248–257).

the plane of the circle. By the measure of a set of great circles on the unit sphere is meant the area of the domain of their poles. Then Crofton's theorem is stated as follows.

THEOREM. *Let Γ be a smooth arc on the unit sphere Σ_0. The measure of the oriented great circles of Σ_0 which meet Γ, each counted a number of times equal to the number of its common points with Γ, is equal to four times the length of Γ.*

Proof. We suppose Γ is defined by a unit vector $e_1(s)$ expressed as a function of its arc length s. Locally (that is, in a certain neighborhood of s), let $e_2(s)$ and $e_3(s)$ be unit vectors depending smoothly on s, such that the scalar products

$$e_i \cdot e_j = \delta_{ij}, \quad 1 \leq i,j \leq 3 \tag{15}$$

and

$$\det(e_1, e_2, e_3) = +1. \tag{16}$$

Then we have

$$\begin{cases} \dfrac{de_1}{ds} = a_2 e_2 + a_3 e_3, \\[2mm] \dfrac{de_2}{ds} = -a_2 e_1 \quad\ + a_1 e_3, \\[2mm] \dfrac{de_3}{ds} = -a_3 e_1 - a_1 e_2. \end{cases} \tag{17}$$

The skew-symmetry of the matrix of the coefficients in the above system of equations follows from differentiation of Equations (15). Since s is the arc length of Γ, we have

$$a_2^2 + a_3^2 = 1, \tag{18}$$

and we put

$$a_2 = \cos\tau(s), \quad a_3 = \sin\tau(s). \tag{19}$$

If an oriented great circle meets Γ at the point $e_1(s)$, its pole is of the form $Y = \cos\theta\, e_2(s) + \sin\theta e_3(s)$, and vice versa. Thus (s,θ) serve as local coordinates in the domain of these poles; we wish to find an expression for the element of area of this domain.

For this purpose, we write

$$dY = (-\sin\theta\, e_2 + \cos\theta\, e_3)(d\theta + a_1\, ds) - e_1(a_2\cos\theta + a_3\sin\theta)\, ds.$$

Since $-\sin\theta\, e_2 + \cos\theta\, e_3$ and e_1 are two unit vectors orthogonal to Y, the element of area of Y is

$$|dA| = |a_2\cos\theta + a_3\sin\theta|\, d\theta\, ds = |\cos(\tau - \theta)|\, d\theta\, ds, \tag{20}$$

where the absolute value at the left-hand side means that the area is calculated in the measure-theoretic sense, with no regard to orientation. To the point Y let $Y^\perp$ be the oriented great circle with Y as its pole, and let $n(Y^\perp)$ be the (arithmetic) number of points common to $Y^\perp$ and Γ. Then the measure μ in our theorem is given by

$$\mu = \int n(Y^{\perp})|dA| = \int_0^{\lambda} ds \int_0^{2\pi} |\cos(\tau - \theta)| \, d\theta,$$

where λ is the length of Γ. As θ ranges from 0 to 2π, the variation of $|\cos(\tau - \theta)|$, for a fixed s, is 4. Hence, we get $\mu = 4\lambda$, which proves Crofton's theorem.

By applying the theorem to each subarc and adding, we see that the theorem remains true when Γ is a sectionally smooth curve on the unit sphere. Actually, the theorem is true for any rectifiable arc on the sphere, but the proof is much longer.

For a closed space curve the tangent indicatrix of which fulfills the conditions of Crofton's theorem, Fenchel's theorem is an easy consequence. In fact, the proof of Fenchel's theorem shows us that the tangent indicatrix of a closed space curve meets every great circle in at least two points—that is, $n(Y^{\perp}) \geq 2$. It follows that its length is

$$\lambda = \int |k| \, ds = \frac{1}{4} \int n(Y^{\perp})|dA| \geq 2\pi,$$

because the total area of the unit sphere is 4π.

Crofton's theorem also leads to the following theorem of Fary and Milnor, which gives a necessary condition on the total curvature of a knot.

THEOREM. *The total curvature of a knot is greater than, or equal to, 4π.*

Since $n(Y^{\perp})$ is the number of relative maxima or minima of the "height function," $Y \cdot X(s)$, it is even. Suppose that the total curvature of a closed space curve C is $< 4\pi$. There exists $Y \in \Sigma_0$, such that $n(Y^{\perp}) = 2$. By a rotation, suppose Y is the point $(0,0,1)$. Then the function $x_3(s)$ has only one maximum and one minimum. These points divide C into two arcs, such that x_3 increases on the one and decreases on the other. Every horizontal plane between the two extremal horizontal planes meets C in exactly two points. If we join them by a segment, all these segments will form a surface which is homeomorphic to a circular disk, which proves that C is not knotted.

For further reading, see:

1. S. S. Chern and R. K. Lashof, "On the total curvature of immersed manifolds," I, *American Journal of Mathematics* 79 (1957), pp. 302–18, and II, *Michigan Mathematical Journal* 5 (1958), pp. 5–12.

2. N. H. Kuiper, "Convex immersions of closed surfaces in E^5," *Comm. Math. Helv.* 35 (1961), pp. 85–92.

On integral geometry compare the article of Santalo in *MAA Studies in Mathematics*, vol. 4 (1957).

5. DEFORMATION OF A SPACE CURVE

It is well-known that a one-one correspondence between two curves under which the arc lengths, the curvatures (when not equal to 0), and the torsions are respectively equal, can only be established by a proper motion. It is natural to study the correspondences under which only s and k are equal. We shall call such a correspondence a deformation of the space curve (in German, *Verwindung*). The most notable result in this direction is a theorem of A. Schur, which formulates the geometrical fact that if an arc is "stretched," the distance between its endpoints becomes longer. Using the name curvature to mean here always its absolute value, we state Schur's theorem as follows.

THEOREM. *Let C be a plane arc with the curvature $k(s)$ which forms a convex curve with its chord, AB. Let C^* be an arc of the same length referred to the same parameter s such that its curvature $k^*(s) \leqq k(s)$. If d^* and d denote the lengths of the chords joining their endpoints, then $d \leqq d^*$. Moreover, the equality sign holds when and only when C and C^* are congruent.*

Proof. Let Γ and Γ^* be the tangent indicatrices of C and C^* respectively, P_1 and P_2 two points on Γ, and P^*_1 and P^*_2 their corresponding points on Γ^*. We denote by $\overset{\frown}{P_1 P_2}$ and $\overset{\frown}{P^*_1 P^*_2}$ their arc lengths and by $\overline{P_1 P_2}$ and $\overline{P^*_1 P^*_2}$ their spherical distances. Then we have

$$\overline{P_1 P_2} \leqq \overset{\frown}{P_1 P_2}, \quad \overline{P^*_1 P^*_2} \leqq \overset{\frown}{P^*_1 P^*_2}.$$

The inequality on the curvature implies

$$\overset{\frown}{P^*_1 P^*_2} \leqq \overset{\frown}{P_1 P_2}. \tag{21}$$

Since C is convex, Γ lies on a great circle, and we have

$$\overline{P_1 P_2} = \overset{\frown}{P_1 P_2},$$

provided that $\overline{P_1 P_2} \leqq \pi$. Now let Q be a point on C at which the tangent is parallel to the chord. Denote by P_0 its image point on Γ. Then the condition $\overline{P_0 P} \leqq \pi$ is satisfied by any point P on Γ, and if P^*_0 denotes the point on Γ^* corresponding to P_0, we have

$$\overline{P^*_0 P^*} \leqq \overline{P_0 P}, \tag{22}$$

from which it follows that

$$\cos \overline{P^*_0 P^*} \geqq \cos \overline{P_0 P}, \tag{23}$$

since the cosine function is a monotone decreasing function of its argument when the latter lies between 0 and π.

Because C is convex, d is equal to the projection of C on its chord:

$$d = \int_0^L \cos \overline{P_0 P} \, ds. \tag{24}$$

On the other hand, we have

$$d^* \geqq \int_0^L \cos \overline{P^*_0 \ P^*} \, ds, \tag{25}$$

for the integral on the right-hand side is equal to the projection of C^*, and hence of the chord joining its endpoints, on the tangent at the point Q^* corresponding to Q. Combining Equations (23), (24), and (25), we get $d^* \geqq d$.

Suppose that $d = d^*$. Then the inequalities in Equations (22), (23), and (25) become equalities, and the chord joining the endpoints A^* and B^* of C^* must be parallel to the tangent at Q^*. In particular, we have

$$\overline{P_0^* P^*} = \overline{P_0 P},$$

which implies that the arcs $A^* Q^*$ and $B^* Q^*$ are plane arcs. On the other hand, we have, by using Equation (21),

$$\overline{P_0^* P^*} \leqq \overparen{P_0^* P^*} \leqq \overparen{P_0 P} = \overline{P_0 P},$$

or

$$\overparen{P_0^* P^*} = \overparen{P_0 P}.$$

Hence, the arcs $A^* Q^*$ and $B^* Q^*$ have the same curvature as AQ and BQ at corresponding points and are therefore respectively congruent.

It remains to prove that the arcs $A^* Q^*$ and $B^* Q^*$ lie in the same plane. Suppose the contrary. They must be tangent at Q^* to the line of intersection of the two distinct planes on which they lie. Since this line is parallel to $A^* B^*$, the only possibility is that it contains A^* and B^*; however, then the tangent to C at Q must also contain the endpoints A and B, which is a contradiction. Hence, C^* is a plane arc and is congruent to C.

Schur's theorem has many applications. For example, it gives a solution of the following minimum problem: Determine the shortest closed curve with a curvature $k(s) \leqq 1/R$, R being a constant. The answer is, of course, a circle.

Remark. The shortest closed curve with curvature $k(s) \leqq 1/R$, R being a constant, is a circle of radius R.

By the corollary to Fenchel's theorem, such a curve has length $2\pi R$. Comparing it with a circle of radius R, we conclude from Schur's theorem (with $d^* = d = 0$) that is must itself be a circle.

As a second application of Schur's theorem, we shall derive a theorem of Schwarz. It is concerned with the lengths of arcs joining two given points having a curvature bounded from the above by a fixed constant. The statement of Schwarz's theorem is as follows:

THEOREM. *Let C be an arc joining two given points A and B, with curvature $k(s) \leqq 1/R$, such that $R \geqq \frac{1}{2} d$, where $d = \overline{AB}$. Let S be a circle of radius R through A and B. Then the length of C is either less than, or equal to, the shorter arc AB or greater than, or equal to, the longer arc AB on S.*

Proof. We remark that the assumption $R \geqq \frac{1}{2} d$ is necessary for the circle S to exist. To prove the theorem, we can assume that the length L of C is less than $2\pi R$; otherwise, there is nothing to prove. We then compare C with an arc of the same length on S having a chord of length d'. The conditions of Schur's theorem are

474 SHIING-SHEN CHERN

satisfied and we get $d' \leqq d$, d being the distance between A and B. Hence, L is either greater than, or equal to, the longer arc of S with the chord AB, or less than, or equal to, the shorter arc of S with the chord AB.

In particular, we can consider arcs joining A and B with curvature of $1/R$, $R \geqq d/2$. The lengths of such arcs have no upper bound, as shown by the example of a helix. They have d as a lower bound, but can be as close to d as possible. Therefore, we have an example of a minimum problem which has no solution.

Finally, we remark that Schur's theorem can be generalized to sectionally smooth curves. We give here a statement of this generalization without proof.

Remark. Let C and C* be two sectionally smooth curves of the same length, such that C forms a simple convex plane curve with its chord. Referred to the arc length s from one endpoint as parameter, let k(s) be the curvature of C at a regular point and a(s) the angle between the oriented tangents at a vertex; denote corresponding quantities for C* by the same notations with asterisks. Let d and d* be the distances between the endpoints of C and C*, respectively. Then, if

$$k^*(s) \leqq k(s) \quad and \quad a^*(s) \leqq a(s),$$

we have $d^* \geqq d$. The equality sign holds if and only if

$$k^*(s) = k(s) \quad and \quad a^*(s) = a(s).$$

The last set of conditions does not necessarily imply that C and C^* are congruent. In fact, there are simple rectilinear polygons in space which have equal sides and equal angles, but are not congruent.

6. THE GAUSS-BONNET FORMULA

We consider the intrinsic Riemannian geometry on a surface M. To simplify calculations and without loss of generality, we suppose the metric to be given in the isothermal parameters u and v:

$$ds^2 = e^{2\lambda(u,v)}(du^2 + dv^2). \tag{26}$$

The element of area is then

$$dA = e^{2\lambda} du\, dv \tag{27}$$

and the area of a domain D is given by the integral

$$A = \iint_D e^{2\lambda} du\, dv. \tag{28}$$

Also, the Gaussian curvature of the surface is

$$K = -e^{-2\lambda}(\lambda_{uu} + \lambda_{vv}). \tag{29}$$

It is well-known that the Riemannian metric defines the parallelism of Levi-Civita. To express it analytically, we write

$$u^1 = u \quad and \quad u^2 = v \tag{30}$$

and

$$ds^2 = \sum g_{ij}\, du^i\, du^j. \tag{31}$$

In this last formula and throughout this paragraph, our small Latin indices will range from 1 to 2 and a summation sign will mean summation over all repeated indices. From g_{ij} we introduce the g^{ij}, according to the equation

$$\sum g_{ij} g^{jk} = \delta_i^k \tag{32}$$

and the Christoffel symbols

$$\begin{cases} \Gamma_{ijk} = \dfrac{1}{2}\left(\dfrac{\partial g_{ij}}{\partial u^k} + \dfrac{\partial g_{jk}}{\partial u^i} - \dfrac{\partial g_{ik}}{\partial u^j} \right) \\[2mm] \Gamma_{ik}^j = \sum g^{jh}\Gamma_{ihk} \end{cases} \tag{33}$$

To a vector with the components ξ^i, the Levi-Civita parallelism defines the "covariant differential"

$$D\xi^i = d\xi^i + \sum \Gamma_{jk}^i\, du^k\, \xi^j. \tag{34}$$

All these equations are well-known in classical Riemannian geometry following the introduction of tensor analysis. The following is a new concept. Suppose the surface M is oriented. Consider the space B of all *unit* tangent vectors of M. This space B is a three-dimensional space, because the set of all unit tangent vectors with the same origin is one-dimensional. (It is called a *fiber space*, meaning that all the unit tangent vectors with origins in a neighborhood form a space which is topologically a product space.) To a unit tangent vector $\xi = (\xi^1, \xi^2)$, let $\eta = (\eta^1, \eta^2)$ be the uniquely determined unit tangent vector, orthogonal to ξ, such that ξ and η form a positive orientation. We introduce the linear differential form

$$\varphi = \sum_{1 \leq i,j \leq 2} g_{ij} D\xi^i \eta^j. \tag{35}$$

Then φ is well-defined in B and is usually called the *connection form*.

Because the vector ξ is a unit vector, we can write its components as follows:

$$\xi^1 = e^{-\lambda}\cos\theta \quad \text{and} \quad \xi^2 = e^{-\lambda}\sin\theta. \tag{36}$$

Then

$$\eta^1 = -e^{-\lambda}\sin\theta \quad \text{and} \quad \eta^2 = e^{-\lambda}\cos\theta. \tag{37}$$

Routine calculation gives

$$\begin{aligned} \Gamma_{11}^1 = \Gamma_{12}^2 = -\Gamma_{22}^1 = \lambda_u, \\ \Gamma_{12}^1 = \Gamma_{22}^2 = -\Gamma_{11}^2 = \lambda_v, \end{aligned} \tag{38}$$

whence the important relation

$$\varphi = d\theta - \lambda_v\, du + \lambda_u\, dv. \tag{39}$$

Its exterior derivative is therefore

$$d\varphi = -K\,dA. \tag{40}$$

Equation (40) is perhaps the most important formula in two-dimensional local Riemannian geometry.

The connection form φ is a differential form in B. We get from φ a differential form in a subset of M, when there is defined on it a field of unit tangent vectors. For example, let C be a smooth curve on M with the arc length s and let $\xi(s)$ be a smooth unit vector field along C. Then $\varphi = \sigma\,ds$, and σ is called the *variation* of ξ along C. The vectors ξ are said to be parallel along C, if $\sigma = 0$. If ξ is everywhere tangent to C, σ is called the geodesic curvature of C. C is a geodesic of M, if along C the unit tangent vectors are parallel, that is, if its geodesic curvature is 0.

Consider a domain D of M, such that there is a unit vector field defined over D, with an isolated singularity at an interior point $p_0 \in D$. Let γ_ε be a circle of geodesic radius ε about p_0. Then, from Equation (39), the limit

$$\frac{1}{2\pi} \lim_{\varepsilon \to 0} \int_{\gamma_\varepsilon} \varphi \tag{41}$$

is an integer, to be called the *index* of the vector field at p_0.

Examples of vector fields with isolated singularities are shown in Figure 4.

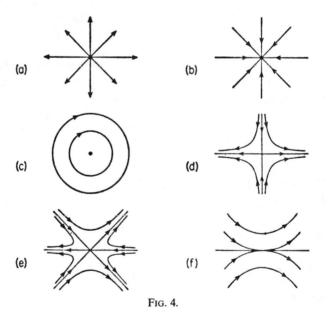

FIG. 4.

These singularities are, respectively, (a) a source or maximum, (b) a sink or minimum, (c) a center, (d) a simple saddle point, (e) a monkey saddle, and (f) a dipole. The indices are, respectively, 1, 1, 1, -1, -2, and 2.

The Gauss-Bonnet formula is the following theorem.

THEOREM. *Let D be a compact oriented domain in M bounded by a sectionally smooth curve C. Then*

$$\int_C k_g\,ds + \int_D K\,dA + \sum_i (\pi - \alpha_i) = 2\pi\chi, \tag{42}$$

where k_g is the geodesic curvature of C, $\pi - \alpha_i$ are the exterior angles at the vertices of C, and χ is the Euler characteristic of D.

Proof. Consider first the case that D belongs to a coordinate domain (u,v) and is bounded by a simple polygon C of n sides, C_i, $1 \leq i \leq n$, with the angles α_i at the vertices. Suppose D is positively oriented. To the points of the arcs C_i we associate the unit tangent vectors to C_i. Thus, to each vertex is associated two vectors at an angle $\pi - \alpha_i$. By the theorem of turning tangents (see Section 1), the total variation of θ as the C_i's are traversed once is $2\pi - \sum(\pi - \alpha_i)$. It follows that

$$\int_C k_g\,ds = 2\pi - \sum_i (\pi - \alpha_i) + \int_C -\lambda_v\,du + \lambda_u\,dv.$$

By Stokes theorem, the last integral is equal to $-\iint_D K\,dA$.

Thus, the formula is proved in this special case.

In the general case, suppose D is subdivided into a union of polygons $D_\lambda, \lambda = 1,\dots,f$, such that (1) each D_λ lies in one coordinate neighborhood and (2) two D_λ have either no point, or one vertex, or a whole side, in common. Moreover, let the D_λ be coherently oriented with D, so that every interior side has different senses induced by the two polygons of which it is a side. Let v and e be the numbers of interior vertices and interior sides in this subdivision of D—i.e., vertices and sides which are not on the boundary, C. The above formula can then be applied to each D_λ. Adding all these relations, we have, because the integrals of geodesic curvature along the interior sides cancel,

$$\int_C k_g\,ds + \iint_D K\,dA = 2\pi f - \sum_{i,\lambda} (\pi - \alpha_{\lambda i}) - \sum_i (\pi - \alpha_i)$$

where α_i are the angles at the vertices of D, while the first sum in the right-hand side is extended over all interior vertices of the subdivision. Since each interior side is on exactly two D_λ and since the sum of interior angles about a vertex is 2π, this sum is equal to

$$-2\pi e + 2\pi v.$$

We call the integer

$$\chi(D) = v - e + f \tag{43}$$

the Euler characteristic of D. Substituting, we get Equation (42). Equation (42) has the consequence that the integer χ is independent of the subdivision.

In particular, if C has no vertex, we have

$$\int_C k_g\,ds + \iint_D K\,dA = 2\pi\chi. \tag{44}$$

Moreover, if D is the whole surface M, we get

$$\iint_S K \, dA = 2\pi\chi. \tag{45}$$

It follows that if $K=0$, the Euler characteristic of M is 0, and M is homeomorphic to a torus. If $K>0$, then $\chi>0$, and S is homeomorphic to a sphere.

The Euler characteristic plays an important role in the study of vector fields on a surface.

Remark. On a closed orientable surface M, the sum of the indices of a vector field with a finite number of singularities, is equal to the Euler characteristic, $\chi(M)$ of M.

Proof. Let p_i, $1 \leqq i \leqq n$, be the singularities of the vector field. Let $\gamma_i(\varepsilon)$ be a circle of radius ε about p_i, and let $\Delta_i(\varepsilon)$ be the disk bounded by $\gamma_i(\varepsilon)$. Integrating K of A over the domain $M - \bigcup_i \Delta_i(\varepsilon)$ and using Equation (40), we get

$$\iint_{M - \cup_i \Delta_i(\varepsilon)} K \, dA = \sum_i \int_{\gamma_i(\varepsilon)} \varphi,$$

where $\gamma_i(\varepsilon)$ is oriented so that it is the boundary of $\Delta_i(\varepsilon)$. The theorem follows by letting $\varepsilon \to 0$.

We wish to give two further applications of the Gauss-Bonnet formula. The first is a theorem of Jacobi. Let $X(s)$ be the coordinate vector of a closed space curve, with the arc length s. Let $T(s)$, $N(s)$, and $B(s)$ be the unit tangent, principal normal, and binormal vectors, respectively. In particular, the curve on the unit sphere with the coordinate vector $N(s)$ is the *principal normal indicatrix*. It has a tangent, wherever

$$k^2 + w^2 \neq 0, \tag{46}$$

where k (when not equal to 0) and w are, respectively, the curvature and torsion of $X(s)$. Jacobi's theorem follows.

THEOREM. *If the principal normal indicatrix of a closed space curve has a tangent everywhere, it divides the unit sphere in two domains of the same area.*

Proof. To prove the theorem, we define τ by the equations

$$k = \sqrt{k^2 + w^2} \, \cos\tau, \quad w = \sqrt{k^2 + w^2} \, \sin\tau. \tag{47}$$

Then we have

$$d(-\cos\tau \, T + \sin\tau \, B) = (\sin\tau \, T + \cos\tau \, B) \, d\tau - \sqrt{k^2 + w^2} \, N \, ds.$$

Hence, if σ is the arc length of $N(s)$, $d\tau/d\sigma$ is the geodesic curvature of $N(s)$ on the unit sphere. Let D be one of the domains bounded by $N(s)$, and A its area. By the Gauss-Bonnet formula, we have, since $K=1$,

$$\int_{N(s)} d\tau + \iint dA = 2\pi.$$

It follows that $A = 2\pi$, and the theorem is proved.

Our second application is Hadamard's theorem on convex surfaces.

, THEOREM. *If the Gaussian curvature of a closed orientable surface in euclidean space is everywhere positive, the surface is convex (that is, it lies at one side of every tangent plane).*

We discussed a similar theorem for curves in Section 1. For surfaces, it is not necessary to suppose that it has no self-intersection.

Proof. It follows from the Gauss-Bonnet formula that the Euler characteristic $\chi(M)$ of the surface M is positive, so that $\chi(M)=2$ and

$$\iint_M K\,dA = 4\pi.$$

Suppose M is oriented. We consider the Gauss mapping

$$g:M\to\Sigma_0 \tag{48}$$

(where Σ_0 is the unit sphere about a fixed point 0), which assigns to every point $p\in M$ the end of the unit vector through 0 parallel to the unit normal vector to M at p. The condition $K>0$ implies that g has everywhere a nonzero functional determinant and is locally one-to-one. It follows that $g(M)$ is an open subset of Σ_0. Since M is compact, $g(M)$ is a compact subset of Σ_0, and hence is also closed. Therefore, g maps onto Σ_0.

Suppose that g is not one-to-one, that is, there exist points p and q of $M, p\neq q$, such that $g(p)=g(q)$. There is then a neighborhood U of q, such that $g(M-U)=\Sigma_0$. Since $\iint_{M-U} K\,dA$ is the area of $g(M-U)$, counted with multiplicities, we will have

$$\iint_{M-U} K\,dA \geqq 4\pi.$$

But

$$\iint_U K\,dA >0,$$

so that

$$\iint_M K\,dA = \iint_U K\,dA + \iint_{M-U} K\,dA >4\pi,$$

which is a contradiction, and Hadamard's theorem is proved.

Hadamard's theorem is true under the weaker hypothesis $K\geqq 0$, but the proof is more difficult; see the article by Chern-Lashof mentioned in Section 4.

For further reading, see:

1. S. S. Chern, "On the curvatura integra in a Riemannian manifold," *Annals of Mathematics*, 46 (1945), pp. 674–84.

2. H. Flanders, "Development of an extended exterior differential calculus," *Transactions of the American Mathematical Society*, 75 (1953), pp. 311–26.

See also Section 8 of Flanders' article in *MAA Studies in Mathematics*, Vol. 4 (1967).

7. UNIQUENESS THEOREMS OF COHN-VOSSEN AND MINKOWSKI

The "rigidity" theorem of Cohn-Vossen can be stated as follows.

THEOREM. *An isometry between two closed convex surfaces is established either by a motion or by a motion and a reflection.*

In other words, such an isometry is always trivial, and the theorem is obviously not true locally. The following proof is the work of G. Herglotz.

Proof. We shall first discuss some notations on surface theory in euclidean space. Let the surface S be defined by expressing its position vector X as a function of two parameters, u and v. These functions are supposed to be continuously differentiable up to the second order. Suppose that X_u and X_v are everywhere linearly independent, and let ξ be the unit normal vector, so that S is oriented. As usual, let

$$I = dX \cdot dX = E\, du^2 + 2F\, du\, dv + G\, dv^2$$
$$II = -dX \cdot d\xi = L\, du^2 + 2M\, du\, dv + N\, dv^2 \tag{49}$$

be the first and second fundamental forms of the surface. Let H and K denote respectively the mean and Gaussian curvatures.

It is sufficient to prove that under the isometry, the second fundamental forms are equal. Assume the local coordinates are such that corresponding points have the same local coordinates. Then E, F, and G are equal for both surfaces, and the same is true of the Christoffel symbols. Let the second surface be S^*, and denote the quantities pertaining to S^* by the same symbols with asterisks. We introduce

$$\lambda = \frac{L}{D}, \quad \mu = \frac{M}{D}, \quad \nu = \frac{N}{D}, \tag{50}$$

where $D = \sqrt{EG - F^2}$. Then the Gaussian curvature is

$$K = \lambda\nu - \mu^2 = \lambda^*\nu^* - \mu^{*2}, \tag{51}$$

and is the same for both surfaces. The mean curvatures are

$$H = \frac{1}{2D}(G\lambda - 2F\mu + E\nu) \quad \text{and}$$
$$H^* = \frac{1}{2D}(G\lambda^* - 2F\mu^* + E\nu^*). \tag{52}$$

We introduce further

$$J = \lambda\nu^* - 2\mu\mu^* + \nu\lambda^*. \tag{53}$$

The proof depends on the following identity:

$$DJ\xi = \frac{\partial}{\partial u}(\nu^* X_u - \mu^* X_v) - \frac{\partial}{\partial v}(\mu^* X_u - \lambda^* X_v). \tag{54}$$

We first notice that the Codazzi equations can be written in terms of λ^*, μ^*, and ν^*

in the form

$$\lambda^*_v - \mu^*_u + \Gamma^2_{22}\lambda^* - 2\Gamma^2_{12}\mu^* + \Gamma^2_{11}\nu^* = 0,$$
$$\mu^*_v - \nu^*_u - \Gamma^1_{22}\lambda^* + 2\Gamma^1_{12}\mu^* - \Gamma^1_{11}\nu^* = 0. \tag{55}$$

We next write the equations of Gauss:

$$X_{uu} - \Gamma^1_{11}X_u - \Gamma^2_{11}X_v - D\lambda\xi = 0,$$
$$X_{uv} - \Gamma^1_{12}X_u - \Gamma^2_{12}X_v - D\mu\xi = 0, \tag{56}$$
$$X_{vv} - \Gamma^1_{22}X_u - \Gamma^2_{22}X_v - D\nu\xi = 0.$$

Multiplying these equations by X_v, $-X_u$, ν^*, $-2\mu^*$, and λ^*, respectively, and adding, we establish Equation (54).

We now write

$$p = Xe_3, \quad y_1 = XX_u, \quad y_2 = XX_v, \tag{57}$$

where the right-hand sides are the scalar products of the vectors in question, so that $p(u,v)$ is the oriented distance from the origin to the tangent plane at $X(u,v)$. Equation (54) gives, after taking scalar product with X,

$$DJp = -\nu^*E + 2\mu^*F - \lambda^*G + (\nu^*y_1 - \mu^*y_2)_u - (\mu^*y_1 - \lambda^*y_2)_v. \tag{58}$$

Let C be a closed curve on S. It divides S into two domains, D_1 and D_2, both having C as boundary. Moreover, if D_1 and D_2 are coherently oriented, C appears as a boundary in opposite senses. To each of these domains, say D_1, we apply Green's theorem, and get

$$\iint_{D_1} Jp\, dA = \iint_{D_1} (-\nu^*E + 2\mu^*F - \lambda^*G)\, du\, dv$$
$$+ \int_C (+\mu^*y_1 - \lambda^*y_2)\, du + (\nu^*y_1 - \mu^*y_2)\, dv. \tag{59}$$

Adding this equation to a similar one for D_2, the line integrals cancel, and we have

$$\iint_S Jp\, dA = \iint_S (-\nu^*E + 2\mu^*F - \lambda^*G)\, du\, dv.$$

By Equation (52),

$$\iint_S Jp\, dA = -2\iint_S H^*\, dA. \tag{60}$$

In particular, this formula is valid when S and S^* are identical, and we have

$$\iint_S 2Kp\, dA = -2\iint_S H\, dA. \tag{61}$$

Subtracting these two equations, we get

$$\iint_S \begin{vmatrix} \lambda^* - \lambda & \mu^* - \mu \\ \mu^* - \mu & \nu^* - \nu \end{vmatrix} p\, dA = 2\iint_S H^*\, dA - 2\iint_S H\, dA. \tag{62}$$

To complete the proof, we need the following elementary lemma:

LEMMA. *Let*

$$ax^2 + 2bxy + cy^2 \quad and \quad a'x^2 + 2b'xy + c'y^2 \tag{63}$$

be two positive definite quadratic forms, with

$$ac - b^2 = a'c' - b'^2. \tag{64}$$

Then

$$\begin{vmatrix} a'-a & b'-b \\ b'-b & c'-c \end{vmatrix} \leqq 0, \tag{65}$$

and the equality sign holds only when the two forms are identical.

As proof, we observe that the statement of the lemma remains unchanged under a linear transformation of the variables. Applying such a linear transformation when necessary, we can assume $b' = b$. Then the left-hand side of Equation (65) becomes

$$(a'-a)(c'-c) = -\frac{c}{a'}(a'-a)^2 \leqq 0,$$

as was to be proved. Moreover, the quantity equals 0 only when we also have $a' = a$ and $c' = c$.

We now choose the origin to be inside S, so that $p > 0$. Then the integrand in the left-hand side of Equation (62) is nonpositive, and it follows that

$$\iint_S H^* dA \leqq \iint_S H dA.$$

Since the relation between S and S^* is symmetrical, we must also have

$$\iint_S H dA \leqq \iint_S H^* dA.$$

Hence,

$$\iint_S H dA = \iint_S H^* dA.$$

It follows that the integral at the left-hand side of Equation (62) is 0, and hence, that

$$\lambda^* = \lambda, \quad \mu^* = \mu, \quad \nu^* = \nu,$$

completing the proof of Cohn-Vossen's theorem.

By Hadamard's theorem, we see that the Gauss map $g: S \rightarrow \Sigma_0$ (see Section 6) is one-to-one for a closed surface with $K > 0$. A point on S can therefore be regarded as a function of its normal vector ξ, and the same is true with any scalar function on S. Minkowski's theorem expresses the unique determination of S when $K(\xi)$ is known.

THEOREM. *Let S be a closed convex surface with Gaussian curvature $K > 0$. The function $K(\xi)$ determines S up to a translation.*

Proof. We shall give a proof of this theorem modeled after the above—that is, by an integral formula [see S. S. Chern, *American Journal of Mathematics* 79 (1957), pp. 949–50]. Let u and v be isothermal parameters on the unit sphere Σ_0, so that we have

$$\xi_u^2 = \xi_v^2 = A > 0 \quad \text{(say)}, \quad \xi_u \xi_v = 0. \tag{66}$$

Through the mapping g^{-1} we regard u and v also as parameters on S. Since ξ_u and ξ_v are orthogonal to ξ and are linearly independent, every vector orthogonal to ξ can be expressed as their linear combination. This fact, taken with the relation $X_u \xi_v = X_v \xi_u$, allows us to write

$$\begin{aligned} -X_u &= a\xi_u + b\xi_v, \\ -X_v &= b\xi_u + c\xi_v. \end{aligned} \tag{67}$$

Forming scalar products of these equations with ξ_u and ξ_v, we have

$$Aa = L, \quad Ab = M, \quad Ac = N. \tag{68}$$

Moreover, taking the vector product of the two relations in Equation (67), we find

$$X_u \times X_v = (ac - b^2)(\xi_u \times \xi_v).$$

But

$$X_u \times X_v = D\xi, \quad \xi_u \times \xi_v = A\xi, \tag{69}$$

so that, combining with Equation (68), we have

$$D = A(ac - b^2) = \frac{KD^2}{A},$$

which gives

$$A = KD, \quad ac - b^2 = \frac{1}{K}. \tag{70}$$

Since $A\,du\,dv$ and $D\,du\,dv$ are, respectively, the volume elements of Σ_0 and S, the first relation in Equation (70) expresses the well-known fact that K is the ratio of these volume elements.

Suppose S^* is another convex surface with the same function, $K(\xi)$. We set up a homeomorphism between S and S^*, so that they have the same normal vector at corresponding points. Then the parameters u and v can be used for both S and S^*, and corresponding points have the same parameter values. We denote by asterisks the vectors and functions for the surface S^*. Since $K = K^*$, we have from Equation (70), $ac - b^2 = a^*c^* - b^{*2}$ and $D = D^*$.

Let

$$p = X \cdot \xi \quad \text{and} \quad p^* = X^* \cdot \xi \tag{71}$$

be the distances from the origin to the tangent planes of the two surfaces. Our basic relation is the identity

$$\begin{aligned} (X, X^*, X_u)_v - (X, X^*, X_v)_u &= A\left\{ 2(ac - b^2)p^* + (-ac^* - a^*c + 2bb^*)p \right\} \\ &= A\left\{ 2(ac - b^2)(p^* - p) + \begin{vmatrix} a - a^* & b - b^* \\ b - b^* & c - c^* \end{vmatrix} p \right\} \end{aligned}$$

which follows immediately from Equations (67), (69), (70), and (71). From it, we find, by Green's theorem, the integral formula

$$\int_{\Sigma_0} \left\{ 2(ac - b^2)(p^* - p) + \begin{vmatrix} a - a^* & b - b^* \\ b - b^* & c - c^* \end{vmatrix} p \right\} A \, du \, dv = 0. \tag{72}$$

By translations if necessary, we can suppose the origin to be inside both surfaces, S and S^*, so that $p > 0$ and $p^* > 0$. Since

$$\begin{pmatrix} a & b \\ b & c \end{pmatrix} \quad \text{and} \quad \begin{pmatrix} a^* & b^* \\ b^* & c^* \end{pmatrix}$$

are positive definite matrices, it follows from our algebraic lemma that

$$\begin{vmatrix} a - a^* & b - b^* \\ b - b^* & c - c^* \end{vmatrix} \leqq 0.$$

Hence,

$$\int_{\Sigma_0} (ac - b^2)(p^* - p) A \, du \, dv \geqq 0. \tag{73}$$

But the same relation is true when S and S^* are interchanged. Hence, the integral at the left-hand side of Equation (73) must be identically 0. It follows from Equation (72) that

$$\int_{\Sigma_0} \begin{vmatrix} a - a^* & b - b^* \\ b - b^* & c - c^* \end{vmatrix} p \, A \, du \, dv = 0,$$

possible only when $a = a^*$, $b = b^*$, and $c = c^*$. The latter implies that

$$X^*_u = X_u \quad \text{and} \quad X^*_v = X_v,$$

which means that S and S^* differ by a translation.

For further reading, see:

1. S. S. Chern, "Integral formulas for hypersurfaces in euclidean space and their applications to uniqueness theorems," *Journal of Mathematics and Mechanics*, 8 (1959), pp. 947–55.

2. T. Otsuki, "Integral formulas for hypersurfaces in a Riemannian manifold and their applications," *Tôhoku Mathematical Journal*, 17 (1965), pp. 335–48.

3. K. Voss, "Differentialgeometrie geschlossener Flächen im euklidischen Raum," *Jahresberichte deutscher Math. Verein.*, 63 (1960–1961), pp. 117–36.

8. BERNSTEIN'S THEOREM
ON MINIMAL SURFACES

A minimal surface is a surface which locally solves the Plateau problem—that is, it is the surface of smallest area bounded by a given closed space curve. Analytically, it is defined by the condition that the mean curvature is identically 0. We suppose the surface to be given by

$$z = f(x, y), \tag{74}$$

where the function $f(x,y)$ is twice continuously differentiable. Then a minimal surface is characterized by the partial differential equation,

$$(1+q^2)r - 2pqs + (1+p^2)t = 0, \tag{75}$$

where

$$p = \frac{\partial f}{\partial x}, \quad q = \frac{\partial f}{\partial y}, \quad r = \frac{\partial^2 f}{\partial x^2}, \quad s = \frac{\partial^2 f}{\partial x \partial y}, \quad t = \frac{\partial^2 f}{\partial y^2}. \tag{76}$$

Equation (75), called the minimal surface equation, is a nonlinear "elliptic" differential equation.

Bernstein's theorem is the following "uniqueness theorem."

THEOREM. *If a minimal surface is defined by Equation (74) for all values of x and y, it is a plane. In other words, the only solution of Equation (75) valid in the whole (x,y)-plane is a linear function.*

Proof. We shall derive this theorem as a corollary of the following theorem of Jörgens [*Math. Annalen* 127 (1954), pp. 130–34].

THEOREM. *Suppose the function $z = f(x,y)$ is a solution of the equation*

$$rt - s^2 = 1, \quad r > 0, \tag{77}$$

for all values of x and y. Then $f(x,y)$ is a quadratic polynomial in x and y.

For fixed (x_0, y_0) and (x_1, y_1), consider the function

$$h(t) = f(x_0 + t(x_1 - x_0), y_0 + t(y_1 - y_0)).$$

We have

$$h'(t) = (x_1 - x_0)p + (y_1 - y_0)q,$$
$$h''(t) = (x_1 - x_0)^2 r + 2(x_1 - x_0)(y_1 - y_0)s + (y_1 - y_0)^2 s \geqq 0,$$

where the arguments in the functions p, q, r, s, t are $x_0 + t(x_1 - x_0)$ and $y_0 + t(y_1 - y_0)$. From the last inequality, it follows that

$$h'(1) \geqq h'(0)$$

or

$$(x_1 - x_0)(p_1 - p_0) + (y_1 - y_0)(q_1 - q_0) \geqq 0. \tag{78}$$

where

$$p_i = p(x_i, y_i) \quad \text{and} \quad q_i = q(x_i, y_i), \quad i = 0, 1. \tag{79}$$

Consider the transformation of Lewy:

$$\xi = \xi(x,y) = x + p(x,y), \quad \eta = \eta(x,y) = y + q(x,y). \tag{80}$$

Setting

$$\xi_i = \xi(x_i, y_i), \quad \eta_i = \eta(x_i, y_i), \quad i = 0, 1, \tag{81}$$

we have, by Equations (78)

$$(\xi_1 - \xi_0)^2 + (\eta_1 - \eta_0)^2 \geq (x_1 - x_0)^2 + (y_1 - y_0)^2. \tag{82}$$

Hence, the mapping

$$(x, y) \rightarrow (\xi, \eta) \tag{83}$$

is distance-increasing.

Moreover, we have

$$\begin{aligned}
\xi_x &= 1 + r, \quad \xi_y = s \\
\eta_x &= s, \quad\quad \eta_y = 1 + t,
\end{aligned} \tag{84}$$

so that

$$\frac{\partial (\xi, \eta)}{\partial (x, y)} = 2 + r + t \geq 2, \tag{85}$$

and the mapping in Equation (83) is locally one-to-one. It follows easily that Equation (83) is a diffeomorphism of the (x, y)-plane onto the (ξ, η)-plane.

We can therefore regard the function $f(x, y)$, which is a solution of Equation (77), as a function in ξ and η. Let

$$F(\xi, \eta) = x - iy - (p - iq), \tag{86}$$

$$\zeta = \xi + i\eta. \tag{87}$$

It can be verified by a computation that $F(\xi, \eta)$ satisfies the Cauchy-Riemann equations, so that $F(\zeta) = F(\xi, \eta)$ is a holomorphic function in ζ. Moreover, we have

$$F'(\zeta) = \frac{t - r + 2is}{2 + r + t} \tag{88}$$

From the last relation, we get

$$1 - |F'(\zeta)|^2 = \frac{4}{2 + r + t} > 0.$$

Thus $F'(\zeta)$ is bounded in the whole ζ-plane. By Liouville's theorem, we have

$$F'(\zeta) = \text{const.}$$

On the other hand, by Equation (88) we have

$$r = \frac{|1 - F'|^2}{1 - |F'|^2}, \quad s = \frac{i(\bar{F}' - F')}{1 - |F'|^2}, \quad t = \frac{|1 + F'|^2}{1 - |F'|^2}. \tag{89}$$

It follows that r, s, and t are all constants, and Jörgens's theorem is proved.

Bernstein's theorem is an easy consequence of Jörgens's theorem. In fact, let

$$W = (1 + p^2 + q^2)^{1/2}. \tag{90}$$

Then the minimal surface equation is equivalent to each of the following equations:

$$\frac{\partial}{\partial x}\frac{-pq}{W} + \frac{\partial}{\partial y}\frac{1+p^2}{W} = 0,$$

$$\frac{\partial}{\partial x}\frac{1+q^2}{W} + \frac{\partial}{\partial y}\frac{-pq}{W} = 0. \tag{91}$$

It follows that there exists a C^2-function, $\varphi(x,y)$, such that

$$\varphi_{xx} = \frac{1}{W}(1+p^2), \quad \varphi_{xy} = \frac{1}{W}pq, \quad \varphi_{yy} = \frac{1}{W}(1+q^2). \tag{92}$$

These partial derivatives satisfy the equation

$$\varphi_{xx}\varphi_{yy} - \varphi_{xy}^2 = 1, \quad \varphi_{xx} > 0.$$

By Jörgens's theorem, φ_{xx}, φ_{xy}, and φ_{yy} are constants. Hence, p and q are constants, and $f(x,y)$ is a linear function. [This proof of Bernstein's theorem is that of J. C. C. Nitsche, *Annals of Mathematics*, 66 (1957), pp. 543–44.]

Minimal surfaces have an extensive literature. See the following expository article:

1. J. C. C. Nitsche, "On new results in the theory of minimal surfaces," *Bulletin of the American Mathematical Society*, 71 (1965), pp. 195–270.

19

NORMAN LEVINSON

Norman Levinson was born on August 11, 1912, in Lynn, Massachusetts. He received the B.S. and M.S. degrees in electrical engineering in 1934 from the Massachusetts Institute of Technology. While a student of electrical engineering he also studied mathematics intensively. Among the graduate courses which he took was the year course given in 1933–34 by Norbert Wiener, *Fourier Series and Integrals*. This represented his initial contact with Wiener which can best be described in his own words as follows: "I became acquainted with Wiener in September 1933, while still a student of electrical engineering, when I enrolled in his graduate course. It was at that time really a seminar course. At that level he was a most stimulating teacher. He would actually carry on his research at the blackboard. As soon as I displayed a slight comprehension of what he was doing, he handed me the manuscript of Paley-Wiener for revision. I found a gap in a proof and proved a lemma to set it right. Wiener thereupon sat down at his typewriter, typed my lemma, affixed my name and sent it off to a journal. A prominent professor does not often act as a secretary for a young student. He convinced me to change my course from electrical engineering to mathematics. He then went to visit my parents, unschooled immigrant working people living in a run-down ghetto community, to assure them about my future in mathematics. He came to see them a number of times during the next five years to reassure them until he finally found a permanent position for me. (In those depression years positions were very scarce)," (see [1]).

Wiener obtained an M.I.T. traveling fellowship for Levinson to spend 1934–35 with G. H. Hardy in Cambridge, England. (Later D. C. Spencer, also an engineering student, was awarded the same M.I.T. fellowship to Cambridge, where he studied with Littlewood.) In 1935 Levinson was awarded the D.Sc. by M.I.T. For the next year and one-half he was a National Research Council Fellow at Princeton University and the Institute for Advanced Study, where he continued his research in gap and density theorems under the nominal supervision of von Neumann. In February 1937 he became an instructor at M.I.T., where he remained except for 1948–49 when he was a Guggenheim Fellow at the Mathematics Institute in Copenhagen and 1967–68 when he was at the University of Tel Aviv. He was named Institute Professor at M.I.T. in 1971. He died October 10, 1975.

After the publication of his Colloquium book, *Gap and Density Theorems*, by the American Mathematical Society in 1940, he shifted his research to differential equations for which he was awarded the Bôcher Prize by the A.M.S. in 1966–68. In 1955, *Theory of Ordinary Differential Equations* by Earl Coddington and Levinson was published. Since then he also published articles in probability, complex linear programming, and analytic number theory.

Norman Levinson was the heart of mathematics at M.I.T., a man who combined creative intellect of the highest order with human compassion and an unremitting dedication to science and to excellence in its pursuit. Throughout the mathematical world, the name M.I.T. and the name of Norman Levinson have been synonymous for many years.

In addition to his research career, Levinson had an extremely successful career as a teacher. He supervised over 35 doctoral students, many of whom moved on to successful careers in mathematics themselves. He was known as a superb lecturer and teacher, [1].

In accepting the award, Professor Levinson expressed himself as very pleased that the Association had honored him with the 1971 Chauvenet Prize. He was particularly gratified when he found that his old teacher G. H. Hardy had been an early recipient of the Prize.

Reference

1. Resolution on the Death of Norman Levinson, Institute Professor and Professor of Mathematics, Massachusetts Institute of Technology.

A MOTIVATED ACCOUNT OF AN ELEMENTARY PROOF OF THE PRIME NUMBER THEOREM

NORMAN LEVINSON, Massachusetts Institute of Technology

1. Introduction. One of the most striking results of mathematics is the prime number theorem first conjectured, independently, by Gauss and Legendre prior to 1800 and proved, independently, by Hadamard and de la Vallée Poussin in 1896. Among the many great mathematicians of the 19th century who did not succeed in proving the theorem were Chebychef and Riemann, both of whom obtained important partial results. Riemann indicated that the prime number theorem was related to the behavior of the zeta function in the complex plane and found many properties of this function which has since borne his name. Riemann's ideas were exploited and augmented in the proofs of Hadamard and de la Vallée Poussin.

In 1949, P. Erdös and A. Selberg, using a formula previously proved by Selberg in an elementary way, jointly succeeded in giving several elementary proofs of the prime number theorem, [3]. While elementary, neither these proofs, nor another one of Selberg [6], are simple.

With the tremendous proliferation of mathematics, many mathematicians no longer study number theory. Therefore it seems worthwhile to give a self-contained and motivated account of an elementary proof of the prime number theorem.

The prime numbers (2, 3, 5, 7, 11, 13, $\cdots$) were known to ancient man and in Euclid there is a proof that they are infinite in number. The number of primes not exceeding x is called $\pi(x)$ and can be represented by

$$(1.1) \qquad \pi(x) = \sum_{p \leq x} 1$$

where the symbol p runs over the sequence of primes in increasing order. The simplest form of Legendre's conjecture was

$$(1.2) \qquad \lim_{x \to \infty} \frac{\pi(x)}{x/\log x} = 1.$$

Gauss' conjecture has turned out to be more profound and was that $\pi(x)$, for large x, is close to $\int_2^x dt/\log t$. He arrived at this by observing from a tabulation of prime numbers that the primes seemed to have an asymptotic density which at

Professor Levinson studied with Norbert Wiener and received the Sc.D. degree in 1935. He was a travelling Fellow at Cambridge University in 1934–35 and an NRC Fellow in 1935–37. He has been on the MIT staff since 1937, except for a year as Guggenheim Fellow at the Mathematics Institute, Copenhagen and a year at the University of Tel Aviv. His main research interests are transforms, entire functions, probability, and differential equations. He is the author of the AMS Colloquium volume, *Gap and Density Theorems*, and (with E. Coddington) *Ordinary Differential Equations*. Professor Levinson received the AMS Bôcher Prize and he is a member of the National Academy of Sciences. *Editor*

x was $1/\log x$. Because of this, for some purposes a better way to find the asymptotic behavior of the primes is to weight each p with $\log p$. This is done in the function

$$(1.3) \qquad \theta(x) = \sum_{p \leq x} \log p.$$

Actually it turns out to be even more convenient to use not $\theta(x)$ but a closely related function $\psi(x)$ as will be seen.

The account that follows begins with the factorization of an integer into the product of powers of primes and proceeds with motivated proofs of the relevant discoveries of the 19th century in sections 2 and 3. This approach is continued in section 4 to prove Selberg's formula, and finally in section 5 where an exposition of proof of Selberg [6] is given as simplified by Wright [4], [9], and further simplified by the author [5].

The elementary proof of the prime number theorem has been extended to give elementary proofs of sharper forms of the theorem with a remainder by Breusch [2], Bombieri [1], Wirsing [8] and others. I am indebted to George B. Thomas for a critical reading of the manuscript.

2. The Chebychef identity and its inversion. Our starting point is that a positive integer can be factored into a product of powers of distinct primes. Thus a positive integer

$$(2.1) \qquad n = p_1^{k_1} p_2^{k_2} \cdots p_m^{k_m},$$

where the p_j, $1 \leq j \leq m$, are distinct primes and each k_j is a positive integer. Because addition is simpler than multiplication a more useful form of (2.1) is

$$(2.2) \qquad \log n = k_1 \log p_1 + k_2 \log p_2 + \cdots + k_m \log p_m.$$

The utility of this formula is very much enhanced by the use of the von Mangoldt symbol $\Lambda(n)$, introduced in 1895, which is defined by

$$(2.3) \qquad \Lambda(n) = \log p \text{ for } n = p^j,$$

where p is a prime number and j is a positive integer, and $\Lambda(n) = 0$ otherwise. Thus $\Lambda(n) \neq 0$ only if n is a power of a prime.

The symbol $\sum_{j|n}$ will be used to denote a sum on j where j runs through all of the positive divisors of the positive integer n. With this notation it will be shown that (2.2) can be written as

$$(2.4) \qquad \log n = \sum_{j|n} \Lambda(j).$$

To prove (2.4) note that because of (2.1) and the definition of $\Lambda(j)$, the only nonzero terms that can appear on the right side of (2.4) are $\log p_1$, $\log p_2$, $\cdots$ $\log$

p_m. Moreover log p_1 appears for $j=p_1$, for $j=p_1^2$, $\cdots$, and for $j=p_1^{k_1}$. Thus log p_1 appears exactly k_1 times; similarly log p_2 appears k_2 times, etc., which shows that (2.4) is a consequence of (2.2). The formula (2.4) is an extremely powerful variant of (2.1) and incorporates the properties of prime numbers which are needed here. The transformation of (2.2) into the form (2.4) is not obvious and historically came relatively late.

The formula (2.4) can be written in the equivalent form

$$(2.5) \qquad \qquad \log n = \sum_{ij=n} \Lambda(j),$$

where i and j are positive integers each of which takes on all possible values satisfying $ij=n$, so that indeed j runs through all positive divisors of n (as does i also).

The number of primes up to x, $\pi(x)$, is closely related to the sum

$$(2.6) \qquad \qquad \psi(x) = \sum_{j \leq x} \Lambda(j).$$

From the definition of $\Lambda(j)$,

$$\psi(x) = \sum_{p \leq x} \log p + \sum_{p^2 \leq x} \log p + \sum_{p^3 \leq x} \log p + \cdots = \theta(x) + \theta(x^{1/2}) + \theta(x^{1/3}) + \cdots$$

The function $\psi(x)$, expressed in the latter form, was already known to Chebychef, who gave a simple proof that the prime number theorem (1.2) is equivalent to

$$(2.7) \qquad \qquad \lim_{x \to \infty} \frac{\psi(x)}{x} = 1.$$

This proof will be given in Lemma 3.4 and (2.7) will be proved in Section 5. (Roughly speaking, $\psi(x)$ acts like $\pi(x) \log x$ for large x because $\psi(x)$ counts each $p \leq x$ with weight log p, (2.3), and because log p is close to log x for "most" of the $p \leq x$. True, $\psi(x)$ also counts log p again for $p \leq x^{1/2}$, for $p \leq x^{1/3}$, etc., but these last are very sparse as will be seen later in the proof.)

To use (2.5) to get information about $\psi(x)$, (2.5) is summed on $n \leq x$ to get $\sum_{n \leq x} \log n = \sum_{n \leq x} \sum_{ij=n} \Lambda(j)$, so that if one defines

$$(2.8) \qquad \qquad T(x) = \sum_{n \leq x} \log n,$$

then

$$(2.9) \qquad \qquad T(x) = \sum_{ij \leq x} \Lambda(j).$$

Because the logarithm is a smooth function, $T(x)$ can be readily appraised for large x, and this will be done in (3.4).

The double sum in (2.9) is taken over those lattice points in the positive quadrant of the (i, j) plane which lie on or below the hyperbola $ij=x$. If the double sum (2.9) is treated as a repeated sum, summing first on j

$$T(x) = \sum_{i \leq x} \sum_{j \leq x/i} \Lambda(j) = \sum_{i \leq x} \psi(x/i).$$

This identity was discovered by Chebychef (1850) and will be rewritten as

$$(2.10) \qquad T(x) = \sum_{n \leq x} \psi(x/n).$$

The Chebychef identity (2.10) is really a transform relationship. It suggests that given a function $F(x)$, defined for $x>1$, one defines a related function $G(x)$ for $x>1$ by

$$(2.11) \quad G(x) = \sum_{n \leq x} F(x/n) = F(x) + F(x/2) + F(x/3) + \cdots F(x/[x]),$$

where as usual $[x]$ is the largest integer not exceeding x. $G(x)$ may be regarded as a transform of $F(x)$. G is seen to be a linear homogeneous function of F. Transform relationships are among the most powerful tools of the mathematician and this one is no exception.

Since $T(x)$ is a comparatively simple function, it is of interest to try to invert the relationship (2.10) to express $\psi(x)$ in terms of T, or in the more general notation, to try to invert (2.11) to find F in terms of G. To solve for F in terms of G, a first modest step would be to eliminate $F(x/2)$ from the right side of (2.11). This is easily done by writing (2.11) with x replaced by $x/2$ to get

$$G\left(\frac{x}{2}\right) = F\left(\frac{x}{2}\right) + F\left(\frac{x}{4}\right) + F\left(\frac{x}{6}\right) + \cdots.$$

Subtracting the above from (2.11) would eliminate $F(x/2)$. This process can be extended at once by writing (2.11) with x replaced by $x/2$, then by $x/3$, etc., to get

$$G(x) \quad = F(x) + F\left(\frac{x}{2}\right) + F\left(\frac{x}{3}\right) + F\left(\frac{x}{4}\right) + F\left(\frac{x}{5}\right) + F\left(\frac{x}{6}\right) + \cdots$$

$$(2.12) \quad G\left(\frac{x}{2}\right) = \qquad\quad F\left(\frac{x}{2}\right) \qquad\quad + F\left(\frac{x}{4}\right) \qquad\qquad + F\left(\frac{x}{6}\right) + \cdots$$

$$G\left(\frac{x}{3}\right) = \qquad\qquad\qquad\quad F\left(\frac{x}{3}\right) \qquad\qquad\qquad\quad + F\left(\frac{x}{6}\right) + \cdots$$

$$G\left(\frac{x}{4}\right) = \qquad\qquad\qquad\qquad\quad F\left(\frac{x}{4}\right) \qquad\qquad\qquad\qquad + \cdots$$

$$
(2.12) \quad
\begin{aligned}
G\left(\frac{x}{5}\right) &= \qquad\qquad\qquad\qquad F\left(\frac{x}{5}\right) \qquad\qquad +\cdots \\
G\left(\frac{x}{6}\right) &= \qquad\qquad\qquad\qquad F\left(\frac{x}{6}\right) +\cdots
\end{aligned}
$$

· ·

If one uses the equations in sequence one can first eliminate $F(x/2)$, then $F(x/3)$, then $F(x/4)$, etc., from the right. For example up to $G(x/6)$ one gets

$$
F(x) = G(x) - G\left(\frac{x}{2}\right) - G\left(\frac{x}{3}\right) - G\left(\frac{x}{5}\right) + G\left(\frac{x}{6}\right) + \cdots .
$$

This suggests, and indeed, because of the diagonal form of the right side of (2.12), actually proves that (2.11) can be inverted by a formula of the type

$$
(2.13) \qquad\qquad F(x) = \sum_{k \le x} \mu(k) G\left(\frac{x}{k}\right),
$$

where the $\mu(k)$ remain to be specified, (Möbius, 1832). To determine the $\mu(k)$ note that by (2.11) $G(x/k) = \sum_{j \le x/k} F(x/jk)$ which in (2.13) gives

$$
F(x) = \sum_{k \le x} \mu(k) \sum_{j \le x/k} F(x/jk) = \sum_{jk \le x} \mu(k) F(x/jk).
$$

If this double sum is summed first on the lattice points on the hyperbolas $jk = n$ and then for n, $1 \le n \le x$,

$$
(2.14) \qquad\qquad F(x) = \sum_{n \le x} F(x/n) \sum_{jk=n} \mu(k).
$$

The equation (2.14) becomes an identity if $\mu(1) = 1$ and, replacing $jk = n$ by $k \mid n$, if

$$
(2.15) \qquad\qquad \sum_{k \mid n} \mu(k) = 0, \qquad n \ge 2.
$$

Setting $n = 2$, 3, 4, etc. successively determines the $\mu(k)$ uniquely. To find the $\mu(k)$ explicitly try the case $n = p$ to get $k = 1$ and $k = p$ which gives $\mu(1) + \mu(p) = 0$ and hence $\mu(p) = -1$. The case $n = p_1 p_2$ gives

$$
u(1) + \mu(p_1) + \mu(p_2) + \mu(p_1 p_2) = 0
$$

and hence $\mu(p_1 p_2) = 1$. Similarly it is easily found that

$$
\mu(p_1 p_2 p_3) = -1, \ \mu(p^2) = 0, \ \mu(p^3) = 0, \cdots, \ \mu(p_1^2 p_2) = 0.
$$

This suggests that

(2.16) $\mu(n) = (-1)^m,$ $n = p_1 p_2 \cdots p_m,$

where $p_1, p_2, \cdots, p_m$ are all distinct primes and

(2.17) $\mu(n) = 0$ if $p^2 \mid n,$

where as usual p is a prime. The function $\mu(n)$ is known as the Möbius function.

It will now be proved that if $\mu(n)$ is defined as in (2.16) and (2.17) above, then (2.15) is indeed valid. We recall that the solution of (2.15) was unique. Because of (2.17) for $n = p_1^{k_1} p_2^{k_2} \cdots p_m^{k_m},$

$$\sum_{j \mid n} \mu(j) = \sum_{j \mid p_1 p_2 \cdots p_m} \mu(j),$$

so only the right side need be treated to prove (2.15). If $m = 1$, (2.15) is true since $\mu(1) = 1$ and $\mu(p_1) = -1$. If $m \geq 2$

(2.18) $\sum_{j \mid p_1 \cdots p_m} \mu(j) = \sum_{k \mid p_1 \cdots p_{m-1}} (\mu(k) + \mu(kp_m)).$

But from (2.16) if k in (2.18) is the product of r primes

$$\mu(kp_m) = (-1)^{r+1} = -\mu(k).$$

Hence each term on the right of (2.18) is zero and so (2.15) is proved. Moreover by (2.16) and (2.17)

(2.19) $|\mu(n)| \leq 1$

(which is the only use we shall make of the material which begins after (2.15) and ends with (2.19)).

Applying the Möbius inversion formula (2.13) to Chebychef's identity (2.10) gives the *inversion formula*

(2.20) $\psi(x) = \sum_{k \leq x} \mu(k) T\left(\dfrac{x}{k}\right).$

Using the definitions of ψ and T this can be written as

$$\sum_{n \leq x} \Lambda(n) = \sum_{k \leq x} \mu(k) \sum_{j \leq x/k} \log j = \sum_{jk \leq x} \mu(k) \log j = \sum_{n \leq x} \sum_{jk=n} \mu(k) \log j$$

$$= \sum_{n \leq x} \sum_{k \mid n} \mu(k) \log n/k.$$

Used for $x = 1, 2, 3, \cdots$ the above proves that

(2.21) $\Lambda(n) = \sum_{k \mid n} \mu(k) \log n/k,$ $n \geq 1,$

which is the inversion formula for (2.4). The referee observes that one could also show (2.21) directly using (2.4) and (2.15).

3. Some elementary results. Although those results concerning prime numbers that follow were all discovered in the 19th century, some were not found until as much as 70 years after the prime number theorem was first conjectured by Legendre and Gauss.

It will be convenient to use the following well-known lemma in which, as usual, $[x]$ is the largest integer not exceeding x.

LEMMA 3.1. *Let $f(t)$ have a continuous derivative, $f'(t)$, for $t \geq 1$. Let c_n, $n \geq 1$, be constants and let $C(u) = \sum_{n \leq u} c_n$. Then*

$$(3.1) \qquad \sum_{n \leq x} c_n f(n) = f(x)C(x) - \int_1^x f'(t)C(t)dt$$

and

$$(3.2) \qquad \sum_{n \leq x} f(n) = \int_1^x f(t)dt + \int_1^x (t - [t])f'(t)dt + f(1) - (x - [x])f(x).$$

Proof. $C(n) - C(n-1) = c_n$ and $C(u) = C([u])$ since $C(u)$ is a step function. Thus if $[x] = N$,

$$\sum_{n \leq x} c_n f(n) = \sum_{n \leq x} (C(n) - C(n-1))f(n)$$

$$= \sum_{n \leq x-1} C(n)(f(n) - f(n+1)) + C(x)f(N)$$

$$= - \sum_{n \leq x-1} C(n) \int_n^{n+1} f'(t)dt + C(x)f(N)$$

$$= - \int_1^N C(t)f'(t)dt + C(x)f(N).$$

Since $C(t)$ is constant on $N \leq t < x$,

$$\int_N^x C(t)f'(t)dt = C(x)(f(x) - f(N)).$$

Adding this to the previous equation and transposing the integral on the left side to the right proves (3.1).

In case $c_n = 1$,   (3.1) becomes

$$\sum_{n \leq x} f(n) = [x]f(x) - \int_1^x f'(t)[t]dt$$

$$= [x]f(x) - \int_1^x f'(t)tdt + \int_1^x f'(t)(t - [t])dt.$$

Integrating the first integral on the right by parts proves (3.2).

It will be convenient to use the following notation. Suppose $f(x)$ is bounded for finite x and that there is a constant K and a $g(x)$ such that for large x

$$|f(x)| \leq Kg(x);$$

then this will be denoted by

(3.3) $$f(x) = O(g(x))$$

and, where convenient, $f(x)$ will be replaced by the right side above.

Applying (3.2) to $f(t) = \log t$ and using $0 \leq t - [t] < 1$ gives

(3.4) $$T(x) = x \log x - x + O(\log x),$$

which is a weak form of Stirling's formula.

LEMMA 3.2. (Chebychef 1850). *For large x*

(3.5) $$\psi(x) < \tfrac{3}{2}x.$$

Proof. Using Chebychef's identity (2.10)

$$T(x) - 2T(x/2) = \psi(x) - \psi(x/2) + \psi(x/3) - \psi(x/4) + \cdots \geq \psi(x) - \psi(x/2)$$

because $\psi(x/(2n-1)) - \psi(x/2n) \geq 0$ since ψ is monotone nondecreasing. Using (3.4), $\psi(x) - \psi(x/2) \leq x \log 2 + K \log x$, $x \geq 2$, for some constant K. Applying the above with x replaced by $x/2^j$,

(3.6) $$\psi\left(\frac{x}{2^j}\right) - \psi\left(\frac{x}{2^{j+1}}\right) \leq \frac{x}{2^j} \log 2 + K \log x$$

so long as $x/2^j \geq 2$ which implies $j < \log x/\log 2$. Recalling that $\psi(t) = 0$, $t < 2$, and adding (3.6) for $0 \leq j < \log x/\log 2$,

$$\psi(x) \leq x \log 2 \left(1 + \frac{1}{2} + \frac{1}{2^2} + \cdots \right) + \frac{\log x}{\log 2} K \log x$$

$$= 2x \log 2 + K \log^2 x/\log 2.$$

Since $\log 2 < .7$ this proves (3.5).

LEMMA 3.3. (Proved 1874 by Mertens in a slightly different form.)

(3.7) $$\sum_{n \leq x} \frac{\Lambda(n)}{n} = \log x + O(1).$$

Proof. In the double sum (2.9), sum first on i and then on j, (the opposite of what was done in the derivation of (2.10)), to get

(3.8)
$$T(x) = \sum_{j \leq x} \Lambda(j) \sum_{i \leq x/j} 1 = \sum_{j \leq x} \Lambda(j) \left[\frac{x}{j}\right]$$

$$= x \sum_{j \leq x} \frac{\Lambda(j)}{j} - \sum_{j \leq x} \Lambda(j) \left(\frac{x}{j} - \left[\frac{x}{j}\right]\right).$$

Moreover

$$(3.9) \qquad 0 \leq \sum_{j \leq x} \Lambda(j)\left(\frac{x}{j} - \left[\frac{x}{j}\right]\right) \leq \sum_{j \leq x} \Lambda(j) = \psi(x) = O(x)$$

by (3.5). Using this and (3.4) in (3.8) proves (3.7).

LEMMA 3.4.

$$(3.10) \qquad \psi(x) = \pi(x) \log x + O\!\left(\frac{x \log \log x}{\log x}\right),$$

so that (2.7) is equivalent to the prime number theorem.

Proof. From its definition (2.6) and from (2.3),

$$(3.11) \qquad \psi(x) = \sum_{p \leq x} \log p + \sum_{p \leq x^{1/2}} \log p + \sum_{p \leq x^{1/3}} \log p + \cdots,$$

where the sums $p \leq x^{1/j}$ above are not zero only if $x^{1/j} \geq 2$ or if $j \leq \log x / \log 2$. Hence

$$\psi(x) \leq \sum_{p \leq x} \log p + \frac{\log x}{\log 2} \sum_{p \leq x^{1/2}} \log p.$$

From the definition (1.1) of $\pi(x)$ this gives

$$\psi(x) \leq \log x \, \pi(x) + \frac{\log x}{\log 2} \pi(x^{1/2}) \log x^{1/2}.$$

Since $\pi(y) \leq y$, the above gives

$$(3.12) \qquad \psi(x) \leq \log x \, \pi(x) + \frac{x^{1/2} \log^2 x}{2 \log 2}.$$

By (3.11)

$$\psi(x) \geq \sum_{x/\log^2 x < p \leq x} \log p \geq \log\left(\frac{x}{\log^2 x}\right) \sum_{x/\log^2 x < p \leq x} 1$$

$$= \log\left(\frac{x}{\log^2 x}\right)\left(\pi(x) - \pi\left(\frac{x}{\log^2 x}\right)\right).$$

Since $\pi(y) \leq y$, this gives

$$\frac{\psi(x)}{\log x - 2 \log \log x} \geq \pi(x) - \frac{x}{\log^2 x},$$

or

$$\pi(x) \log x \leq \psi(x) \frac{\log x}{\log x - 2 \log \log x} + \frac{x}{\log x}$$

$$= \psi(x) + \psi(x) \frac{2 \log \log x}{\log x - 2 \log \log x} + \frac{x}{\log x} \cdot$$

Using (3.5) and $2 \log \log x < (\log x)/4$ for large x,

$$(3.13) \qquad \pi(x) \log x \leqq \psi(x) + \frac{4x \log \log x}{\log x} + \frac{x}{\log x},$$

which with (3.12) proves the lemma.

It will be useful later to apply (3.2) to $f(t) = 1/t$.

LEMMA 3.5.

$$(3.14) \qquad \sum_{n \leq x} 1/n = \log x + \gamma + O(1/x),$$

where γ is a constant (Euler's constant).

Proof. Applying (3.2) to $f(t) = 1/t$

$$\sum_{n \leq x} \frac{1}{n} = \log x - \frac{x - [x]}{x} + 1 - \int_1^x \frac{t - \lfloor t \rfloor}{t^2} \, dt.$$

If

$$(3.15) \qquad \gamma = 1 - \int_1^\infty \frac{t - [t]}{t^2} \, dt,$$

then $\sum_{n \leq x} 1/n = \log x + \gamma + H$, where

$$H = \int_x^\infty \frac{t - [t]}{t^2} \, dt - \frac{x - [x]}{x} = O\left(\frac{1}{x}\right)$$

since $0 \leq t - [t] < 1$, which proves (3.14).

REMARK. From (3.15), $0 < \gamma < 1$.

4. Selberg's elementary inequality. The Möbius inversion formula (2.20) which expresses ψ in terms of T will now be used in an attempt to find how $\psi(x)$ behaves for large x. The computation will be simplified if it is possible to find a simple $F(x)$, say $\tilde{F}(x)$, with a transform $\tilde{G}(x)$ which is close to $T(x)$. In that case subtract the Möbius inversion formula for $\tilde{F}$ (2.13), from that for ψ:

$$(4.1) \qquad \psi(x) - \tilde{F}(x) = \sum_{k \leq x} \mu(k) \left(T\left(\frac{x}{k}\right) - \tilde{G}\left(\frac{x}{k}\right) \right);$$

if the right side could be shown to be small, then $\psi(x)$ would be close to $\tilde{F}(x)$.

If it were proved that $\psi(x)/x \to 1$, then $\psi(x)$ would be close to x for large x. This suggests one try $\tilde{F}(x) = F_0(x) = x$. Hence $G_0(x) = \sum_{n \leq x} F_0(x/n) = x \sum_{n \leq x} n^{-1}$

which by (3.14) becomes $G_0(x) = x \log x + \gamma x + O(1)$. This is not close enough to $T(x)$ as given by (3.4). As a refinement let $\tilde{F}(x) = F_1(x) = x - C$ where C is a constant. (There are many choices other than C that would work here.) Then

$$G_1(x) = x \sum_{n \leq x} \frac{1}{n} - C \sum_{n \leq x} 1 = x \log x + \gamma x + O(1) - C[x]$$
$$= x \log x - (C - \gamma)x + O(1).$$

Hence if $C = 1 + \gamma$ then by (3.4)

(4.2) $T(x) - G_1(x) = O(\log x)$

which is comparatively small. Using (4.1) with $\tilde{F} = x - C$

(4.3) $\psi(x) - x + C = \sum_{k \leq x} \mu(k) \left(T\left(\frac{x}{k}\right) - G_1\left(\frac{x}{k}\right) \right).$

Even if the right side of (4.2) were in the stronger form $O(1)$ (which is false), the fact that (by (2.19)) $|\mu(k)| \leq 1$ would imply only that the right side of (4.3) is $O(x)$. Thus the inversion formula (4.3) gives, *not* the prime number theorem, but at most the much weaker result

(4.4) $\psi(x) = O(x),$

(already proved more simply in Lemma 3.2). Actually (4.3) does give (4.4) as the following crude argument shows.

Since the logarithm grows more slowly than any positive algebraic power, $\log x = O(x^{1/2})$. Thus, for example, (4.2) implies the much weaker result

(4.5) $T(x) - G_1(x) = O(x^{1/2}).$

Using this and $|\mu(k)| \leq 1$, there is a constant K such that the right side of (4.3) is dominated by

(4.6)
$$Kx^{1/2} \sum_{k \leq x} k^{-1/2} < Kx^{1/2}\left(1 + \sum_{2 \leq k \leq x} \int_{k-1}^{k} u^{-1/2}du\right)$$
$$\leq Kx^{1/2}\left(1 + \int_{1}^{x} u^{-1/2}du\right) = O(x)$$

which does in fact prove (4.4).

Thus the Möbius inversion of Chebychef's formula yields only the crude result (4.4), and herein lies the reason for the long delay in the discovery of an elementary proof of the prime number theorem.

Note that *the crude result* (4.5) *serves just as well as the much more refined* (4.2) in appraising the right side of (4.3). This suggests the following idea.

In the Möbius inversion formula

(4.7) $F(x) = \sum_{k \leq x} \mu(k) G\left(\frac{x}{k}\right)$

(where for us $F=\psi-x+C$ and $G=T-G_1$), we can increase the terms in the sum on the right side somewhat since doing so will not change the crude appraisal $O(x)$, (4.6), for this side. On the other hand, a judicious increase of the terms on the right side might possibly replace $F(x)$ (and hence $\psi(x)$) on the left side by some growing function multiplied by $F(x)$, which would then make the appraisal $O(x)$ for the right side useful.

A little experimentation shows that the simplest case to compute explicitly is where the right side of (4.7) is replaced by

$$(4.8) \qquad J(x) = \sum_{k\leq x} \mu(k) \log \frac{x}{k} \, G\left(\frac{x}{k}\right).$$

This must now be computed in terms of F. From the definition of G,

$$J(x) = \sum_{k\leq x} \mu(k) \log \frac{x}{k} \sum_{j\leq x/k} F\left(\frac{x}{jk}\right)$$

$$= \sum_{jk\leq x} \mu(k) \log \frac{x}{k} F\left(\frac{x}{jk}\right) = \sum_{n\leq x} F\left(\frac{x}{n}\right) \sum_{jk=n} \mu(k) \log \frac{x}{k}$$

$$= \sum_{n\leq x} F\left(\frac{x}{n}\right) \sum_{k|n} \mu(k) \log \frac{x}{k}.$$

Using $\log x/k = \log x/n + \log n/k$

$$J(x) = \sum_{n\leq x} F\left(\frac{x}{n}\right) \log \frac{x}{n} \sum_{k|n} \mu(k) + \sum_{n\leq x} F\left(\frac{x}{n}\right) \sum_{k|n} \mu(k) \log \frac{n}{k}.$$

By (2.15) and (2.21) this becomes $J(x) = F(x) \log x + \sum_{n\leq x} F(x/n)\Lambda(n)$. With (4.8) this gives

$$(4.9) \qquad F(x) \log x + \sum_{n\leq x} F\left(\frac{x}{n}\right) \Lambda(n) = \sum_{k\leq x} \mu(k) \log \frac{x}{k} G\left(\frac{x}{k}\right),$$

and this is the Tatuzawa-Iseki identity [7] which leads easily to the inequality of Atle Selberg. Indeed by (4.2)

$$\log x(T(x) - G_1(x)) = O(\log^2 x) = O(x^{1/2})$$

and hence as already shown in (4.6)

$$\sum_{k\leq x} \mu(k) \log \frac{x}{k} \left(T\left(\frac{x}{k}\right) - G_1\left(\frac{x}{k}\right)\right) = O(x).$$

Thus (4.9) with $F(x)=\psi(x)-x+C$ becomes

$$(4.10) \qquad (\psi(x) - x) \log x + \sum_{n\leq x} \left(\psi\left(\frac{x}{n}\right) - \frac{x}{n}\right)\Lambda(n) = O(x),$$

where use is made of (4.4) to incorporate $C\psi(x)$ together with $C \log x$ in $O(x)$. (4.10) is a form of the famous inequality of Atle Selberg [6].

Because of Lemma 3.3, (4.10) can be written as

$$(4.11) \qquad \psi(x) \log x + \sum_{n \leq x} \Lambda(n)\psi(x/n) = 2x \log x + O(x).$$

With $c_n = \Lambda(n)$, (3.1) and (3.5) yield

$$(4.12) \qquad \sum_{n \leq x} \Lambda(n) \log n = \psi(x) \log x - \int_1^x \frac{\psi(t)}{t} \, dt = \psi(x) \log x + O(x).$$

Also

$$(4.13) \qquad \sum_{j \leq x} \Lambda(j)\psi\left(\frac{x}{j}\right) = \sum_{j \leq x} \Lambda(j) \sum_{k \leq x/j} \Lambda(k) = \sum_{jk \leq x} \Lambda(j)\Lambda(k).$$

Thus if

$$(4.14) \qquad \Lambda_2(n) = \Lambda(n) \log n + \sum_{jk=n} \Lambda(j)\Lambda(k),$$

then (4.12) and (4.13) in (4.11) yield $\sum_{n \leq x} \Lambda_2(n) = 2x \log x + O(x)$ as an equivalent to (4.11). By (3.4) $\sum_{n \leq x} \log n = x \log x + O(x)$. Combining the above two inequalities,

$$(4.15) \qquad Q(n) = \sum_{k \leq n} (\Lambda_2(k) - 2 \log k) = O(n), \qquad n \geq 2, \text{ and } Q(1) = 0.$$

5. Proof of the prime number theorem. If $R(x) = \psi(x) - x, x \geq 2$, and $R(x) = 0$, $x < 2$, then (4.10) becomes

$$(5.1) \qquad R(x) \log x + \sum \Lambda(n) R(x/n) = O(x),$$

where the summation is self terminating since $R(x/n) = 0$ for $n > x/2$. The goal (2.7) takes the form

$$(5.2) \qquad \lim_{x \to \infty} \frac{R(x)}{x} = 0.$$

The derivation of (5.2) from (5.1) is complicated because the weights $\Lambda(n)$ in the weighted sum in (5.1) depend on the location of the prime numbers which is just what we are trying to find. Because of this complication no easy derivation of (5.2) from (5.1) has been found.

The proof that follows uses several smoothing operations on (5.1) to get a more tractable inequality. Most of these smoothings involve a loss of information, and the objective is to smooth for tractability but not to degrade (5.1) completely.

First $R(x)$ will be replaced by the smoother

$$(5.3) \qquad S(y) = \int_2^y \frac{R(x)}{x}\, dx, \qquad y \geq ;2$$

$S(y) = 0$, $y < 2$. Fortunately it is easy to show, as will be done later, that (5.2) is implied if we can prove

$$(5.4) \qquad \lim_{y \to \infty} \frac{S(y)}{y} = 0.$$

LEMMA 5.1. *There exists a constant c such that*

$$(5.5) \qquad |S(y)| \leq cy \qquad y \geq 2$$

and

$$(5.6) \qquad |S(y_2) - S(y_1)| \leq c |y_1 - y_2|.$$

Moreover a consequence of (5.1) is

$$(5.7) \qquad S(y) \log y + \sum \Lambda(j) S\left(\frac{y}{j}\right) = O(y).$$

Proof. From (3.5), $-x \leq \psi(x) - x \leq \tfrac{1}{2}x$ for large x. Hence

$$(5.8) \qquad \limsup_{x \to \infty} \frac{|R(x)|}{x} \leq 1$$

and, since $|R(x)|$ is bounded for finite x, there must exist a constant c such that

$$(5.9) \qquad |R(x)| \leq cx, \qquad x \geq 2.$$

By (5.3) $S'(y) = R(y)/y$ except at $y = p^i$ where $R(y)$ is discontinuous. By (5.9) then

$$(5.10) \qquad |S'(y)| \leq c, \qquad y \neq p^j.$$

Hence, first for the case where the interval $y_1 < y < y_2$ contains no p^i, (5.6) is true. However since $S(y)$ is continuous, the fact that the magnitude of a sum is less than or equal to the sum of the magnitudes, allows (5.6) to be extended for all y_1 and y_2. The condition (5.6) is known as a Lipschitz condition. The result (5.5) follows from (5.6) with $y_1 = 2$.

Since $||a| - |b|| \leq |a - b|$, (5.6) yields

$$(5.11) \qquad ||S(y_2)| - |S(y_1)|| \leq c |y_2 - y_1|.$$

To prove (5.7), divide (5.1) by x and integrate to get

$$(5.12) \qquad \int_2^y \frac{R(x)}{x} \log x\, dx + \sum \Lambda(n) \int_2^y R\left(\frac{x}{n}\right) \frac{dx}{x} = O(y).$$

Integrating the first term by parts

$$\int_2^y \frac{R(x)}{x} \log x \, dx = \log y \, S(y) - \int_2^y \frac{S(x)}{x} \, dx = \log y \, S(y) + O(y)$$

by (5.5). Also if $\xi = x/n$

$$\int_2^y R\left(\frac{x}{n}\right) \frac{dx}{x} = \int_2^{y/n} \frac{R(\xi)}{\xi} \, d\xi = S\left(\frac{y}{n}\right).$$

These in (5.12) prove (5.7).

To make the weighted sum in (5.7) more tractable, the density of the set of points where $S(y/j)$ actually appears in the sum will be increased by iterating (5.7).

LEMMA 5.2. *With* $\Lambda_2(n) = \Lambda(n) + \sum_{ij=n} \Lambda(i)\Lambda(j)$ *as in* (4.14) *and* K_1 *a constant*

$$(5.13) \qquad \log^2 y \, |\, S(y)\,| \leq \sum \Lambda_2(m) \, |\, S(y/m)\,| + K_1 y \log y.$$

Proof. Replace y in (5.7) by y/k, multiply by $\Lambda(k)$, and sum for $k \leq y$ to get

$$\sum \Lambda(k) S\left(\frac{y}{k}\right) \log \frac{y}{k} + \sum \sum \Lambda(k)\Lambda(j) S\left(\frac{y}{jk}\right) = O(y) \sum_{k \leq y} \frac{\Lambda(k)}{k} .$$

Setting $jk = m$ in the second sum and summing on m, and setting $\log y/k = \log y - \log k$ in the first sum and replacing this latter k by m,

$$\log y \sum \Lambda(k) S\left(\frac{y}{k}\right) - \sum_{m \leq y} S\left(\frac{y}{m}\right) \left\{ \Lambda(m) \log m - \sum_{jk=m} \Lambda(j)\Lambda(k) \right\} = O(y \log y)$$

where (3.7) is used to get the right side. The first sum above is now replaced by use of (5.7) to give

$$S(y) \log^2 y = - \sum S\left(\frac{y}{m}\right) \left\{ \Lambda(m) \log m - \sum_{jk=m} \Lambda(j)\Lambda(k) \right\} + O(y \log y).$$

Replacing all terms in the sum on the right by their magnitude gives (5.13).

The inequality (4.15) suggests that on the average $\Lambda_2(m)$ acts like $2 \log m$. A weighted sum with weights $2 \log m$ is quite tractable and this suggests modifying (5.13) by replacing $\Lambda_2(m)$ by $2 \log m$.

LEMMA 5.3. *There is a constant* K_2 *such that*

$$(5.14) \qquad \log^2 y \, |\, S(y)\,| \leq 2 \sum |\, S(y/m)\,| \, \log m + K_2 y \log y.$$

Proof.

$$(5.15) \qquad \sum |\, S(y/m)\,| \, \Lambda_2(m) = 2 \sum_{m \leq y} |\, S(y/m)\,| \, \log m + J(y)$$

where, since by (4.15), $\Lambda_2(m) - 2 \log m = Q(m) - Q(m-1)$,

$$J(y) = \sum_{m \leq y} (Q(m) - Q(m-1)) \, | \, S(y/m) \, |$$

$$= \sum_{m \leq y} Q(m) \, | \, S(y/m) \, | \; - \; \sum_{m \leq y} Q(m) \, | \, S(y/(m+1)) \, |$$

$$= \sum_{2 \leq m \leq y} Q(m)(\, | \, S(y/m) \, | \; - \; | \, S(y/(m+1)) \, | \,)$$

since $S(y) = 0$, $y < 2$. Using (4.15) and (5.11) there is a constant K_3 such that

$$J(y) \leq K_3 \sum_{2 \leq m \leq y} m \left(\frac{y}{m} - \frac{y}{m+1} \right)$$

$$= K_3 y \sum_{2 \leq m \leq y} \frac{1}{m+1} < K_3 y \int_1^y \frac{dv}{v} = K_3 y \log y.$$

This and (5.15) now prove that (5.14) is a consequence of (5.13).

There is a further simplification in replacing the sum in (5.14) by an integral.

LEMMA 5.4. *There is a constant K_4 such that*

(5.16) $$\log^2 y \, | \, S(y) \, | \; \leq 2 \int_2^y | \, S(y/u) \, | \; \log u \, du + K_4 y \log y.$$

Proof. Since $\log u$ is increasing

$$\log m \, | \, S(y/m) \, | \leq \int_m^{m+1} \log u \, | \, S(y/m) \, | \, du.$$

On the right use $| \, S(y/m) \, | \leq | \, S(y/u) \, | + | \, S(y/m) - S(y/u) \, |$ to get

$$\log m \, | \, S(y/m) \, | \; \leq \int_m^{m+1} \log u \, | \, S(y/u) \, | \, du + J_m$$

(5.17)

$$J_m = \int_m^{m+1} \log u \, | \, S(y/m) - S(y/u) \, | \, du.$$

Using (5.6)

$$J_m \leq c \left(\frac{y}{m} - \frac{y}{m+1} \right) \int_m^{m+1} \log u \, du \leq \frac{cy \log (m+1)}{m(m+1)}.$$

Since $\log (m+1) \leq m$, the above in (5.17) gives

$$\log m \left| S \left(\frac{y}{m} \right) \right| \leq \int_m^{m+1} \log u \left| S \left(\frac{y}{u} \right) \right| dy + \frac{cy}{m+1}.$$

Using this in (5.14) now gives (5.16) with $K_4 = K_2 + c$.

The inequality (5.16) assumes a simpler form with an exponential change of variable. Replace u by $v = \log y/u$. Also let $x = \log y$. Then (5.16) becomes

$$(5.18) \qquad x^2 \left| S(e^x) \right| \leq 2 \int_0^{x - \log 2} \left| S(e^v) \right| (x - v) e^{(x-v)} dv + K_4 x e^x.$$

If

$$(5.19) \qquad W(x) = e^{-x} S(e^x)$$

then (5.18) becomes

$$(5\ 20) \qquad \left| W(x) \right| \leq \frac{2}{x^2} \int_0^x (x - v) \left| W(v) \right| dv + \frac{K_4}{x}.$$

This inequality contains valuable information since it says in effect that $\left| W(x) \right|$ is dominated by a weighted average of $\left| W \right|$. This has as a consequence the following lemma. (Note that γ below is not Euler's constant.)

LEMMA 5.5. *Let*

$$(5.21) \qquad \alpha = \limsup_{x \to \infty} \left| W(x) \right|, \qquad \gamma = \limsup_{x \to \infty} \frac{1}{x} \int_0^x \left| W(\xi) \right| d\xi;$$

then $\alpha \leq 1$ and

$$(5.22) \qquad\qquad\qquad\qquad \alpha \leq \gamma.$$

REMARK. Recalling (5.4) and (5.19), our goal now is $\alpha = 0$.

Proof. That $\alpha \leq 1$ follows from (5.19) and the fact that (5.8) and (5.3) imply that

$$(5.23) \qquad\qquad\qquad \limsup_{y \to \infty} \frac{\left| S(y) \right|}{y} \leq 1.$$

The key result $\gamma \geq \alpha$ will be proved by use of (5.20) and *this is the only use that is made of* Lemmas 5.2, 5.3 and 5.4. Note that (5.20) can be written as

$$(5.24) \qquad \left| W(x) \right| \leq \frac{2}{x^2} \int_0^x u \, du \left(\frac{1}{u} \int_0^u \left| W(v) \right| dv \right) + \frac{K_4}{x}$$

as can be verified by inverting the order of integration. But

$$\frac{2}{x^2} \int_0^x u \, du = 1$$

and hence the integral on the right of (5.24) is simply a weighted average of $(1/u) \int_0^u \left| W(v) \right| dv = (1/u) \int_0^u e^{-v} \left| S(e^v) \right| dv \leq c$ by (5.5). Hence for any fixed x_1 and $x > x_1$,

$$I(x) = \frac{2}{x^2} \int_0^x u \, du \left(\frac{1}{u} \int_0^u | W(v) | \, dv \right)$$

(5.25)

$$\leq \frac{2c}{x^2} \int_0^{x_1} u \, du + \frac{2}{x^2} \int_{x_1}^x u \, du \left(\frac{1}{u} \int_0^u | W(v) | \, dv \right).$$

Given $\epsilon > 0$, for sufficiently large x_1,

$$\frac{1}{u} \int_0^u | W(v) | \, dv < \gamma + \epsilon \qquad u \geq x_1$$

from the definition of γ. Hence (5.25) gives

$$I(x) \leq \frac{c x_1^2}{x^2} + (\gamma + \epsilon) \left(1 - \frac{x_1^2}{x^2} \right).$$

Thus for large x, (5.24) yields

$$| W(x) | \leq \gamma + \epsilon + \frac{c x_1^2}{x^2} + \frac{K_4}{x}.$$

Letting $x \to \infty$, $\alpha \leq \gamma + \epsilon$, and since this is true for all $\epsilon > 0$ it implies (5.22).

Two more facts are required about W to prove that $\alpha = 0$.

LEMMA 5.5. *If $k = 2c$ then*

(5.26)
$$| W(x_2) - W(x_1) | \leq k | x_2 - x_1 |,$$

and hence

(5.27)
$$|| W(x_2) | - | W(x_1) || \leq k | x_2 - x_1 |.$$

Proof. Since $W(x) = e^{-x} S(e^x)$,

$$| W'(x) | \leq e^{-x} | S(e^x) | + | S'(e^x) | \qquad x \neq j \log p.$$

Hence by (5.5) and (5.10), $| W'(x) | \leq 2c = k$ for $x \neq j \log p$. This leads to (5.26) just as (5.10) led to (5.6).

LEMMA 5.7. *If $W(v) \neq 0$ for $v_1 < v < v_2$, then there exists a number M such that*

(5.28)
$$\int_{v_1}^{v_2} | W(v) | \, dv \leq M, \qquad W(v) \neq 0, \qquad v_1 < v < v_2.$$

Proof. From (3.1) letting $c_n = \Lambda(n)$ and $f(n) = 1/n$, (3.7) implies

$$\int_2^x \frac{\psi(t)}{t^2} \, dt = \log x + O(1),$$

or since $R(t) = \psi(t) - t$,

(5.29)
$$\int_2^x \frac{R(t)}{t^2}\, dt = O(1).$$

But

$$\int_2^x \frac{S(y)}{y^2}\, dy = \int_2^x \frac{dy}{y^2}\int_2^y \frac{R(t)}{t}\, dt = \int_2^x \frac{R(t)}{t}\left(\int_t^x \frac{dy}{y^2}\right) dt$$

$$= \int_2^x \frac{R(t)}{t^2}\, dt - \frac{1}{x}\int_2^x \frac{R(t)}{t}\, dt.$$

Using (5.29) and (5.9), $\int_2^z (S(y)/y^2)dy = O(1)$, or letting $y = e^u$, $x = e^v$,

$$\int_{\log 2}^v W(u)du = O(1).$$

Writing this for $v = v_1$ and $v = v_2$ and subtracting, the resulting integral is bounded and hence there is a constant M such that

$$\left|\int_{v_1}^{v_2} W(u)du\right| \le M.$$

But if $W(u) \ne 0$, $v_1 < u < v_2$, this can be written as (5.28). Since M can be increased if convenient it can be assumed $Mk > 1$.

LEMMA 5.8. *A function $W(x)$ subject to the three conditions (5.22), (5.27) and (5.28) must in fact have $\alpha = 0$.*

Proof. Choose $\beta > \alpha$. Then from the definition of α there exists an x_β such that

(5.30)
$$|W(x)| \le \beta \qquad x \ge x_\beta.$$

If $W(x) \ne 0$ for all large x it follows from (5.28) that $\gamma = 0$ and hence that $\alpha = 0$. Suppose then that $W(x)$ has arbitrarily large zeros. Let a and b be successive zeros of $W(x)$ for $x > x_\beta$.

CASE 1. $b - a \ge 2M/\beta$. By (5.28), since $W(x) \ne 0$, $a < x < b$,

$$\int_a^b |W(x)| \le M \le \tfrac{1}{2}(b - a)\beta.$$

(Hence the average of $|W|$ on (a, b) is less than $\tfrac{1}{2}\beta$.)

CASE 2. $b - a \le 2\beta/k$. In this case it follows from (5.27) that if the graph of $|W(x)|$ rises as rapidly as possible going right from $x = a$ and left from $x = b$, it cannot lie above a triangle with altitude $k(b-a)/2 \le \beta$ and hence

$$\int_a^b |W(x)|\, dx \le \tfrac{1}{2}(b - a)\beta.$$

CASE 3. $2\beta/k < b - a < 2M/\beta$. Reasoning as in Case 2 for a distance β/k from each endpoint and using (5.30) otherwise,

$$\int_a^b |W(x)| \leq \frac{\beta^2}{k} + \left(b - a - \frac{2\beta}{k}\right)\beta$$

(5.31)
$$= (b-a)\beta\left(1 - \frac{\beta}{k(b-a)}\right) \leq (b-a)\beta\left(1 - \frac{\beta^2}{2Mk}\right)$$

$$< (b-a)\beta\left(1 - \frac{\alpha^2}{2Mk}\right).$$

Since $Mk > 1$ and since $\alpha \leq 1$, (5.31) is valid in Cases 1 and 2 also. If x_1 is the first zero of $W(x)$ to the right of x_β and $\bar{x}$ the largest zero to the left of y, then (5.31) and (5.28) imply that

$$\int_0^y |W(x)| \, dx \leq \int_0^{x_1} |W(x)| \, dx + (\bar{x} - x_1)\beta\left(1 - \frac{\alpha^2}{2Mk}\right) + M.$$

Dividing by y and noting that $\bar{x} \leq y$,

$$\frac{1}{y}\int_0^y |W(x)| \, dx \leq \frac{1}{y}\int_0^{x_1} |W(x)| \, dx + \beta\left(1 - \frac{\alpha^2}{2Mk}\right) + \frac{M}{y} \cdot$$

Letting $y \to \infty$, $\gamma \leq \beta(1 - \alpha^2/2Mk)$, and since $\gamma \geq \alpha$,

$$\alpha \leq \beta\left(1 - \frac{\alpha^2}{2Mk}\right).$$

Since this holds for all $\beta > \alpha$ it must hold for $\beta = \alpha$. Hence $\alpha^3 \leq 0$, and since $\alpha \geq 0$, this implies $\alpha = 0$. Since $W(x) = e^{-x}S(e^x)$, this implies that $|S(y)|/y \to 0$ as $y \to \infty$. Hence if given $\epsilon > 0$, if y is large enough,

$$|S(y)| \leq \tfrac{1}{3}\epsilon^2 y.$$

Thus $S(y(1+\epsilon)) - S(y) \leq \tfrac{1}{3}\epsilon^2(y(1+\epsilon) + y) < \epsilon^2 y$, or

$$\int_y^{y(1+\epsilon)} \frac{R(u)}{u} \, du \leq \epsilon^2 y.$$

Since $R(u) = \psi(u) - u$ and ψ is nondecreasing,

$$\frac{\psi(y)}{y(1+\epsilon)}\int_y^{y(1+\epsilon)} du - \int_y^{y(1+\epsilon)} du \leq \epsilon^2 y.$$

Hence $\psi(y)/y \leq (1+\epsilon)^2$. Similarly $S(y) - S(y(1-\epsilon)) \geq -\epsilon^2 y$ for large enough y leads to $\psi(y)/y \geq (1-\epsilon)^2$. Since ϵ is arbitrary this proves (2.7).

Supported in part by the Office of Naval Research and by the National Science Foundation, NSF GP 7477.

References

1. E. Bombieri, Sulle formule di A. Selberg generalizzate per classi di funzioni aritmetiche e le applicazioni al problema del resto nel "Primzahlsatz", Riv. Mat. Univ. Parma, (2) 3 (1962) 393–440.

2. R. Breusch, An elementary proof of the prime number theorem with remainder term, Pacific J. Math., 10 (1960) 487–497.

3. P. Erdös, On a new method in elementary number theory which leads to an elementary proof of the prime number theorem, Proc. Nat. Acad. Sci. U. S. A., 35 (1949) 374–384.

4. G. H. Hardy and E. M. Wright, An Introduction to the Theory of Numbers, Oxford University Press, New York, 1960.

5. N. Levinson, On the elementary proof of the prime number theorem, Proc. Edinburgh Math. Soc., (2) 15 (1966) 141–146.

6. A. Selberg, An elementary proof of the prime number theorem, Ann. of Math., (2) 50 (1949) 305–313.

7. T. Tatuzawa and K. Iseki, On Selberg's elementary proof of the prime number theorem, Proc. Japan Acad., 27 (1951) 340–342.

8. E. Wirsing, Elementare Beweise des Primzahlsatzes mit Restglied, I, J. Reine Angew. Math., 211 (1962) 205–214.

9. E. M. Wright, The elementary proof of the prime number theorem, Proc. Roy. Soc. Edinburgh. Sect. A., 63 (1951) 257–267.

20

JEAN FRANÇOIS TRÈVES

Professor Trèves was born on April 23, 1930, in Brussels, Belgium. He received the first and second baccalaurate degrees in Paris in 1949 and 1950, his *licence en science* and his Ph.D. at the Sorbonne in 1953 and 1958. From 1958 to 1961, he was assistant professor at the University of California, Berkeley, from 1961 to 1964 an associate professor at Yeshiva University, and from 1964 to 1970 a professor at Purdue University. Since 1970, he has been a professor at Rutgers University.

Professor Trèves was an Alfred P. Sloan Fellow in 1960–62 and 1962–64. From June to November 1961 he was under the auspices of the Organization of American States at the *Instituto de Matematica Pura e Aplicada* in Rio de Janeiro, Brazil; in September 1965, he was a Visiting Professor at the Tata Institute of Fundamental Research in Bombay, India, and from 1965 to 1967, and again, during the academic year 1974–75 he was a Visiting Professor at the Sorbonne in Paris.

Professor Trèves' significant contributions to various branches of analysis, but, in particular, to partial differential equations and functional analysis, are contained in his numerous publications.

In accepting the award, Professor Trèves stressed that he was very much honored and thankful for having been awarded the 1972 Chauvenet Prize. He added that, because of the apparent increasing technicality of mathematical research, it is becoming ever more difficult to exchange information between mathematicians working in different fields—even in the same field. He felt this to be a worrisome situation, which makes expository talks and articles more necessary than ever.

ON LOCAL SOLVABILITY OF LINEAR PARTIAL DIFFERENTIAL EQUATIONS*

JEAN FRANÇOIS TRÈVES, Purdue University

The title indicates more or less what the talk is going to be about. It is going to be about the problem which is probably the most primitive in partial differential equations theory, namely to know whether an equation does, or does not, have a solution. Even this is meant in the most primitive terms. I would like to begin by explaining what the terms are.

As you all know, the really difficult analysis these days, and perhaps always, is the *global analysis*. Well, the problem that I am going to discuss is purely *local*—in the strictest possible sense: we would like to find out if a linear partial differential equation, with coefficients as smooth as you wish, admits locally a solution. Obviously, in this connection, *negative* results are very important: and negative results about local solvability have global implications. But of course positive results have also their importance. Let us state precisely what is the problem. The partial differential equation under study will be

$$Pu = f \tag{1}$$

and we would like to know whether for given f, defined in the neighborhood of some point, there are solutions u. This is really too vague, so that the first thing we shall do is to make it a little more precise. Let us say that the *differential operator* P is defined in an open set Ω of the Euclidean space R^n and that we wish to solve the equation in Ω; suppose that the right-hand side f is very regular, say $f \in \mathcal{C}^\infty(\Omega)$. Then you would like to know whether there are solutions (defined in Ω). Now, the experience we have acquired since 1950 in the field of linear partial differential equations tells us that if you pose the problem in this way, you will encounter very serious difficulties of *global* nature. After all, Ω is a manifold, in general it possesses a boundary, or points at infinity, and the fact is that the behaviour of f at the boundary, or at infinity, may have an influence on the answer to our question. We want to avoid this, for this is much too difficult a problem for us to handle at the present time. We shall therefore assume that f has the best possible behaviour at the boundary, which means vanishes identically in the neighborhood of the boundary. In other words, f has *compact* support in Ω. Let me point out that local solvability theorems have been known for a long time. The most famous of them, which even the algebraists know, is the theorem of Cauchy-Kovalevska. If Ω is small enough, if P has analytic coefficients, if f is analytic, and if the standard conditions on characteristics are satisfied, then there always is an analytic solution.

* An address delivered before the Cincinnati meeting of the American Mathematical Society by invitation of the Committee to Select Hour Speakers for Western Sectional Meetings, April 19, 1969; received by the editors December 3, 1969.

AMS Subject Classifications. Primary 3501, 3520.

Key Words and Phrases. Partial differential equations, local solvability, distribution, infinitely differentiable functions, compact support, Cauchy problem, analytic functionals, pseudodifferential operators.

The kind of data I shall consider are of a completely different nature: simply by requiring that they have compact support, I exclude the analytic case. In a sense one considers that the analytic case is separate and we want to study the $\mathcal{C}^\infty$ case. In summary we try to solve the equation $Pu = f$ for arbitrary *right-hand sides f*, $\mathcal{C}^\infty$ with compact support in Ω. Now the question is as to what kind of solution I am willing to consider. Well, to start with, I would like to have the best kind, say $\mathcal{C}^\infty$ solutions. But this turns out to be a difficult request to satisfy, at least for the time being—so that we shall ask for any solution at all, were it to be of the worst kind. What are the worst possible solutions? This depends on your viewpoint. In today's lecture I will consider that distributions are bad enough. Thus we shall seek solutions which are distributions in Ω. Let me mention that you could substantially enlarge the inventory of possible solutions, for instance by allowing what are called *Sâto hyperfunctions* or, when the coefficients of the differential operator P are analytic, by considering *analytic functionals*. Later on, I hope to say a few words about these aspects.

We have not gone, yet, far enough along the road to simplification: one further step remains to be taken. For the problem, as we have stated it so far, is not local enough. The set Ω has been kept fixed, and this is too rigid. We shall allow ourselves to choose Ω at will—provided that it ranges over the collection of open neighborhoods of a given point x_0. We consider a point x_0 of R^n and we ask the following question: *Is there an open set Ω containing x_0 such that, given any function $f \in \mathcal{C}^\infty$ with compact support contained in Ω, there is a distribution u in Ω satisfying the partial differential equation $Pu = f$ in Ω?* If the answer to this question is yes, we shall say that the equation $Pu = f$ is *locally solvable* at x_0. The problem is then to find, if possible, necessary and sufficient conditions in order that a given linear partial differential equation be locally solvable at a given point.

Now that I have stated the problem, let me say that the talk will mostly be historical. But some of it will be very recent history, because in the last four months the problem, or at least an important part of it, has been nearing complete solution, and it is the progress towards its solution that I would like to describe.

The problem discussed here originated after Hans Lewy, ca. 1956, exhibited his now famous example of a linear partial differential equation in the three-dimensional space which is *not* locally solvable at any point of R^3. Lewy's operator is

$$L = \frac{\partial}{\partial x_1} + i\frac{\partial}{\partial x_2} + i(x_1 + ix_2)\frac{\partial}{\partial x_3}.$$

(Allow me to insert a personal anecdote: in 1955 I was given the following thesis problem: prove that every linear partial differential equation with smooth coefficients, not vanishing identically at some point, is locally solvable at that point. My thesis director was, and still is, a leading analyst; his suggestion simply shows that, at that time, nobody had any inkling of the structure underlying the local solvability problem, as it is now gradually revealed.)

Before Lewy's example had been discovered, there were many *positive* cases which were known. From the viewpoint of later history, the most interesting was presented in Hörmander's thesis (1955); I shall describe it soon. But even prior to this, there were many known cases of local solvability. Let me describe rapidly what was known, on the subject of local solvability, prior to Hörmander. First of all, every equation belonging to any one of the three "classical types" was known to be locally solvable at any point of its domain of definition. By the classical types I mean the elliptic equations, like Laplace's, the parabolic ones, like the heat equation, the hyperbolic, like the wave equation. For decades the study of partial differential equations had been essentially limited to these, and very nice and very important results about them had been obtained. In particular, as I have said, they all are locally solvable at any point. It is worth underlining that the solvability is (in general) strictly local: in general, elliptic equations are not globally solvable; parabolic and hyperbolic equations, under reasonable hypotheses, fare slightly better and are usually solvable in every open set with compact closure.

There was another important class of linear partial differential equations for which solvability was known, as a matter of fact, solvability in any *bounded* open subset of the Euclidean space, and these were the equations with constant coefficients. In particular they are all locally solvable. Thus the relevant case, in the study of local solvability, is the case of variable coefficients. Of course, you realize that local solvability is an intrinsic property, coordinates independent, whereas the property of having constant coefficients depends on the coordinates in which the differential operator is expressed. If you change the coordinates, the coefficients will not usually remain constant.

We come now to the "positive" local solvability theorem in Hörmander's thesis. Consider a linear partial differential operator in n variables, which I must, for the first time, write down explicitly:

$$P(x,D) = \sum_{|\alpha| \leqslant m} c_\alpha(x) D^\alpha,$$

where I have used the standard notation: $\alpha = (\alpha_1, \ldots, \alpha_n)$, $|\alpha| = \alpha_1 + \cdots + \alpha_n$,

$$D^\alpha = D_1^{\alpha_1} \cdots D_n^{\alpha_n}, \quad D_j = -\sqrt{(-1)} \partial/\partial x_j \quad (j = 1, \ldots, n).$$

The coefficients $c_\alpha(x)$ are complex-valued $\mathcal{C}^\infty$ functions (in fact, Hörmander's result is true under much weaker regularity assumptions).

The condition in Hörmander's theorem bears only on the principal part of the operator, i.e., on its leading terms: today one would say on its *principal symbol*:

$$P_m(x,\xi) = \sum_{|\alpha| = m} c_\alpha(x) \xi_1^{\alpha_1} \cdots \xi_n^{\alpha_n} \quad (x \in \Omega, \xi \in R^n, \xi \neq 0).$$

This is a homogeneous polynomial of degree m with respect to the variables ξ (with coefficients depending on x).

It has been a standard practice, in the study of linear partial differential equations, to drop the lower order terms. But when is it permitted to do so? To this question there is a kind of philosophical answer, it is not a theorem but more like a conjecture that is to be checked in each instance: *you may neglect the lower order terms whenever the real characteristics are simple.* This means that the cone of zeros of the equation

$$P_m(x,\xi) = 0 \tag{2}$$

(called the *characteristic* equation) has no singularities in R^n (we exclude the origin: at any rate the proper set-up for studying (2) is the *projective* space). This can be rephrased by saying that, for all $\xi \in R^n, \xi \neq 0$,

$$\mathrm{grad}_\xi P_m(x,\xi) \neq 0.$$

Indeed, by Euler's homogeneity formula, whenever $\mathrm{grad}_\xi P_m(x,\xi)$ vanishes so does $P_m(x,\xi)$ and such a zero is perforce multiple. In all this x is kept fixed (but arbitrary).

There are cases where you cannot drop the lower order terms. The best known examples are the heat operator

$$\partial^2/\partial x^2 - \partial/\partial t \tag{3}$$

and the Schrödinger operator

$$\partial^2/\partial x^2 - (1/i)(\partial/\partial t). \tag{4}$$

In both cases the principal symbol is ξ^2; the characteristics are double. It is clear that consideration of the principal symbol alone will not give you sufficient information; for instance it will not enable you to distinguish between the heat equation and Schrödinger's. It is worthwhile pointing out, however, that consideration of the principal symbol or of some symbol which can be called principal (by means of a modified homogeneity) has been an almost universal practice in the study of PDE's. This is true about the heat and the Schrödinger equations: it is only that one replaces the homogeneity with respect to all variables (as manifested in $P_m(x,\xi)$) by a *separate* homogeneity, e.g., of degree 2 in x and degree 1 in t as in (3) and (4). Keeping this in mind one sees easily that most results about linear PDE's are obtained by dropping the lower order terms. One important exception is the theory of linear PDE's with *constant coefficients* (and there are a few more, of lesser importance).

It should also be pointed out that the principal symbol $P_m(x,\xi)$ is a kind of "invariant" attached to the equation (1): it is coordinates invariant in the sense that it is a well-defined function on the cotangent bundle. It is the only *simple* "invariant" attached to the equation (if one excepts the zero-order terms), the others are more complicated and have not yet been much used, to my knowledge.

Let us return to the solvability result in Hörmander's thesis. As I said the hypotheses concerned only the principal symbol. In keeping with the general principle stated above, the first assumption was that *the real characteristics*, if there were any, *be simple*. Hörmander put a label on differential operators which have this property, he said they are *of principal type*; the label has stuck. The second assumption was, in a sense, more strange: *the coefficients $c_\alpha(x)$ in the principal part* (hence, $|\alpha| = m$) *had to be real*. Under these two hypotheses, there is local solvability —at any point in whose neighborhood they hold.

The solvability is strictly local; the exceptions to global solvability are trivial: the simplest is probably the rotation operator in $R^2\backslash\{0\}$,

$$x_1(\partial/\partial x_2) - x_2(\partial/\partial x_1).$$

It satisfies all of Hörmander's conditions, but it is certainly not globally solvable in any annular domain surrounding the origin.

Now, if we go back to Lewy's operator, we see that it has first order. Any first order operator whose principal part does not vanish identically at any point has simple real characteristics, since the gradient with respect to ξ of its principal symbol is constant with respect to ξ. Of course, the coefficients in Lewy's operator are not real: otherwise there would be a contradiction. On the other hand, it comes very close to having constant coefficients. Remember, however, that constant coefficients PDE's are locally (even globally) solvable. In Lewy's operator all coefficients are constant except one, and that one is linear in x_1, x_2. Note also that it is closely related to the Cauchy-Riemann operator—which, of course, is solvable. In a sense, Lewy's operator is barely on the edge of nonsolvability. But on the other hand, it has that remarkable property of not being locally solvable at *any* point.

The negative result of Lewy and the positive one of Hörmander, mentioned above, called for some kind of general explanation. The beginning of such an explanation was provided by Hörmander in 1959 with a necessary condition of local solvability. This work, as much as the work of Lewy, is crucial for the subsequent history—for two reasons. First of all, the nature of the necessary condition, found by Hörmander, is remarkable and opens the way to much further investigation. Second, the techniques used in its proof were very original, very striking, and are still the essential techniques one uses, with some improvements and modifications. In the necessary and sufficient conditions known today, the proof of the necessity follows the pattern given by Hörmander in 1959.

Let me state the condition. Clearly, both the positive result of 1955, which required the coefficients of the principal part to be real, and Lewy's example show that the nonreal nature of these coefficients is essential if we are to have nonsolvability. Let me add that the proof of the positive result used a priori estimates in the space L^2; these could be established because one had control of the commutator $[P_m(x,D), \bar{P}_m(x,D)] = P_m(x,D)\bar{P}_m(x,D) - \bar{P}_m(x,D)P_m(x,D)$ ($\bar{P}_m(x,\xi)$ is the homogeneous polynomial of degree m in ξ obtaining by replacing each coefficient of $P_m(x,\xi)$ by its complex conjugate). When the coefficients of P_m are real, the above commutator is identically zero, but later work of Hörmander, concerning what he called *principally normal* equations, confirmed the impression that control of the commutator leads to very good L^2 estimates. Now, $[P_m(x,D), \bar{P}_m(x,D)]$ is a differential operator of order $2m-1$; familiarity with the derivation of the estimates and the fact that the real characteristics of P are simple (which enables one to neglect lower order terms) strongly suggest that what is relevant, in the commutator, is its principal part, $C_{2m-1}(x,D)$. Indeed, Hörmander's necessary condition states that if P is locally solvable at the point x, then, for all $\xi \in R_n, \xi \neq 0$,

$$P_m(x,\xi) = 0 \Rightarrow C_{2m-1}(x,\xi) = 0.$$

This is the necessary condition of local solvability. When one encounters it for the first time, I am certain that it must seem almost as mysterious as Lewy's example. Of course, it is not satisfied, at any point $x = (x^1, x^2, x^3)$, by Lewy's operator: for it was the behaviour of this operator that Hörmander wanted to "explain." The simplest operator which does not satisfy it (at every point of the line $x^1 = 0$) is

$$\partial/\partial x^1 + \sqrt{(-1)}x^1(\partial/\partial x^2) \tag{5}$$

(for points where $x^1 \neq 0$, (5), viewed as an operator in R^2, is elliptic, hence locally solvable).

In a sense, Hörmander's necessary condition is not really satisfactory: for one thing, it lacks a conceptual basis; for another, one would like to know how far it is from being sufficient. At any rate, the next natural step was to seek necessary and sufficient conditions for local solvability. In 1962 Louis Nirenberg and I decided to tackle this problem; it looked rather formidable and we set ourselves very modest goals. We decided to look at a *single first order* linear partial differential equation, at least as a beginning. Note that such an equation will necessarily have simple characteristics—provided that the coefficients of its principal part do not vanish simultaneously at some point.

We had a starting point, namely Hörmander's necessary condition, exemplified, in particularly simple form, by (5). That condition led quite naturally to studying the commutators, not only the first one, $[L_0, \bar{L}_0]$ (I am now denoting by L_0 the principal part of the first order linear partial differential operator under study), but all the successive ones,

$$\left[L_0, \left[L_0, \bar{L}_0\right]\right], \quad \left[L_0, \left[L_0, \left[L_0, \bar{L}_0\right]\right]\right], \text{ etc.,}$$

and also, perhaps, those where a number of L_0 have been replaced by $\bar{L}_0$. Whatever conjecture one could come up to, it could always be tested on the operators

$$\frac{\partial}{\partial x^1} + \sqrt{(-1)}(x^1)^k \frac{\partial}{\partial x^2}, \qquad k = 0, 1, 2, \ldots. \tag{6}$$

What did we know about these operators? When $k = 0$, it is essentially the Cauchy-Riemann operator, everywhere solvable; when $k = 1$, we get (5), which is not solvable in any open set intersecting the line $x^1 = 0$. Of course, when $k > 1$, Hörmander's necessary condition would not provide any more information. But an easy modification of its proof shows that (6) is not solvable in any such open set when k is *odd*. On the other hand, when k is even, (6) is locally solvable (at any point)—for a very simple reason: namely that one can write explicitly, by a modification of Cauchy's formula, a solution to the corresponding equation (1). I will give, a little later, a different proof of this fact, a proof which ought to throw some light, I hope, on the question of solvability.

The facts known about (6) raise the question of the relation between the number k and the various commutators of L_0 and $\bar{L}_0$. It is convenient, at this point, to modify a little bit our approach and instead of looking at the complex vector fields L_0 and $\bar{L}_0$, to look at the real vector fields $A = \operatorname{Re} L_0, B = \operatorname{Im} L_0$. In the case of (6),

$$A = \partial/\partial x^1, \quad B = (x^1)^k \partial/\partial x^2.$$

Let us then set $C_1 = [A, B], C_{p+1} = [A, C_p] (p = 1, 2, \ldots)$; these are real vector fields; let us denote by $C_p(x, \xi)$ the symbol of C_p. In the case of (6),

$$C_p(x, \xi) = \frac{k!}{(k-p)!} (x^1)^{k-p} (i\xi_2) \qquad \text{if } p \leqslant k,$$
$$= 0 \qquad\qquad\qquad\qquad \text{if } p > k.$$

Consider a point (x, ξ) such that $x^1 = 0, \xi_1 = 0, \xi_2 \neq 0$. At any such point, the symbols of A, B and of every C_p for $p < k$ vanish; $C_k(x, \xi)$ does not. This provides a characterization of the integer k which seems not to depend on the peculiarities of the example (6). The concept can easily be extended to higher order equations. Indeed, let $A(x, D)$ (resp. $B(x, D)$) denote the real (resp. imaginary) part of $P_m(x, D)$. We may form the successive commutators or, rather, the principal parts of these commutators:

$$C_1(x, D) = \text{principal part of } \big[A(x, D), B(x, D) \big],$$
$$C_p(x, D) = \text{principal part of } \big[A(x, D), C_{p-1}(x, D) \big] \qquad \text{for } p > 1.$$

We look now at the symbols of all these operators, in (x, ξ)-space (ξ must always be different from zero), and with every point (x, ξ) we associate a number $k_0(x, \xi)$: this is the least integer k such that $C_k(x, \xi) \neq 0$. We agree that $k_0(x, \xi)$ is taken to be *zero* whenever $P_m(x, \xi) \neq 0$. Then, in analogy with what happens in the case (6), we shall conjecture that the parity of all these numbers $k_0(x, \xi)$ determines whether there is local solvability in the (open) set where the point x ranges.

Well, this approach cannot possibly lead to the full answer. Indeed it has several serious defects: for one, it assigns a privileged role to the real part $A(x, D)$, and this is inadmissible, since A and B can be interchanged by multiplying P_m by $\sqrt{(-1)}$; such a multiplication certainly cannot affect solvability. Also the approach will completely fail if *all* the commutators happen to vanish at some point (x, ξ), as in the example

$$\partial/\partial x^1 + \sqrt{(-1)} \exp\big(-1/(x^1)^2 \big) \partial/\partial x^2. \tag{7}$$

There is a way around the latter difficulty; this is a very important observation, which will also allow us to eliminate the first defect we mentioned. It has to do with the most classical part of PDE theory, the part that goes back to Lagrange, Hamilton, Jacobi. When dealing with functions such as $A(x, \xi), B(x, \xi)$, etc., defined (and smooth) in the (x, ξ)-space (in the cotangent bundle as we say now), one can introduce the Hamilton-Jacobi equations which, for A, e.g., read

$$dx/dt = \operatorname{grad}_\xi A(x, \xi),$$
$$d\xi/dt = -\operatorname{grad}_x A(x, \xi). \tag{8}$$

I am taking A here, but I could as well consider B, C_p, etc. Note however that in our examples (5), (6), (7) the gradient with respect to ξ of $A(x,\xi)$ did not vanish at any point (x,ξ) under consideration. Our basic hypothesis, that P is of principal type, means that given any point $(x_0, \xi^0), \xi^0 \neq 0$, either $\text{grad}_\xi A$ or $\text{grad}_\xi B$ does not vanish at that point, or that none does. We may assume that $\text{grad}_\xi A(x_0, \xi^0) \neq 0$. In this case, the (unique) solution $(x(t), \xi(t))$ of (8) such that $x(0) = x_0, \xi(0) = \xi^0$ describes a true curve in (x,ξ)-space whose projection on the x-space is, in fact, a curve. Let us denote by $\Gamma_A(x_0, \xi^0)$ the integral curve of (8) through (x_0, ξ^0) in the (x,ξ)-space; it is often called the *bicharacteristic strip* of A through that point. It follows at once from the form of (8) that the function $A(x,\xi)$ must be constant along such a curve. Thus, if we assume that $A(x_0, \xi^0) = 0$, we will have $A = 0$ on the whole of $\Gamma_A(x_0, \xi^0)$. In this case we shall refer to $\Gamma_A(x_0, \xi^0)$ as the *null bicharacteristic strip* of A through (x_0, ξ^0).

The relevance of these bicharacteristic strips from our viewpoint is due to the following fact: an immediate computation shows that the symbol of the first "commutator" $C_1(x, D)$, the principal part of $[A(x,D), B(x,D)]$ is given by

$$C_1(x,\xi) = \sum_{j=1}^{n} \left\{ \frac{\partial A}{\partial \xi_j}(x,\xi) \frac{\partial B}{\partial x^j}(x,\xi) - \frac{\partial A}{\partial x^j}(x,\xi) \frac{\partial B}{\partial \xi_j}(x,\xi) \right\},$$

that is,

$$C_1(x,\xi) = (dB/dt)(x,\xi),$$

where d/dt denotes the differentiation in the direction tangential to the bicharacteristic strip of A through the point (x,ξ). By iteration,

$$C_p = d^p B / dt^p.$$

Let us then suppose that $P_m(x_0, \xi^0) = 0$, that is, that both A and B vanish at (x_0, ξ^0). Let us look at the function $B(x,\xi)$ restricted to the null bicharacteristic strip of A through (x_0, ξ^0), $\Gamma_A(x_0, \xi^0)$. Hörmander's necessary condition can be restated by saying that if the equation $Pu = f$ is locally solvable at x_0 then the first derivative of B along $\Gamma_A(x_0, \xi^0)$ must vanish at (x_0, ξ^0). In the case (6) where, say $\xi^0 = (0, 1)$ and $x_0^1 = 0$, we see that there is local solvability if the first derivative of B along $\Gamma_A(x_0, \xi^0)$ which does not vanish is of *even* order, and that there is no solvability if this derivative is of *odd* order. But there is a way of rephrasing this property which does not rely on the fact that the zero of $B(x,\xi)$ along $\Gamma_A(x_0, \xi^0)$ is of finite order, and which therefore enables us to extend the property to equations such as (7): namely, that

(9) $B(x,\xi)$ *does not change sign at* (x_0, ξ^0) *along the null bicharacteristic strip of* $A(x,\xi)$ *through that point.*

We could now make the conjecture that local solvability of the equation $Pu = f$ in the open set Ω is equivalent with the fact that (9) is valid for all $x_0 \in \Omega$ and all $\xi^0 \in R^n, \xi^0 \neq 0$—if it were not for the fact that the statement (9) is blatantly dissymmetric in A and B. We shall therefore symmetrize in the following fashion:

(10) *Let* $x_0 \in \Omega$, $\xi^0 \in R^n, \xi^0 \neq 0$, *be such that* $P_m(x_0, \xi^0) = 0$. *Let* z *be any complex number such that*

$$\mathrm{grad}_\xi \, \mathrm{Re}(zP_m) \neq 0 \quad at \quad (x_0, \xi^0).$$

Then the function $\mathrm{Im}(zP_m)(x, \xi)$ *does not change sign at* (x_0, ξ^0) *along the null bicharacteristic strip of* $\mathrm{Re}(zP_m)(x, \xi)$ *through this point.*

Clearly, solvability of Equation (1) is unchanged if we substitute $zP, z \neq 0$, for P. Condition (10) may seem somewhat awkward, as it involves all complex numbers $z \neq 0$. But the remarkable fact is that, essentially,* it is true for all numbers $z \neq 0$ as soon as it is true for at least one of them. More precisely:

THEOREM 1. *Let* $q(x, \xi)$ *be a complex-valued function of* (x, ξ), *defined,* C^∞ *and nowhere vanishing in some open neighborhood of* (x_0, ξ^0). *Let us suppose that neither* $\mathrm{grad}_\xi A$ *nor* $\mathrm{grad}_\xi \mathrm{Re}(qP_m)$ *vanish at* (x_0, ξ^0). *Then, if Property (9) holds, the analogous property, where* $P_m(x, \xi)$ *is replaced by* $q(x, \xi)P_m(x, \xi)$, *also does.*

This was proved in the first order case (i.e., $m = 1$) in [6]; it is proved for arbitrary m in [7] under the additional assumption that any change of sign B along a null bicharacteristic strip of A must occur at a zero of finite order of B (along that strip).† Let me add that these proofs are by no means trivial.

If we take Theorem 1 into account and keep in mind that, at any point (x, ξ), either $\mathrm{grad}_\xi A(x, \xi)$ or $\mathrm{grad}_\xi B(x, \xi)$ are $\neq 0$, we see that the validity of Property (10) for all $x_0 \in \Omega, \xi^0 \in R^n, \xi^0 \neq 0$, can be rephrased as follows:

(11) $\mathrm{Re} P_m(x, \xi)$ *does not change sign along any null bicharacteristic of* $\mathrm{Im} P_m(x, \xi)$ *(lying over* Ω) *and* $\mathrm{Im} P_m(x, \xi)$ *does not change sign along any null bicharacteristic strip of* $\mathrm{Re} P_m(x, \xi)$.

We have tacitly agreed not to speak of a null bicharacteristic strip of a function of (x, ξ) unless the gradient with respect to ξ of this function does not vanish anywhere on the strip.

We may now formulate the main conjecture:

CONJECTURE. *The equation* $Pu = f$ *is locally solvable at every point of* Ω *if and only if* (11) *holds.*

This is the conjecture which Nirenberg and I were led to make, in 1962. Note that it is compatible with the positive result in Hörmander's thesis, where the coefficients of $P_m(x, \xi)$ had to be real, i.e., $\mathrm{Im}(i^m P_m(x, \xi)) \equiv 0$. Soon after, we were able to prove that it indeed held for *first order* differential equations—with a "minor" qualification: the necessity of Condition (11) was established only in the case where, along the null bicharacteristic strips of A, the changes of sign of the function B occurred at zeros of finite order (as a matter of fact, we could prove

---

* Not "essentially" but "truly"; see next footnote †.

† In [7] Theorem 1 is proved in full generality: the hypothesis about the changes of sign of B is removed.

necessity even if some changes of sign occurred at zeros of infinite order, but not in full generality: this is still an open question). This is not to say that B could not have zeros of infinite order, but only that B could not change sign at such a zero; for instance, B could very well vanish identically on a null bicharacteristic strip of A, as in the operator

$$\partial/\partial x^1 + \sqrt{(-1)}x^2\partial/\partial x^2.$$

Of course, the case of analytic coefficients was completely settled; in this case, Condition (11) is necessary and sufficient—without any qualification.

This was the state of affairs in 1963. Since then, until recently, nothing much has happened. We could not extend our result to higher order equations. We could not handle the lower order terms. This was due to a certain feature in our treatment of the first order case. I should say that the obstruction was on the side of the sufficiency proof. As far as necessity of (11) was concerned we had the feeling that it was, and would remain within our reach, that all we had to do was, some day, to sit down and muddle through the technicalities, and we would get what we expected. But not so with sufficiency! Recent experience has confirmed that these were sound feelings.

I would like to explain what the obstruction was, for it had an important implication concerning first order equations. The proof of sufficiency was based on an estimate of a quite unusual nature, and we could not see, we still cannot see, how to extend it to higher order equations. Let me describe the estimate we did obtain, and also the one we would have liked to obtain. In fact I shall begin by the latter. Let us denote by P' the *formal transpose* of P. This is the operator derived from P by integration by parts:

$$\int (Pu)v\,dx = \int u(P'v)\,dx, \quad u,v \in C_c^\infty(\Omega).$$

Now what we would have liked to obtain is an estimate

$$\int |u|^2\,dx \leqslant C\int |P'u|^2\,dx, \quad u \in C_c^\infty(\Omega). \tag{12}$$

Here Ω is some open set. A very elementary lemma about Hilbert spaces (or else a straightforward application of the Hahn-Banach theorem) shows that, if (12) holds, then to every $f \in L^2(\Omega)$ there is $u \in L^2(\Omega)$ such that $Pu = f$ (in Ω, in the distribution sense).

Thus (12) is what one needs if he is to prove solvability in the "L^2 sense." As a matter of fact, we would be very happy if we could have even more: if we could choose the constant C in (12) so that it decreases and tends to zero together with the diameter of Ω. In such a case one can always perturb the zero order term in P (or in P') as widely as he wishes, for the smallness of C will take care of that. In other words, estimates like (12), with $C \rightsquigarrow 0$ with diam Ω, are stable under zero order perturbations. This stability is crucial if we are to try to extend such an estimate to higher order equations; transposed into this context it would mean that we could neglect lower order terms.

As I said (12) is not at all what we proved; what we proved was the estimate

$$\int |u|^2 dx \leqslant C \int |\text{grad}(P'u)|^2 dx, \quad u \in C_c^\infty(\Omega). \tag{13}$$

Now, here, even when the constant $C \searrow 0$ with diam Ω, one cannot perturb the zero order terms in P' and deduce a similar estimate; such a perturbation introduces a quantity of the form

$$\text{const.} \int |\text{grad}\, u|^2 dx$$

on the right-hand side of (13), and there is no way of absorbing it into the left-hand side. Of course, (13) is valid (under Assumption (11)) regardless of what the zero order terms in P are; but as far as the method of proof is concerned, you cannot neglect these zero order terms, you have to handle them with special tricks. This is what precludes extension to higher order case since we could not possibly hope to extend those tricks so as to handle lower order, but not any more zero order, terms.

Moreover an estimate such as (13) does not imply L^2 solvability. What it does imply is that, to every $f \in L^2(\Omega)$ there is $u \in H^{-1}(\Omega)$ satisfying the equation $Pu = f$ in Ω; $H^{-1}(\Omega)$ is the (Sobolev) space of distributions in Ω which can be expressed as finite sums of first order partial derivatives of functions belonging to $L^2(\Omega)$.

It is now time to describe more recent results. Last December, 1968, I was able to show that Property (11) is sufficient for local solvability in $\Omega \subset R^2$. More precisely:

THEOREM 2. *In the case of two independent variables* $(n = 2)$, *if Property* (11) *holds, the equation* $Pu = f$ *is locally solvable at every point of* Ω.

Conversely, suppose that, for some $x_0 \in \Omega$ *the following is true*:

(14) *There is* $\xi^0 \neq 0$ *such that* $P_m(x_0, \xi^0) = 0$, $d_\xi \text{Re}\, P_m(x_0, \xi^0) \neq 0$, *and such that, along the null bicharacteristic strip of* $\text{Re}\, P_m$ *through* (x_0, ξ^0), $\text{Im}\, P_m$ *has a zero of finite odd order at that point*.

Then the equation $Pu = f$ *is not locally solvable at* x_0.

But now something more is true, for the estimate used to prove the existence of solutions is of the kind (12), not of the kind (13). More precisely, since we deal with a differential operator of order m, the estimate reads:

$$\sum_{|\alpha| \leqslant m-1} \int |D^\alpha u|^2 dx \leqslant \epsilon \int |P'u|^2 dx, \quad u \in C_c^\infty(\Omega), \tag{15}$$

where $\epsilon \searrow 0$ when diam $\Omega \searrow 0$. This is optimal—since we know it is optimal for *hyperbolic* equations (any better estimate, i.e. where derivatives of order $> m - 1$ of u would be dominated, would imply hypoellipticity). It is not very difficult to derive from (15) that to every $f \in L^2(\Omega)$ there is $u \in H^{m-1}(\Omega)$ such that $Pu = f$ in Ω; $H^{m-1}(\Omega)$ is the Sobolev space of functions whose distributions derivatives of order $\leqslant m - 1$ belong to $L^2(\Omega)$.

Differential operators in two independent variables, if they are of principal type, are quite special. Indeed, their principal symbol $P_m(x, \xi)$ is a polynomial in the two variables (ξ_1, ξ_2), homogeneous of degree m, hence can be factorized,

$$P_m(x,\xi) = a(x) \prod_{j=1}^{m} (\xi_1 - \lambda_j(x)\xi_2),$$

and all the properties of P (at least, those of interest to us here) can be translated in terms of the *characteristic roots* $\lambda_j(x)$. These are functions of x alone; ξ does not enter. Because of this, Condition (11) can be rephrased, in this case, in the following manner:

(11′) *For each $j = 1, \ldots, m$, the function* $\mathrm{Im}\,\lambda_j$ *does not change sign along the integral curves of the vector field*

$$\partial/\partial x^1 - (\mathrm{Re}\,\lambda_j(x))\partial/\partial x^2.$$

Suppose for a moment that the coefficients of $P_m(x,\xi)$ are analytic. Then the changes of sign of $\mathrm{Im}\,\lambda_j$ must perforce occur at points belonging to a proper analytic subset of Ω. Even in the C^∞ case, they certainly could not occur everywhere. Nonsolvability is therefore a "rare" event. This shows that an example such as Lewy's, where nonsolvability occurs at every point, can only be encountered if the number of independent variables is > 2.

What is now known when there are $n > 2$ independent variables? Well, in the first order case, we have now the best possible result. Indeed, I have been able to prove an estimate of the kind (12), here again with a constant $C \rightsquigarrow 0$ with diam Ω. Thus:

THEOREM 3. *If P is first order linear partial differential operator with C^∞ coefficients in Ω, having Property (11), every point $x_0 \in \Omega$ has an open neighborhood $U(x_0) \subset \Omega$ such that*

$$PL^2(U(x_0)) \supset L^2(U(x_0)).$$

This is an improvement on the 1962 result concerning the existence of solutions to first order PDE's. The proof of Theorem 3 uses Theorem 2 via the observation that, locally, any first order linear PDE, of the form*

$$\frac{\partial u}{\partial t} + \sqrt{(-1)} \sum_{j=1}^{n} b_j(x,t)\frac{\partial u}{\partial x^j} = f,$$

having Property (11), can be transformed by a change of variables into a first order PDE in *two* independent variables. The unfortunate thing is that such changes of variables are to be performed in the complement of the set

$$\{x; b_j(x,t) = 0, \quad j = 1, \ldots, n, \quad (x,t) \in \Omega\},$$

and their domain of validity tends to shrink as we approach this set. The

---

* It is easy to see that every first order linear PDE with C^∞ coefficients, without any zero order term, which does vanish identically at any point, can be brought into that form—at least locally and up to a nowhere vanishing factor—by a change of coordinates.

derivatives of the cut-off functions which are used tend, in absolute value, to $+\infty$ as the domains of the local transformations shrink and this creates some difficulty when one patches together the L^2-estimates which have been established in each domain individually. The derivatives must be kept under control and this is achieved by means of a lemma which resembles very much a classical lemma of Whitney used to extend to the whole space of differentiable functions defined in a closed set.

At last I come to PDE's of principal type, in any number of independent variables, of any order. Very recently, Louis Nirenberg and I have shown the necessity of Condition (11), under the usual restrictions. Precisely we proved:

THEOREM 4. *Suppose that for some point* $x_0 \in \Omega$, *Property* (14) *holds. Then the equation* $Pu = f$ *is not locally solvable at* x_0.

As a matter of fact, we prove this theorem when P is a pseudodifferential operator of order m (which is then an arbitrary real number), under the assumption that the homogeneity degrees, with respect to ξ, of the various terms in the symbol $P(x, \xi)$ of P are $m, m-1, m-2, \ldots, m-k, \ldots$. One must then modify the statement of Condition (14) as follows:

(14′) *There is* $\xi^0 \neq 0$, *such that* $P_m(x_0, \xi^0) = 0, d_\xi \operatorname{Re} P_m(x_0, \xi^0) \neq 0$, *and such that, along the null bicharacteristic strip of* $\operatorname{Re} P_m$ *through* (x_0, ξ^0), $\operatorname{Im} P_m$ *has a zero of finite odd order at that point and changes sign there from minus to plus.*

This property is invariant under multiplication of $P_m(x, \xi)$ by a nowhere vanishing function $q(x, \xi)$ as considered in Theorem 1. This is shown in [7]. In the pseudodifferential case one must also suitably modify the concept of local solvability.

We have learned from Professor Olga Oleĭnik that one of her Ph.D. students, Y. V. Egorov, has proved a result which appears to be stronger than Theorem 4. I must say that I am not extremely surprised that others should have also proved this kind of theorem, as it is really a question of reworking, with a little bit of skill, the original proof of the necessary condition of Hörmander of 1959.

Theorems 2, 3, 4 represent the results known at the present time on the subject of local solvability of linear PDE's in the sense of functions and distributions.*

I would like to end this lecture by a few remarks about solvability in a more general sense than function or distribution solvability. As a starting point for these remarks I wish to take a fresh look at the operators (6) which I now write

$$\partial/\partial t + it^k \partial/\partial x, \quad i = \sqrt{-1}.$$

After performing a Fourier transformation in the variable x, the associated equa-

---

* During the summer of 1969, L. Nirenberg and I were able to prove our conjecture (p. 520) for all PDE's of principal type, of any order, in any number of variables, if they have *analytic* coefficients. The results obtained are the best possible, as we prove an estimate of the kind (15). On the other hand, Y. V. Egorov claims to have obtained very general sufficient conditions for the validity of the so-called *subelliptic estimates*, which imply local solvability. But it is not clear to us, at this time, what his conditions are (added on the proofs).

tion reads

$$\hat{u}_t + t^k \xi \hat{u} = \hat{f}, \quad f \in (C_c^\infty)_{x,t} \tag{16}$$

where the "hats" denote Fourier transformation in x (on the Fourier transforms-side the variable is denoted by ξ). We may solve (16) in two different ways:

$$\hat{u}(\xi,t) = \int_{-\infty}^{t} \hat{f}(\xi,s) \exp\left(-\frac{t^{k+1}-s^{k+1}}{k+1}\xi\right) ds, \tag{17}$$

$$\hat{u}(\xi,t) = -\int_{t}^{+\infty} \hat{f}(\xi,s) \exp\left(-\frac{t^{k+1}-s^{k+1}}{k+1}\xi\right) ds. \tag{18}$$

If we want to return to the original equation we must then perform an inverse Fourier transformation, reverting from the variable ξ to x. But we will end up with a distribution solution if and only if we start from a solution of (16) which is *tempered* with respect to ξ, that is to say (roughly speaking), which, as a function of ξ, does not grow at infinity faster than some polynomial. The question is then whether such a solution of (16) exists. Suppose that k is *even*. Then we could select (17) as the solution for $\xi > 0$ and (18) when $\xi < 0$ (the solution does not have to be a continuous function of ξ but only a tempered distribution in ξ!). It is clear that, with these choices, the exponent

$$-(t^{k+1}-s^{k+1})\xi/(k+1)$$

in each integral, remains bounded. We have thus defined a solution $\hat{u}$ of (16) which is an L^∞ function of ξ. But what if k is *odd*? It is clear that this procedure will not work. By virtue of the nonexistence theorems stated earlier we know that no procedure will work. On the other hand, we have at our disposal lots of solutions of (16), and they have very nice expressions; one feels that they should be useful in some computations at least, and that we should be able to define and use their inverse Fourier transforms. This is indeed the case, but these inverse Fourier transforms will, of course, not be distributions—they will be objects of a fairly general nature, known under the name of analytic functionals. These are linear functionals on spaces of holomorphic functions, exactly like distributions are linear functionals on the space of C^∞ functions with compact support. One of the essential differences between the former and the latter is that there exist partitions of unity consisting of C_c^∞ functions (subordinated to any locally finite open covering of the base space) but there are no such partitions consisting of holomorphic functions. This difference has, of course, extremely deep consequences; one of these is that one can define the support of a distribution but not the support of an analytic functional. One has the right to say that a given distribution takes these values in a given open set and these different values in a different open set, but nothing of the sort may be asserted about analytic functionals. In summary, distributions are *localizable* whereas analytic functionals are not. This does not mean that the latter are useless, it only means that they do not have the properties which go with localization. As a matter of fact, a great deal can be said, and has been said, about analytic functionals (see e.g. the works of A. Martineau, in particular [5]), mainly through their "Fourier" transforms, which happen to be called Fourier-Borel transforms (sometimes Fourier-Laplace, sometimes Borel-Laplace transforms). Let Ω be an open subset of the complex n-dimensional space C^n intersecting the real space R^n and let $\Omega_R = \Omega \cap R^n$; then any L^1 function $f(x)$ with compact support in Ω_R defines a unique analytic functional on Ω through the formula

$$\langle f,h \rangle = \int f(x)h(x)\,dx,$$

where h is an arbitrary holomorphic function in Ω. Similarly any distribution with compact support in Ω_R defines a unique analytic functional in Ω. Thus distributions with compact support can be identified with certain analytic functionals (but of course not every analytic functional can be thus defined by a distribution with compact support—otherwise our remarks above about localization would be some-what irrelevant). Keeping this in mind, I would like to state a theorem about solvability in spaces of analytic functionals. In fact, it concerns local solvability of the *Cauchy problem* posed for a differential operator of order m with analytic coefficients (only analytic coefficients can act as multipliers on analytic function-als). In the present case we must distinguish between space variables, $x = (x^1,\dots,x^n)$, and a time variable; we assume that the differential operator under study has an expression

$$P = D_t^m - \sum_{j=1}^{m} c_j(x,t,D_x)D_t^{m-j},$$

where each $c_j(x,t,D_x)$ is a differential operator of order $\leqslant j$ with respect to x with coefficients which are analytic functions of (x,t), say in a neighborhood U of the origin in R^{n+1}. We consider the Cauchy problem for P relative to the hyperplane $t = 0$, which is the problem of finding a solution u to

$$Pu = f \quad \text{in } U, \tag{19}$$

$$D_t^k u = g_k \quad \text{in } U \cap \{(x,t); t = 0\}, \qquad k = 0,\dots,m-1. \tag{20}$$

The "objects" g_k depend only on x, they are called the *Cauchy data*; f will be a function of t, say a continuous function of t, valued in a space of analytic functionals with respect to the variables x; and we shall also take the g_k to be analytic functionals with respect to x. Possibly after some shrinking of the neigh-borhood U, it can be (roughly) stated that the problem (19)–(20) has a unique solution which is a C^1 function of t with values in the space of analytic functionals with respect to x. For the precise statement, see [15, §11]. The proof of this theorem is easy, it is a dual form of the Cauchy-Kowalewski theorem. I should add that much more than existence and uniqueness of the solution u can be asserted: the solution u can be expressed in terms of the data, f and the g_k, by means of an integral formula having a kernel which is an analytic function of the intervening variables. On this subject see [15, Chapter II], also [16]. At any rate, we see that when f and the g_k are C^∞ functions of x and t with compact support or distributions of x and t with compact support (f does not have to be a function with respect to t in order that we have a solution u to (19)–(20), f can be only a distribution in t, but then u will also have to be a distribution in t), the equation (19) admits a solution—always! but of course it is not going to be a distribution, only a distribution with respect to t, but an analytic functional with respect to x.

It should be mentioned that we should not expect a theorem such as the one just stated to hold for differential operators with C^∞ coefficients—not only because these coefficients would not multiply analytic functionals, for we would be willing to try to replace analytic functionals with something else. The obstruction here is the *uniqueness* part in the statement: indeed it is known that there is no uniqueness in the (noncharacteristic) Cauchy problem for certain equations with C^∞ coefficients—even when the solutions happen to be C^∞ functions of x and t with compact support! Examples of such equations have been constructed by Paul Cohen and A. Pliš.

There are objects, that is, generalized functions, which are more general than distributions and, in a certain sense, less general than analytic functionals—and which manage to retain the localization properties: they are the, so-called, Sâto hyperfunctions. Very roughly speaking they are boundary values (carried by pieces of the real space R^n) of holomorphic functions (defined in complex neighborhoods of those "pieces" of R^n); they have attracted the attention of theoretical physicists as they lend themselves to certain operations which are not allowed in distribution theory. Also they have provided solutions to PDE *with constant coefficients* in cases where no distribution solution exists (see [1]). Thus it is only natural to ask the question as to whether one has to go so far as analytic functionals in order to attain solvability (at least in the case of equations with analytic coefficients) and whether Sâto hyperfunctions, with their localization properties, would not suffice. Alas, this does not seem to be the case. A student of Laurent Schwartz, Pierre Schapira, has shown that, at least for first order PDE with analytic coefficients, nonsolvability in the distribution sense implies nonsolvability in the space of Sâto hyperfunctions (see [11], [12]). I am convinced that what lies at the root of nonsolvability of PDE's of principal type is not our insistence on considering distributions, it is not just compliance with fashion; it is our insistence in having solutions which are *localizable* generalized functions.

References

1. R. Harvey, Hyperfunctions and linear partial differential equations, Proc. Nat. Acad. Sci. U.S.A. 55 (1966), 1042–1046. MR 34#495.

2. L. Hörmander, On the theory of general partial differential operators, Acta Math. 94 (1955), 161–248. MR 17, #853.

3. ———, Differential equations without solutions, Math. Ann. 140 (1960), 169–173. MR 26#5279.

4. H. Lewy, An example of a smooth linear partial differential equation without solution, Ann. of Math. (2) 66 (1957), 155–158. MR 19, 551.

5. A. Martineau, Équations différentielles d'ordre infini, Bull. Soc. Math. France 95 (1967), 109–154.

6. L. Nirenberg and F. Trèves, Solvability of a first order linear partial differential equation, Comm. Pure Appl. Math. 16 (1963), 331–351, MR 29 #348.

7. ———, On local solvability of linear partial differential equations. Part I: Necessary conditions, Comm. Pure Appl. Math. (to appear).

8. ———, On local solvability of linear partial differential equations. Part II: Sufficient conditions, Comm. Pure Appl. Math. (to appear).

9. ———, Conditions nécessaires de résolubilité locale des équations pseudodifférentielles, C. R. Acad. Sci. Paris 269 (1969), 774–777.

10. ———, Conditions suffisantes de résolubilité locale des équations aux dérivées partielles linéaires, C. R. Acad. Sci. Paris 269 (1969), 853–856.

11. P. Schapira, Une équation aux dérivées partielles sans solutions dans l'espace des hyperfonctions, C. R. Acad. Sci. Paris Sér. A–B 265 (1967), A665–A667. MR 36#4112.

12. ———, Solutions hyperfonctions des équations aux dérivées partielles du premier ordre, Bull. Soc. Math. France 97 (1969), 243–255.

13. F. Trèves, On the local solvability of linear partial differential equations in two independent variables, Amer. J. Math. (to appear).

14. ———, Local solvability in L^2 of first order linear PDE's, Amer. J. Math. (to appear).

15. ———, Ovcyannikov theorem and hyperdifferential operators, Notas de Matematica, no. 46, Inst. Mat. Pura Apl. Con. Nac. Pesquisas, Rio de Janeiro, 1968.

16. ———, Hyperdifferential operators in complex space, Bull. Soc. Math. France 97 (1969), 193–223.

21

CARL DOUGLAS OLDS

Professor Olds was born on May 11, 1912, in Wanganui, New Zealand. He received all his degrees at Stanford University, the A.B., 1936, the A.M., 1937, and the Ph.D. in 1943 under Professor J. V. Uspensky. From 1935 to 1940 and in the summer of 1942, he was an acting instructor at Stanford University, and from 1940 to 1945 an assistant professor at Purdue University. Since 1945, he has been at California State University, San Jose, advancing through the ranks to full professor.

Professor Olds has served the mathematical community extensively. His service to the Mathematical Association of America included the acting chairmanship of the Northern California Section for part of 1951, Secretary-Treasurer of that Section from 1952 to 1955, and Sectional Governor for the period 1956–58. He served as first editor of the *Mathematical Log*, the official publication of Mu Alpha Theta, the national high school and junior college mathematics club. He was awarded the Mu Alpha Theta service plaque in 1966.

Professor Olds' skill as a teacher was recognized by the award to him of a California State College Distinguished Teaching Award for the academic year 1965–66.

Professor Olds' substantial contributions to various parts of number theory are contained in his many publications in a great variety of periodicals. He is also the author of the book *Continued Fractions*, published as part of the New Mathematics Library of Random House in 1963.

In accepting the award, Professor Olds indicated how very pleased and honored he felt. He added that in the past, and especially during the last few years, the Editors of the *Monthly* have done a fine job in encouraging expository writing. He thought that more mathematicians would write such articles if they realized that expository articles do not have to be long, do not have to cover an entire field of study, and do not need to include every reference that exists on a subject.

THE SIMPLE CONTINUED FRACTION EXPANSION OF e

C. D. OLDS, San Jose State College, California

1. Introduction. Students, and others, have asked where they can find a readable account of the simple continued expansion of e, namely, the beautiful result due to Euler [1] that

(1)
$$e = 2 + \frac{1}{1} + \frac{1}{2} + \frac{1}{1} + \frac{1}{1} + \frac{1}{4} + \frac{1}{1} + \frac{1}{1} + \frac{1}{6} + \cdots$$
$$= \langle 2, 1, 2, 1, 1, 4, 1, 1, 6, 1, \cdots \rangle.$$

Simple continued fractions have the form

(2)
$$a_1 + \cfrac{1}{a_2 + \cfrac{1}{a_3 + \cfrac{1}{a_4 + \cdot }}} ,$$

where a_1 is usually a positive or negative integer (but could be zero), and where the terms $a_2, a_3, a_4, \cdots$ are positive integers. It is convenient to write (2) in the form

$$a_1 + \frac{1}{a_2} + \frac{1}{a_3} + \frac{1}{a_4} + \cdots ,$$

with the "+" signs after the first one lowered to indicate the "step-down" process in forming a continued fraction; or simply to represent it by the symbol $\langle a_1, a_2, a_3, \cdots \rangle$. To understand the continued fraction part of this expository paper one need only read, say, a few pages of Chapter 7 in Niven and Zuckerman [4]; noting, in particular, that the convergents of the continued fraction (2), namely,

(3)
$$c_1 = \frac{a_1}{1} = \frac{p_1}{q_1}, \; c_2 = a_1 + \frac{1}{a_2} = \frac{a_1 a_2 + 1}{a_2} = \frac{p_2}{q_2}, \cdots ,$$

can be calculated, successively, from the equations

(4)
$$p_i = a_i p_{i-1} + p_{i-2},$$
$$q_i = a_i q_{i-1} + q_{i-2},$$

for $i = 1, 2, 3, 4, \cdots$, provided the *undefined* terms which occur are assigned the values $p_{-1} = 0$, $p_0 = 1$, $q_{-1} = 1$, $q_0 = 0$.

Euler derived the expansion (1) by converting the infinite series expansion of e into a continued fraction, an effective method when it succeeds. One does not always end up with a *simple* continued fraction. See Wall [6, p. 17].

Since Euler's time, mathematicians such as Lambert, Gauss, Liouville, Hurwitz, Stieltjes, to mention only a few, established continued fractions as a field worthy of independent study, with applications to many branches of mathematics. It comes as no surprise, then, that the expansion (1), and similar ones, are special cases of later developments of the subject.

However, for historical reasons, we present Hermite's derivation of (1). The main ideas are contained in his famous paper [2] in which he gave the first proof that e is a transcendental number. Hermite needed approximations to e and its integral powers, and, as a matter of fact, those he used were not convergents to their respective continued fractions. What follows, then, is not a mere translation of what Hermite wrote, but, rather, a re-working of his ideas, with changes and additions to make a self-contained exposition starting with the integral (5), given below, and ending with (1).

2. Hermite's Method. Hermite starts with an integral

(5)
$$\int e^{-rx} f(x)\, dx,$$

where $r \neq 0$ is an arbitrary constant, and where $f(x)$ is a polynomial of degree $n = 2m$. Repeated integration by parts transforms (5) into the form

$$\int e^{-rx} f(x)\, dx = -\frac{1}{r} e^{-rx} f(x) + \frac{1}{r} \int f^{(1)}(x) e^{-rx} dx$$

(6)
$$= -\frac{1}{r} e^{-rx} f(x) - \frac{1}{r^2} e^{-rx} f^{(1)}(x) + \frac{1}{r^2} \int f^{(2)}(x) e^{-rx} dx$$

$$\cdots \cdots \cdots \cdots \cdots \cdots \cdots \cdots \cdots \cdots \cdots$$

$$= - e^{-rx}\left(\frac{1}{r} f(x) + \frac{1}{r^2} f^{(1)}(x) + \cdots + \frac{1}{r^{2m+1}} f^{(2m)}(x)\right)$$

$$= - e^{-rx} \Phi(x).$$

Hence,

(7)
$$\int_0^1 e^{-rx} f(x)\, dx = - e^{-rx} \Phi(x)\big]_0^1 = \Phi(0) - e^{-r} \Phi(1).$$

In (7) let $f(x) = x^m (x-1)^m$, then

$$f^{(j)}(0) = f^{(j)}(1) = 0, \qquad j = 0, 1, 2, \cdots, m-1,$$

consequently, the expression $\Phi(x)$ in (6) reduces for $x = 0$ and $x = 1$, respectively, to

(8)
$$\Phi(0) = \sum_{j=m}^{2m} \frac{f^{(j)}(0)}{r^{j+1}}, \qquad \Phi(1) = \sum_{j=m}^{2m} \frac{f^{(j)}(1)}{r^{j+1}}.$$

On the other hand, Taylor's expansion about $x=0$ shows that

$$(9) \qquad f(x) = \sum_{j=m}^{2m} \frac{f^{(j)}(0)}{j!} x^j = \sum_{j=m}^{2m} \alpha_j x^j,$$

where $f^{(j)}(0) = j!\alpha_j$. Since $f(x)$ is a polynomial with integral coefficients, the α_j's are integers. Similarly, writing $f(x) = [(x-1)+1]^m(x-1)^m$, Taylor's expansion about $x=1$, shows that

$$(10) \qquad f(x) = \sum_{j=m}^{2m} \frac{f^{(j)}(1)}{j!} (x-1)^j = \sum_{j=m}^{2m} \beta_j(x-1)^j,$$

where $f^{(j)}(1) = j!\beta_j$, the β_j's being integers. Hence, the equations (8) take the form

$$(11) \qquad \Phi(0) = \sum_{j=m}^{2m} \frac{j!\alpha_j}{r^{j+1}} = \frac{m!M_m(r)}{r^{2m+1}},$$

$$\Phi(1) = \sum_{j=m}^{2m} \frac{j!\beta_j}{r^{j+1}} = \frac{m!N_m(r)}{r^{2m+1}},$$

where $M_m(r)$ and $N_m(r)$ are polynomials of degree m in r with integral coefficients. Using (11), with $f(x) = x^m(x-1)^m$, we rewrite (7) in the form

$$(12) \qquad e^r M_m(r) - N_m(r) = \frac{r^{2m+1}e^r}{m!} J_m = V_m e^r,$$

where

$$(13) \qquad J_m = \int_0^1 e^{-rx} x^m(x-1)^m dx, \qquad V_m = \frac{r^{2m+1}}{m!} J_m.$$

Setting $m=0$, $f(x)=1$, $\Phi(x)=r^{-1}$, and so $\Phi(0)=r^{-1}$, $M_0(r)=1$, $N_0(r)=1$. Similarly, for $m=1$, $f(x)=x^2-x$, and so $M_1(r)=2-r$, $N_1(r)=2+r$.

The crucial part of Hermite's development hinges on a relationship between J_m, J_{m-1}, and J_{m-2}, $m\geq 2$, of the form $J_m+aJ_{m-1}+bJ_{m-2}=0$. To this end, integration by parts shows that

$$(14) \qquad J_m = \frac{m}{r}\int_0^1 e^{-rx}x^{m-1}(x-1)^m dx + \frac{m}{r}\int_0^1 e^{-rx}x^m(x-1)^{m-1}dx.$$

In the first integral above replace $(x-1)^m$ by $(x-1)^{m-1}(x-1)$ to obtain

$$(15) \qquad \int_0^1 e^{-rx}x^m(x-1)^{m-1}dx = \frac{r}{2m}J_m + \frac{1}{2}J_{m-1},$$

and with m replaced by $m-1$,

$$(16) \qquad \int_0^1 e^{-rx}x^{m-1}(x-1)^{m-2} = \frac{r}{2(m-1)}J_{m-1} + \frac{1}{2}J_{m-2}.$$

A second integration by parts shows that

$$(17) \qquad \int_0^1 e^{-rx} x^m (x-1)^{m-1} dx = \frac{m}{r} J_{m-1} + \frac{m-1}{r} \int_0^1 e^{-rx} x^m (x-1)^{m-2} dx.$$

On the other hand, since $(x-1)^{m-1} = x(x-1)^{m-2} - (x-1)^{m-2}$, it follows that

$$(18) \qquad J_{m-1} = \int_0^1 e^{-rx} x^m (x-1)^{m-2} dx - \int_0^1 e^{-rx} x^{m-1} (x-1)^{m-2} dx.$$

Using (16), and then solving (18) for the first integral on the right, we get

$$(19) \qquad \int_0^1 e^{-rx} x^m (x-1)^{m-2} dx = J_{m-1} + \frac{r}{2(m-1)} J_{m-1} + \frac{1}{2} J_{m-2}.$$

Now, substitute (19) into (17), and follow this by equating the right side of (15) with the new form of (17). The result, after simplifications, is

$$(20) \qquad r^2 J_m - 2m(2m-1) J_{m-1} - m(m-1) J_{m-2} = 0.$$

Using the relationship between V_m and J_m given in (13), we can replace (20) by

$$(21) \qquad V_m - (4m-2) V_{m-1} - r^2 V_{m-2} = 0.$$

Next, calculate V_m, V_{m-1}, V_{m-2} from (12) and substitute into (21) to obtain

$$(22) \qquad \begin{aligned} e^r [M_m(r) - (4m-2) M_{m-1}(r) - r^2 M_{m-2}(r)] \\ - [N_m(r) - (4m-2) N_{m-1}(r) - r^2 N_{m-2}(r)] = 0, \end{aligned}$$

and since e^r is irrational this demands that

$$(23) \qquad \begin{aligned} M_m(r) - (4m-2) M_{m-1}(r) - r^2 M_{m-2}(r) = 0, \\ N_m(r) - (4m-2) N_{m-1}(r) - r^2 N_{m-2}(r) = 0. \end{aligned}$$

From (11) we know that $M_m(r)$ and $N_m(r)$ are polynomials of degree m in r. Hence, if we replace r by $2/k$, where k is a positive integer, then these polynomials can be written in the form

$$(24) \qquad M_m(2/k) = S_m/k^m, \qquad N_m(2/k) = R_m/k^m,$$

where S_m and R_m are integers. Next, substitute the values of M_m and N_m from (24) into (12), use (13) to get

$$e^{2/k} S_m - R_m = \frac{2^{2m+1} e^{2/k}}{k^{m+1} m!} \int_0^1 e^{-2x/k} x^m (x-1)^m dx.$$

Since $|x(x-1)| \leq 1/4$ if $0 \leq x \leq 1$, and since $\int_0^1 e^{-2x/k} dx < k/2$, it follows easily that

$$(25) \qquad |e^{2/k} S_m - R_m| < \frac{e^{2/k}}{k^m m!}.$$

Again, using (24) the relations (23) can be replaced by

(26)
$$S_m = (4m - 2)kS_{m-1} + 4S_{m-2},$$
$$R_m = (4m - 2)kR_{m-1} + 4R_{m-2}.$$

If, for convenience, we let

(27)
$$S_m + R_m = 2^{m+1}\,T_m,$$
$$S_m - R_m = -\,2^{m+1}\,Z_m,$$

then the recurrence relations (26) can be replaced by

(28)
$$T_m = (2m - 1)kT_{m-1} + T_{m-2},$$
$$Z_m = (2m - 1)kZ_{m-1} + Z_{m-2},$$

where, in particular,

$$T_0 = 1,\ T_1 = k,\ T_2 = 3k^2 + 1,\ T_3 = 15k^3 + 6k,\ \cdots,$$
$$Z_0 = 0,\ Z_1 = 1,\ Z_2 = 3k,\ Z_3 = 15k^2 + 1,\ \cdots.$$

Referring to equations (4), this shows that $T_0/Z_0,\ T_1/Z_1,\ T_2/Z_2,\ \cdots$, are the convergents to the simple continued fraction $\langle k,\ 3k,\ 5k,\ 7k,\ \cdots \rangle$. Moreover, from (27) we have

$$S_m = 2^m(T_m - Z_m),\ R_m = 2^m(T_m + Z_m),$$

which transforms the inequality (25) into the form

$$\left|\,(e^{2/k} - 1)T_m - (e^{2/k} + 1)Z_m\,\right| < \frac{e^{k/2}}{(2k)^m m!},$$

and after dividing both sides by $Z_m(e^{2/k} - 1)$, and noting by the recursion formula (28) that Z_m increases as m increases, we see that

$$\left|\frac{e^{2/k} + 1}{e^{2/k} - 1} - \frac{T_m}{Z_m}\right| < \frac{e^{2/k}}{(2k)^m m! Z_m(e^{2/k} - 1)} \to 0$$

as $m \to \infty$. We are justified, then, in writing as a simple continued fraction

(29)
$$\frac{e^{2/k} + 1}{e^{2/k} - 1} = k + \frac{1}{3k} + \frac{1}{5k} + \frac{1}{7k} + \cdots = \langle k,\ 3k,\ 5k,\ 7k, \ldots \rangle.$$

Subtracting 1 from both sides of (29) gives

(30)
$$\frac{2}{e^{2/k} - 1} = (k - 1) + \frac{1}{3k} + \frac{1}{5k} + \frac{1}{7k} + \cdots.$$

This can give excellent approximations to e. For example, if $k = 2$, then from

(30) we get

(31)
$$\frac{e-1}{2} = \cfrac{1}{1+\cfrac{1}{6}+\cfrac{1}{10}+\ \cdots}\ .$$

Using the 7th convergent to (31), we have $(e-1)/2 \approx 342762/398959$, which shows that $e \approx 2.718281828458 \cdots$, in error by only one unit in the last place.

From (31) we see that

(32)
$$e = 1 + \cfrac{2}{1+\cfrac{1}{6}+\cfrac{1}{10}+\cfrac{1}{14}+\ \cdots}\ ,$$

and it is quite easy to transform this into a simple continued fraction by the use of two transformations. Write

$$\cfrac{2}{a+\cfrac{1}{b+\cfrac{1}{c+\ \cdot}}} = \cfrac{2}{a+\cfrac{1}{b+\cfrac{1}{y}}}\ .$$

Then it follows easily that if a is even,

$$\cfrac{2}{a+\cfrac{1}{b+\cfrac{1}{y}}} = \cfrac{1}{\cfrac{a}{2}+\cfrac{1}{2b+\cfrac{2}{y}}}\ ,$$

and if a is odd, that

$$\cfrac{2}{a+\cfrac{1}{b+\cfrac{1}{y}}} = \cfrac{1}{\cfrac{(a-1)}{2}+\cfrac{1}{1+\cfrac{1}{1+\cfrac{2}{(b-1)+\cfrac{1}{y}}}}}\ .$$

Thus, the continued fraction (32) transforms easily into Euler's result:

(33)
$$e = 2 + \cfrac{1}{1}+\cfrac{1}{2}+\cfrac{1}{1}+\cfrac{1}{1}+\cfrac{1}{4}+\ \cdots = \langle 2, \overline{1, 2n, 1} \rangle_{n=1}^{\infty}$$

where the "bar" indicated the periodic part of the fraction.

A similar discussion would show that

$$(34) \qquad\qquad e^2 = \langle 7, \overline{3n - 1, 1, 1, 3n, 6(2n +1)} \rangle^{\infty}_{n=1} \; ;$$

and that for $k \geqq 3$ and odd

$$(35) \quad e^{2/k} = \overline{\langle 1, \tfrac{1}{2}[(6n + 1)k - 1], 6k(2n + 1), \tfrac{1}{2}[(6n + 5)k - 1], 1 \rangle}^{\infty}_{n=1} .$$

3. Conclusion. The continued fraction (29) was published by Lambert in 1761. For historical data on continued fractions see Perron [5]. For the technique for proving e transcendental, see Chapter 2 and 9 in Niven [3].

What gave Hermite the idea to start with the integral (5)? A partial answer can be found by thumbing through his previous papers to see his great skill in handling various difficult integrals involving transcendental functions. One soon senses that he must already have had in mind all the basic techniques needed to prove that e is transcendental.

References

1. L. Euler, Introductio analysin infinitorium, Lausanne 1 (1748), Chapter 18.
2. C. Hermite, Compt. Rend. Acad. Sci. Paris, 77 (1873) 18–24, 74–79, 285–293.
3. Ivan Niven, Irrational Numbers, Wiley, New York, 1956.
4. I. Niven and H. S. Zuckerman, An Introduction to the Theory of Numbers, Wiley, New York, 1960.
5. Oskar Perron, Die Lehre von den Kettenbruchen, Teubner, Leipzig, 1929.
6. H. S. Wall, Analytic Theory of Continued Fractions, Van Nostrand, Princeton, N. J., 1948.

22

PETER D. LAX

Peter D. Lax was born on May 1, 1926, in Hungary. He received his A.B. degree in 1947 and his Ph.D. in 1949 under Professor K. O. Friedrichs, both degrees at New York University. He joined the staff at New York University in 1949 as assistant professor, advancing rapidly through the ranks and becoming Director of the Courant Institute of Mathematical Sciences at New York University in 1972, after having served as a director of the A.E.C. Computing and Applied Mathematics Center at the Courant Institute of Mathematical Sciences.

He has been a frequent summer visitor at Stanford, a member of the staff of the Los Alamos Scientific Laboratories in 1950, a consultant to the Radiation Laboratory in California, and a Fulbright Lecturer in the summer of 1958.

In the summer of 1968, Professor Lax gave the SIAM von Neumann Lecture; in the spring of 1972, he gave the Weyl Lectures at the Institute for Advanced Study at Princeton, and in the summer of 1973 he was the Hedrick Lecturer of the Association at the Dartmouth meeting.

Professor Lax is a member of the National Academy of Sciences. He was Vice-President of the American Mathematical Society in 1969–71. He has served the Association in many capacities: as a member of the Joint Committee (with the AMS) on the Graduate Program in Mathematics; a member of the Panel of CUPM on Pre-graduate Training in 1962–64, also its Subcommittee on Applied Mathematics and a member of the CUPM Panel on Computing in 1968–69. He was an Associate Editor of the *Monthly* for the period 1969–73, a member of the Advisory Committee on Individual Lecture Films since 1964, and was appointed to the Committee on Earle Raymond Hedrick Lectures in 1973. The same year, he was elected a member of the Board of Governors.

He has twice received the Association's Lester R. Ford Award: in 1966 for his paper *Numerical Solution of Partial Differential Equations*, in the *Monthly*, 72 (1965) Part II (Slaught Paper No. 10) 74–84, and in 1973 for the same paper for which he won the Chauvenet Prize. He also received the Norbert Wiener Prize in 1975, awarded jointly by the American Mathematical Society and Society for Industrial and Applied Mathematics.

Professor Lax's interest in mathematics has centered on partial differential equations, linear and nonlinear functions of mathematical physics, computing, and functional analysis whenever needed. His many significant contributions to these areas of mathematics are contained in his nearly sixty publications, including the book published with R. Phillips *Scattering Theory*, Academic Press, 1967, and the monograph *Hyberbolic Systems of Conservation Laws and the Mathematical Theory of Shock Waves,* Conference Board of the Mathematical Sciences, Monograph No. 11, SIAM, 1973.

In his acceptance, Professor Lax expressed his gratitude for the honor of having been awarded the 1974 Chauvenet Prize.

He expressed his indebtedness to James Glimm for his fertile ideas on the decay of shock waves. He declared himself extremely fortunate in having teachers who were outstanding expositors: Miss R. Peter, author of the charming popular book, *Playing with Infinity*; Professors Polya and Szego of *Aufgaben und Lehrsätze* fame (now happily available in an English enlarged edition) and, most of all, to Richard Courant—a master expositor in both pure and applied mathematics.

THE FORMATION AND DECAY OF SHOCK WAVES

PETER D. LAX, Courant Institute, New York University

1. Introduction. The theory of propagation of shock waves is one of a small class of mathematical topics whose basic problems are easy to explain but hard to resolve. This article is a brief introduction to the subject: we shall describe the origin of the governing equations, some of the striking phenomena, and a few of the mathematical tools used to analyse them.

2. What is a conservation law? A conservation law asserts that the change in the total amount of a physical entity contained in any region G of space is due to the **flux** of that entity across the boundary of G. In particular, the rate of change is

$$(2.1) \qquad \frac{d}{dt} \int_G u\,dx = - \int_{\partial G} f \cdot n\,dS,$$

where u measures the **density** of the physical entity under discussion, and the vector f describes its flux; n is the outward normal to the boundary ∂G of G. If u and f are differentiable functions, we can, on the left, perform the differentiation under the integral sign and on the right apply the divergence theorem. We obtain

$$\int_G \{u_t + \operatorname{div} f\}\,dx = 0.$$

This relation is assumed to be valid for every domain G. Letting G shrink to a point and dividing by the volume of G we get the differential form of the conservation law:

$$(2.2) \qquad u_t + \operatorname{div} f = 0.$$

To complete the theory we need some law relating f to u. E.g., Newton's law of cooling asserts that the flux of heat is proportional to the negative gradient of u, where u is temperature; in this case $f = -h \operatorname{grad} u$, h positive, so (2.2) becomes

$$u_t - h\Delta u = 0, \qquad \Delta = \operatorname{div} \operatorname{grad}.$$

In this example f depends on the derivatives of u; in what follows *we assume that f depends on u alone*. More precisely, we shall be looking at systems of conservation laws

$$(2.3) \qquad u_t^j + \operatorname{div} f^j = 0, \qquad j = 1, \cdots, n,$$

where each f^j is a function of all the $u^1, \cdots, u^n$, and a nonlinear function at that.

Peter Lax received his Ph.D. at New York University under K. Friedrichs and has spent most of his academic career at New York University, where he is presently a professor. He is a frequent summer visitor at Stanford and the Los Alamos Scientific Lab. His research contributions in partial differential equations, linear and non-linear problems of mathematical physics, computing, and functional analysis have had a profound impact. He was a Fullbright lecturer in 1958, he is a Vice-President of the AMS, he is an elected member of the National Academy of Sciences, he was an AMS Gibbs lecturer, and he received an MAA Lester Ford Award. He is co-author with R. Phillips of *Scattering Theory* (Academic Press, 1967). *Editor.*

Many equations of mathematical physics are of this form, in particular, those governing the flow of a nonviscous, compressible fluid.

We shall concern ourselves with the **initial value problem** for systems of form (2.3); that is, given the value of each u^j at $t = 0$ as function of x, determine u^j as function of x and t for all $t > 0$.

3. The theory of a single nonlinear conservation law. In this section we shall study conservation laws for a single quantity u dependent on only one space variable x; in this case f has only one component:

(3.1) $$u_t + f_x = 0,$$

where f is some nonlinear function of u. Denoting

(3.2) $$\frac{df}{du} = a(u)$$

we can write (3.1) in the form

(3.3) $$u_t + a(u)u_x = 0$$

which asserts that u is constant along trajectories $x = x(t)$ which propagate with speed a:

(3.4) $$\frac{dx}{dt} = a.$$

For this reason a is called the **signal speed**; the trajectories, satisfying (3.4), are called **characteristics**. Note that if f is a nonlinear function of u, both signal speed and characteristics depend on the solution u.

The constancy of u along characteristics combined with (3.4) shows that the characteristics propagate with constant speed; so they are straight lines. This leads to the following geometric solution of the initial value problem

$$u(x, 0) = u_0(x).$$

Draw straight lines issuing from points y of the x-axis, with slope $1/u_0(y)$ (see Fig. 1).

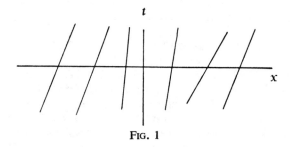

FIG. 1

As we shall show, if u_0 is a C^1 function, these lines simply cover a neighborhood of the x-axis; since the value of u along the line issuing from the point y is $u_0(y)$, $u(x,t)$ is uniquely determined near the x-axis.

An analytical form of this construction goes like this (see Fig. 2)

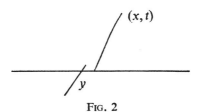

FIG. 2

Let (x, t) be any point, y the intersection of the characteristic through x, t with the x-axis. Then $u = u(x,t)$ satisfies

(3.5)
$$u = u_0(y), \quad y = x - t\,a(u).$$

Assume u_0 differentiable; then, according to the implicit function theorem, (3.5) can be solved for u as a differentiable function of x and t for t small enough, and

(3.6)
$$u_t = -\frac{u_0'a}{1 + u_0'a_u t} \qquad u_x = \frac{u_0'}{1 + u_0'a_u t}.$$

Substituting (3.6) into (3.3) we see immediately that u defined by (3.5) satisfies (3.3).

Let's assume that equation (3.3) is **genuinely nonlinear**, i.e., that $a_u \neq 0$ for all u, say

(3.7)
$$a_u > 0$$

Then if u_0' is ≥ 0 for all x, u_t and u_x as given by formulas (3.6) remain bounded for all $t > 0$; on the other hand, if u_0' is < 0 at some point, both u_t and u_x tend to ∞ as $1 + u_0' a_u(u_0)t$ approaches zero. Both these facts can be deduced from the geometric form of the solution contained in Figure 1·

In the first case, when $u_0(x)$ is an increasing function of x, the characteristics issuing from the x-axis diverge in the positive t direction, so that the characteristics simply cover the whole half-plane $t > 0$. In the second case there are two points y_1 and y_2 such that $y_1 < y_2$, and $u_1 = u_0(y_1) > u_0(y_2) = u_2$; then by (3.7) also $a_1 = a(u_1) > a(u_2) = a_2$ so that the characteristics issuing from these points intersect at time

$$t = \frac{y_2 - y_1}{a_1 - a_2}.$$

At the point of intersection, u has to take on the value u_1 and u_2 both, an impossibility (see Fig. 3).

Both the geometric and the analytic argument prove beyond the shadow of a

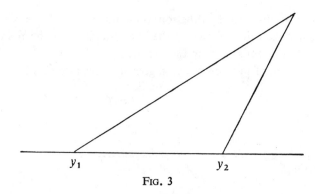

Fig. 3

doubt that if the initial value u_0 is not an increasing function of x then *no continuous function $u(x, t)$ exists for all $t > 0$ with initial value u_0 which solves equation (3.3) in the ordinary sense!*

What happens after continuous solutions cease to exist? After all, the world does not come to an end. For an answer, we turn to experiments with compressible fluids: these clearly show the appearance of discontinuities in solutions. We begin our study of discontinuous solutions with the simplest kind, those satisfying (3.1) in the ordinary sense on each side of a smooth curve $x = y(t)$ across which u is discontinuous. We shall denote by u_l and u_r the values of u on the left and right sides respectively of $x = y(t)$. Choose a and b so that the curve y intersects the interval $a \leqq x \leqq b$ at time t (see Fig. 4).

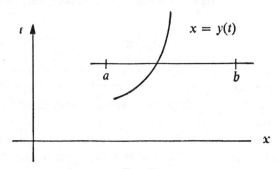

Fig. 4

Denoting by $I(t)$ the quantity $I(t) = \int_a^b u(x, t)\, dx = \int_a^y + \int_y^b$, we have

$$(3.8) \qquad \frac{dI}{dt} = \int_a^y u_t dx + u_l s + \int_y^b u\, dx - u_r s,$$

where we have used the abbreviation

$$(3.9) \qquad\qquad s = \frac{dy}{dt}$$

for the speed with which the discontinuity propagates. Since on either side of the discontinuity (3.1) is satisfied we may set $u_t = -f_x$ in the integrals in (3.8); after carrying out the integration we obtain $dI/dt = f_a - f_l + u_l s - f_b + f_r - u_r s$; here we have used the handy abbreviations

$$f(u_l) = f_l, \qquad f(u_r) = f_r,$$
$$f(u(a)) = f_a, \qquad f(u(b)) = f_b.$$

The conservation law asserts that $dI/dt = f_a - f_b$. Combining this with the above relation we deduce the **jump condition**

(3.10) $$s[u] = [f],$$

where $[u] = u_r - u_l$ and $[f] = f_r - f_l$ denote the jump in u and in f across y.

We show now in an example that previously unsolvable initial value problems can be solved for all t with the aid of discontinuous solutions. Take

(3.11) $$f(u) = \tfrac{1}{2}u^2,$$

$$u_0(x) = \begin{cases} 1 & \text{for } x \leq 0 \\ 1 - x & \text{for } 0 \leq x \leq 1 \\ 0 & \text{for } 1 \leq x. \end{cases}$$

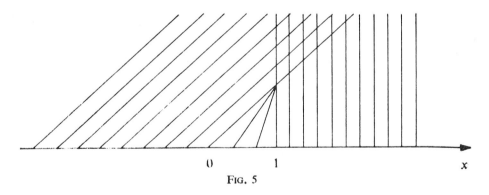

FIG. 5

The geometric solution is single valued for $t \leq 1$ but double valued thereafter (see Fig. 5). Now we define for $t \geq 1$

$$u(x,t) = \begin{cases} 1 & \text{for } x < (1+t)/2 \\ 0 & \text{for } (1+t)/2 < x. \end{cases}$$

The discontinuity starts at $(1,1)$; it separates the state $u_l = 1$ on the left from the state $u_r = 0$ on the right; the speed of propagation was chosen according to the jump condition (3.10), with $f(u) = \tfrac{1}{2}u^2$:

$$s = \frac{0 - \frac{1}{2}}{0 - 1} = \frac{1}{2}.$$

Introducing generalized solutions makes it possible to solve intial value problems which could not be solved within the class of genuine solutions. At the same time there is the danger that the enlarged class of solutions is so large that there are several generalized solutions with the same initial data. The following example shows that this anxiety is well founded:

$$u_0(x) = \begin{cases} 0 \text{ for } x < 0 \\ 1 \text{ for } 0 < x. \end{cases}$$

The geometric solution

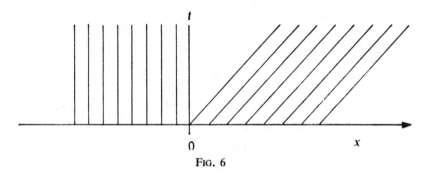

FIG. 6

is single valued for $t > 0$ (see Fig. 6) but does not determine the value of u in the wedge $0 < x < t$. We could fill this gap in the fashion of the previous example and set

(3.12)
$$u(x, t) = \begin{cases} 0 \text{ for } x < t/2 \\ 1 \text{ for } t/2 < x. \end{cases}$$

The speed of propagation was so chosen that the jump condition (3.10) is satisfied. On the other hand the function

(3.12)' $u(x, t) = x/t, \qquad 0 \leqq x \leqq t$

satisfies the differential equation (3.3) with $a(u) = u$, and joins continuously the rest of the solution determined geometrically. Clearly only one of these solutions can have physical meaning; the question is which?

We reject the discontinuous solution (3.12) for failure to satisfy the following criterion:

The characteristics starting on either side of the discontinuity curve when continued in the direction of positive t intersect the line of discontinuity. This will be the case if

(3.13) $a(u_l) > s > a(u_r).$

Under condition (3.7) for a this means that

(3.14) $u_l > u_r.$

Clearly this condition is violated in the solution given by (3.12).

The analysis at the beginning of this section shows that signals propagate along characteristics. Condition (3.13) allows each point of the discontinuity to be reached by characteristics on both sides, so that the shock is influenced by the initial data of the solution; this constitutes one justification of Condition (3.13). Another justification can be based on characterising the physically meaningful solutions as limits, when u tends to zero, of the viscous equation

$$u_t + f(u)_x = \mu u_{xx}, \qquad \mu > 0.$$

Yet another justification can be based on the theory of entropy. We shall not go into this interesting matter any deeper here, but merely record the gratifying fact that when $a(u)$ is a monotonic function of u, condition (3.13) is restrictive enough to make the solution of the initial value problem unique, yet it is broad enough to allow the construction of a solution for all time $t > 0$, having as initial value any integrable function u_0. True, the concept of solution has to be generalized beyond simple discontinuities: a bounded measurable function $u(x,t)$ is said to satisfy the conservation law (3.1) in the sense of distributions, if for all continuously differentiable test functions $\phi(x,t)$, with support in $t > 0$,

(3.15) $$\int \int [\phi_t u + \phi_x f(u)]dxdt = 0.$$

It is easy to verify that for the previously considered class of piecewise continuous solutions condition (3.15) is equivalent with the jump condition (3.10).

For merely bounded, measurable solutions u_l and u_r in condition (3.13) have to be interpreted as follows:

$$u_l = \lim_{y \to x, \ y < x} \inf u(y,t),$$

$$u_r = \lim_{y \to x, \ x < y} \sup u(y,t).$$

For the main existence theorem we refer the reader to [8] and [13], and for uniqueness to [1], [14], and [16].

It turns out that when $a(u)$ is not monotonic, condition (3.13) is not sufficient to guarantee unique determination of solutions by their initial data. A replacement for this condition has been found by Oleinik; this condition, together with the existence and uniqueness theorem is described in [15]; other interesting discussions of this condition are contained in [4], [6], and [16].

4. The decay of solutions. Existence and uniqueness of solutions is not the end but merely the beginning of a theory of differential equations. The really interesting questions concern the behavior of solutions.

Here we shall study the asymptotic behavior for large time of solutions of conservation laws of form (3.1) which satisfy condition (3.14); we assume that $a(u)$ is an *increasing* function of u.

As remarked in Section 3, any differentiable solution u is constant along characteristics

$$(4.1) \qquad \frac{dx}{dt} = a(u) = f'(u).$$

Let $x_1(t)$ and $x_2(t)$ be a pair of characteristics, $0 \leq t \leq T$. Then there is a whole one-parameter family of characteristics connecting the points of the interval $[x_1(0), x_2(0)]$, $t = 0$ with points of the interval $[x_1(T), x_2(T)]$, $t = T$; since u is constant along these characteristics, $u(x, 0)$ *on the first interval and* $u(x, T)$ *on the second interval are equivariant*, i.e., they take on the same values in the same order. Since equivariant functions have the same total increasing and decreasing variations, we conclude that *the total increasing and decreasing variations of a differentiable solution between any pair of characteristics are conserved.*

Denote by $D(t)$ the width of the strip bounded by x_1 and x_2:

$$(4.2) \qquad D(t) = x_2(t) - x_1(t) > 0.$$

Differentiating (4.2) with respect to t and using (4.1), we get

$$(4.3) \qquad \frac{d}{dt} D(t) = \frac{dx_2}{dt} - \frac{dx_1}{dt} = a(u_2) - a(u_1).$$

Integrating with respect to t we get

$$(4.4) \qquad D(T) = D(0) + [a(u_2) - a(u_1)]T.$$

Suppose there is a shock y present in u between the characteristics x_1 and x_2 (see Fig. 7). Since according to condition (3.13) characteristics on either side of a shock run into the shock, there exist for any given time T two characteristics y_1 and y_2 which intersect the shock y at exactly time T. Assuming that there are no other shocks present we conclude that the increasing variation of u on $(x_1(t), y_1(t))$, as well as on $(x_2(t), y_2(t))$, is independent of t. According to condition (3.14), u decreases across shocks, so the increasing variation of u along $[z_1(T), x_2(T)]$ equals the sum of the increasing variations of u along $[x_1(0), y_1(0)]$ and along $[y_2(0), x_2(0)]$. This sum is in general less than the increasing variation of u along $[x_1(0), x_2(0)]$, therefore we conclude that if shocks are present, *the total increasing variation of u between two characteristics decreased with time.*

We give now a quantitative estimate of this decrease. Let I_0 be any interval of the x-axis; we subdivide it into subintervals $[y_{j-1}, y_j]$, $j = 1, \cdots, n$ in such a way

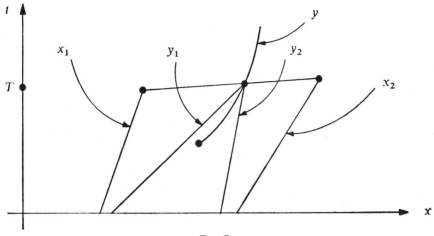

FIG. 7

that $u(x, 0)$ is alternately increasing and decreasing on the intervals (we here assumed for simplicity that u_0 is piecewise monotonic). We denote by $y_j(t)$ the characteristic issuing from the jth point y_j, with the understanding that if $y_j(t)$ runs into a shock, $y_j(t)$ is continued as that shock.

It is easy to show that for any $t > 0$, $u(x, t)$ is alternately increasing and decreasing on the intervals $(y_{j-1}(t), y_j(t))$. Since a is an increasing function of u, and since according to (3.14) u decreases across shocks, the total increasing variation $A^+(T)$ of $a(u)$ across the interval $I(T) = [y_0(T), y_n(T)]$ is

$$(4.5) \qquad \sum_{j \text{ odd}} a(u_j(T)) - a(u_{j-1}(T)) = A^+(T),$$

where $u_{j-1}(T)$ denotes the value of u on the right edge of $y_{j-1}(T)$, $u_j(T)$ denotes the value of u on the left edge of $y_j(T)$; in case $y_{j-1}(T)$ and $y_j(T)$ are the same, the jth term in (4.5) is zero. Suppose $y_{j-1}(T)$ and $y_j(T)$ are shocks; then there exist characteristics $x_{j-1}(t)$ and $x_j(t)$ which start at $t = 0$ inside (y_{j-1}, y_j) and which at $t = T$ run into $y_{j-1}(T)$ and $y_j(T)$ respectively. The value of u along $x_j(t)$ is $u_j(T)$.

Denote $x_j(t) - x_{j-1}(t)$ by $D_j(t)$; according to (4.4)

$$D_j(T) = D_j(0) + [a(u_j) - a(u_{j-1})]T.$$

Summing over j odd and using (4.5) we get

$$(4.6) \qquad \sum D_j(T) = \sum D_j(0) + A^+(T)T.$$

Since the intervals $[x_{j-1}(T), x_j(T)] = [y_{j-1}(T), y_j(T)]$ are disjoint and lie in $I(T)$, their total length cannot exceed the length $L(t)$ of $I(T)$; so we deduce from (4.6) that

$$(4.7) \qquad A^+(T) \leqq \frac{L(T)}{T},$$

where $A^+(T)$ is the total increasing variation of $a(u)$ along $I(T)$.

Let $u(x,t)$ be a solution of (3.1), possibly discontinuous, whose initial values are bounded, and zero outside a finite interval I_0. Since signals propagate with finite speed, for every t the solution $u(x,t)$ is zero outside some finite x-interval $I(t)$. Denote by $v(t)$ and $w(t)$ the values of u at the left and right endpoints of $I(t)$ respectively. Since the endpoints may lie on shocks, these values need not be zero, however it follows from (3.14) that

$$(4.8) \qquad v(t) \leqq 0, \qquad 0 \leqq w(t).$$

Denote by s_{left} and s_{right} the speed with which the shocks at the endpoints propagate; according to the jump relation (3.10).

$$(4.9) \qquad s_{\text{left}} = \frac{f(v) - f(0)}{v}, \quad s_{\text{right}} = \frac{f(w) - f(0)}{w}.$$

Since a is an increasing function of u, $f(u)$ is convex. It follows from the mean value theorem that the difference quotient of f over an interval is not less than f' at the left endpoint, and not greater than f' at the right endpoint of that interval. So it follows from (4.8) that

$$(4.10) \qquad a(v) \leqq \frac{f(v) - f(0)}{v}, \quad \frac{f(w) - f(0)}{w} \leqq a(w).$$

At this point we assume that a is strictly increasing, i.e., that for some positive number k

$$(4.11) \qquad 0 < k \leqq a';$$

here we abbreviate d/du by prime. It follows that inequalities (4.10) are strict; combining these with (4.9) we can put them into this form

$$(4.12) \qquad s_{\text{right}} - s_{\text{left}} \leqq \theta[a(w) - a(v)],$$

where θ is < 1.

Denote the length of $I(t)$ by $L(t)$; since s_{left} and s_{right} are the speeds with which the endpoints of I move,

$$(4.13) \qquad \frac{dL}{dt} = s_{\text{right}} - s_{\text{left}}.$$

Substituting the inequalities (4.12) into (4.13) we get

$$\frac{dL}{dt} \leqq \theta[a(w) - a(v)].$$

Since by (4.8) $v < w$, $a(w) - a(v)$ is bounded by the total increasing variation $A^+(t)$ of $a(u)$ over $I(t)$:

$$(4.14) \qquad a(w) - a(v) \leqq A^+(t).$$

Combining the last two inequalities we get

$$\frac{dL}{dt} \leq \theta A^+(t).$$

Using inequality (4.7) we get

$$\frac{dL}{dt} \leq \frac{\theta}{t} L(t);$$

and multiplying by $t^{-\theta}$ we deduce that

$$\frac{d}{dt}(t^{-\theta}L) \leq 0.$$

Thus $t^{-\theta}L(t)$ is a decreasing function of time; in particular

(4.15) $$L(t) \leq t^{\theta}L(1) \qquad \text{for } t > 1.$$

Substituting this into the right side of (4.7) we get

$$A^+(t) \leq t^{\theta-1}L(1).$$

Since $\theta < 1$, this shows that $A^+(t) \to 0$ as $t \to \infty$.

It follows from the strictly increasing character (4.11) of $a(u)$ that the total increasing variation of u along $I(t)$ is bounded by $A^+(t)/k$. Since u is ≤ 0 at the left endpoint of $I(t)$ and ≥ 0 at the right endpoint, it follows that likewise *the maximum $m(t)$ of $u(x,t)$ over $I(t)$ is bounded by $A^+(t)/k$;*

(4.16) $$m(t) \leq A^+(t)/k.$$

Combining this with the above estimate for A^+ we get that $m(t) \leq \text{const } t^{\theta-1}$ which shows that *the maximum of u at time t tends to zero like $t^{\theta-1}$.*

This result is somewhat crude; a more detailed analysis will furnish a more precise result. (A different derivation was given by Barbara Quinn in her dissertation at New York University, 1970.) We start by expressing $f(r)$, $f(w)$ in (4.9) by their Taylor expansions; we get

$$s_{\text{left}} = f'(0) + \frac{1}{2}f''(0)v + \frac{1}{6}f'''(\bar{v})v^2,$$

(4.17)

$$s_{\text{right}} = f'(0) + \frac{1}{2}f''(0)w + \frac{1}{6}f'''(\bar{w})w^2,$$

where $v < \bar{v} < 0$, $0 < \bar{w} < w$.

Denote by K an upper bound for f'''; since m is an upper bound for $|v|$ and w, it follows that

$$s_{\text{left}} \geq f'(0) + \frac{1}{2}\left[f''(0) + \frac{K}{3}m\right]v$$

$$s_{\text{right}} \leq f'(0) + \frac{1}{2}\left[f''(0) + \frac{K}{3}m\right]w.$$

Substituting this into (4.13) we get

(4.18)　　　　　$$\frac{dL}{dt} \leq \frac{1}{2}\left[f''(0) + \frac{K}{3}m\right](w - v).$$

It follows from (4.11) and (4.14) that

(4.19)　　　　　$$w - v \leq \frac{a(w) - a(v)}{k} \leq \frac{A^+(t)}{k}.$$

The constant k in (4.11) has to be a lower bound of $a' = f''(u)$ for $|u| \leq m$; in particular we can take

(4.20)　　　　　$$k = f''(0) - Km.$$

Substituting this into (4.19) and then into (4.18) we get that for m small enough

(4.21)　　　　　$$\frac{dL}{dt} \leq \frac{1}{2}\left[\frac{f''(0) + K/3\,m}{f''(0) - Km}\right]A^+ \leq \left(\frac{1}{2} + Hm\right)A^+.$$

We substitute into (4.21) estimate (4.16) for m, and then estimate (4.7) for A^+; we obtain the following inequality:

(4.22)　　　　　$$\frac{dL}{dt} \leq \left(\frac{1}{2} + \frac{H}{k}\frac{L}{t}\right)\frac{L}{t}.$$

Introduce a new variable J by $L = J\sqrt{t}$; (4.22) becomes

$$\sqrt{t}\,\frac{dJ}{dt} \leq \frac{H}{k}\frac{J^2}{t}.$$

Dividing by $\sqrt{t}\,J^2$ we get, after integrating from T to $t > T$, that

$$\frac{1}{J(T)} - \frac{1}{J(t)} \leq \frac{H}{2k}\left(\frac{1}{\sqrt{T}} - \frac{1}{\sqrt{t}}\right),$$

which implies that

(4.23)　　　　　$$\frac{1}{J(T)} - \frac{H}{2k\sqrt{T}} \leq \frac{1}{J(t)}.$$

According to (4.15), $L(T)/T = J(T)/\sqrt{T}$ tends to 0 as $T \to \infty$; this implies that for T large enough, the left side of (4.23) is positive. Then (4.23) furnishes an upper bound for $J(t)$ for all $t > T$. The boundedness of $J(t)$ implies that $L(t)$ is $O(\sqrt{t})$ as $t \to \infty$. Combining this with the estimates (4.7) and (4.16) we reach the following conclusion.

THEOREM 4.1.   *Let u be a possibly discontinuous solution of the conservation law $u_t + f_x = 0$, where f is three times differentiable and strictly convex. Suppose that all discontinuities of u satisfy (3.13), and that $u(x,0)$ has compact support. Then*
   (a) *the length of the support of $u(x,t)$ is $O(\sqrt{t})$,*
   (b) $\text{Max}_x |u(x,t)| = O(1/\sqrt{t})$.

It turns out that this result is rather precise: Using an explicit formula one can show, see [9], that the length of the support of u divided by $\sqrt{t}$ tends to a limit, and so does $\sqrt{t}\,\text{Max}\,|u|$.

We turn now to solutions which are **periodic** in x:

$$u(x + p, t) = u(x, t).$$

We take $I(T)$ to be any interval of length p at time T. According to our basic estimate (4.7), the increasing variation of $a(u)$ per period is $\leq p/T$. It follows then from (4.11) that the increasing variation per period of u itself does not exceed p/kT. Since u is periodic, its decreasing and increasing variations are equal, and serves as bound for the oscillation of u, in particular for the deviation of u from its mean value per period.

For a periodic solution $u(x,t)$, the flux f at $(0,t)$ equals the flux at (p,t); thus the total flux into an interval of length p is zero, and so the mean value of u,

$$\bar{u} = \frac{1}{p} \int_0^p u(x,t)dx,$$

is independent of t. We summarize our results as follows:

THEOREM 4.2.   *Let $u(x,t)$ be a possibly discontinuous solution of $u_t + f_x = 0$, f strictly convex, $f'' > k > 0$. Suppose that all discontinuities of u satisfy (3.13) and that u is periodic in x with period p. Then*
   (a) *The total variation of u at time t does not exceed $2p/kt$,*
   (b)

(4.24) $$|u(x,t) - \bar{u}| \leq 1/kt,$$

where $\bar{u}$ is the mean value of u.

Again it can be shown that (4.24) is sharp, i.e., that

(4.25) $$\lim_{t\to\infty} t \max_x |u(x,t) - \bar{u}| = k = f''(\bar{u}).$$

The surprising, almost paradoxical feature of inequality (4.24) is that it holds uniformly for all solutions with period p; it is independent of the amplitude of the initial disturbance. All that the initial amplitude can influence is the time when the asymptotic estimate (4.24) becomes accurate: The *larger* the initial amplitude, the *sooner* (4.25) converges. This is in sharp contrast to the linear case where the asymptotic amplitude of a signal for large time is proportional to its initial amplitude, but the time it takes to reach the asymptotic shape is independent of the initial amplitude.

Let $u_1(x)$ be an initial function which is zero outside the interval $[0, p]$, and define $u_2(x)$ to be equal $u_1(x)$ in $[0, p]$, and periodic (see Fig. 8).

According to Theorem 4.1, $u_1(x, t)$ decays like $1/\sqrt{t}$; $u_2(x, t)$ on the other hand is periodic,[1] so its asymptotic behavior is governed by Theorem 4.2: $u_2(x, t)$ decays like $1/t$. So we have the paradoxical result that u_2, which represents a much larger initial disturbance than u_1, nevertheless decays faster than u_1.

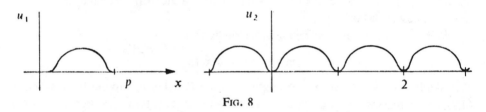

FIG. 8

5. Systems of conservation laws.

Models which are at all realistic are governed by a whole system of conservation laws, rather than by a single one. The value of what we have learned about single equations lies in the light this knowledge sheds on systems. It turns out that the main phenomena we have found: the breakdown of continuous solutions, the necessity of imposing an entropy-like condition to distinguish those discontinuous solutions which are physically realizable from those which are not, and the decay of solutions as $t \to \infty$, have their counterparts for systems. That is not to say that the theory is as far advanced for systems as it is for single equations; on the contrary, what we have is a sea of conjectures, confined partly by the shores of numerical computations, with a few islands of solidly proved mathematical facts.

What are the proven facts about systems? In [10] the author has shown that solutions of 2×2 systems of conservation laws break down after a finite time, unless the initial data satisfy a monotonicity condition. In [9], an analogue of the entropy condition (3.13) is described, and a condition for genuine nonlinearity is given. In [15], Oleinik gives a uniqueness theorem for solutions of systems of two conservation laws of which one is linear. In [2], Glimm solves the initial value problem for systems, for initial data with small oscillation. In [5], Johnson and Smoller solve the initial value problem for initial data which satisfy a certain monotonicity condition, for 2×2 systems which satisfy a certain convexity-like condition. The only existence

---

[1] Solutions whose initial values are periodic are periodic for all t; this follows from the uniqueness theorem that solutions which are equal at $t = 0$ are equal for all $t > 0$.

theorem with no restrictions on the initial data is due to Nishida, [12], and works only for the system

$$u_t + v_x = 0, \qquad v_t - \left(\frac{1}{u}\right)_x = 0.$$

In [3], Glimm and the author prove the decay of solutions with small oscillation of 2×2 systems. The method described in Section 4 is taken from that paper.

For those who wish to work in this field I recommend Glimm's paper [2]. It contains a wealth of ideas, such as the use of an approximation scheme containing a sequence of random parameters; the scheme is shown to converge for almost all values of the parameters. Glimm also introduces novel, nonlocally defined functionals; the estimate of the growth and decay of these functionals plays a crucial role in the existence theorem.

This article is an expanded version of an invited address delivered at the January 1970 meeting of the MAA at San Antonio, Texas. Other versions of this talk were given at Oregon State University, Corvallis; Texas Tech. University, Lubbock, and at Brown University. The talk is partly based on the joint paper [3] with James Glimm.

References

1. A. Douglis, An ordering principle, Comm. Pure Appl. Math., 12 (1959) 87.

2. J. Glimm, Solutions in the large for nonlinear hyperbolic systems of equations, Comm. Pure Appl. Math., 18 (1965) 697–715.

3. J. Glimm, and P. D. Lax, Decay of solutions of systems of nonlinear hyperbolic conservation laws, Mem. Amer. Math. Soc., No. 101 (1970).

4. E. Hopf, The partial differential equation $u_t + uu_x = \mu u_{xx}$, Comm. Pure Appl. Math., 3 (1950) 201–230.

5. J. L. Johnson, and J. Smoller, Global solutions for an extended class of hyperbolic systems of conservation laws, Arch. Rational Mech. Anal., 32 (1969) 169–189.

6. S. Iv. Krushkov, Results on the character of continuity of solutions of parabolic equations and some of their applications, Mat. Zametki, 6 (1969) 97–108.

7. ———, First order quasi-linear equations in several independent variables, Math. USSR Sbornik, 10 (1970) No. 2.

8. P. D. Lax, Weak solutions of nonlinear hyperbolic equations and their numerical computation, Comm. Pure Appl. Math., 7 (1954) 159–193.

9. ———, Hyperbolic systems of conservation laws, II, Comm. Pure Appl. Math., 10 (1957) 537–566.

10. ———, Development of singularities of solutions of nonlinear hyperbolic partial differential equations, J. Mathematical Phys., 5 (1964) 611–613.

11. ———, On a notion of entropy, Proc. of Symposium at the University of Wisconsin, 1971, ed. E. Zarantonello.

12. T. Nishida, Global solutions for an initial value problem of a quasilinear hyperbolic system, Proc. Japan Acad., 44 (1968) 642–646.

13. O. A. Oleinik, Discontinuous solutions of nonlinear differential equations, Uspehi Mat. Nauk, (1957) 3–73, English Translation in Amer. Math Soc. Trans., Ser. 2, No. 26, pp. 95–172.

14. ———, On the uniqueness of the generalized solution of the Cauchy problem for a nonlinear system of equations occurring in mechanics, Uspehi Mat. Nauk, 78 (1957) 169–176.

15. ———, Uspehi Mat. Nauk (N. S.), 14 (1959) 165–170.

16. B. Quinn, Solutions with shocks, an example of an L_1-contractive semigroup, Comm. Pure Appl. Math., 24 (1971).

23

MARTIN D. DAVIS AND REUBEN HERSH

Professor Davis was born on March 8, 1928, in New York City. He received his B.S. degree at the City College of New York in 1948, his M.A. in 1949, and Ph.D. in 1950, both at Princeton University. He began his academic career in 1950 as a Research Instructor in Mathematics at the University of Illinois; in 1952 he joined the School of Mathematics at the Institute for Advanced Study in Princeton. From 1954 on, he advanced rapidly through the ranks at various institutions beginning with the University of California at Davis and ending with his current position, held as of 1965, as Professor of Mathematics at the Courant Institute of Mathematical Sciences.

Professor Davis' research interests are in mathematical logic, and, in particular, recursive functions, Diophantine decision problems and mechanical theorem-proving. His many contributions to these areas of mathematics, which have received a great deal of attention, are contained in his nearly thirty publications including several books and the paper *Hilbert's Tenth Problem is Unsolvable*, which appeared in the *Monthly*, 80 (1973) 233–269, for which he was awarded a 1974 Lester R. Ford Award by the Association and a LeRoy P. Steele Prize by the American Mathematical Society.

Professor Hersh was born in 1927 in New York City. He received his B.A. degree in English Literature at Harvard University in 1946, and with the aid of the G.I. Bill of Rights, earned a certificate in machine-shop practice at Machine and Metal Trades High School in Manhattan in 1952. He received his Ph.D. in mathematics at New York University under Professor P. D. Lax in 1962. He was an assistant professor at Fairleigh Dickinson University in 1962, an instructor at Stanford University from 1962 to 1964, and has been at the University of New Mexico since that time, advancing to full professor in 1970.

He was a participant in the Battelle Mathematics Physics Rencontres in Seattle in 1968, a visiting member of the Courant Institute of Mathematical Sciences in 1970–71, and directed the Rocky Mountain Mathematics Consortium's 1972 Symposium in Edmonton on "Stochastic Differential Equations."

Professor Hersh's mathematical research has centered on partial differential equations, or "random evolutions." The article for which he was awarded the Chauvenet Prize is the fifth he has written for the *Scientific American*, with various co-authors, on recent discoveries in pure mathematics.

In their acceptances, Professor Davis and Professor Hersh expressed deep gratitude for the honor of having been awarded the 1975 Chauvenet Prize.

HILBERT'S 10TH PROBLEM

MARTIN DAVIS, Courant Institute of Mathematical Sciences, New York University, AND
REUBEN HERSH, University of New Mexico

Can a procedure be devised that will indicate if there are solutions to a Diophantine equation (an equation where whole-number solutions are sought)? This question on a famous list has now been answered.

"We hear within us the perpetual call: there is the problem. Seek its solution. You can find it by pure reason, for in mathematics there is no *ignorabimus* [We shall not know]." So did David Hilbert address the Second International Congress of Mathematicians in Paris on August 8, 1900, greeting the new century by presenting a list of 23 major problems to challenge future mathematicians. Some of Hilbert's problems are still unsolved. Others have inspired generations of mathematical investigators and have led to major new mathematical theories. The most recently conquered of Hilbert's problems is the 10th, which was solved in 1970 by the 22-year-old Russian mathematician Yuri Matyasevich.

David Hilbert was born in Königsberg in 1862 and was professor at the University of Göttingen from 1895 until his death in 1943. After the death of Henri Poincaré in 1912 he was generally regarded as being the foremost mathematician of his time. He made fundamental contributions in several fields, but he is perhaps best remembered for his development of the abstract method as a powerful tool in mathematics.

Hilbert's 10th problem is easily described. It has to do with the simplest and most basic mathematical activity: solving equations. The equations to be solved are polynomial equations, that is, equations such as $x^2 - 3xy = 5$, which are formed by adding and multiplying constants and variables and by using whole-number exponents. Moreover, Hilbert specified that the equations must use only integers, that is, positive or negative whole numbers. No irrational or imaginary numbers or even fractions are allowed in either the equations or their solutions. Problems of this type are called Diophantine equations after Diophantus of Alexandria, who wrote a book on the subject in the third century.

Hilbert's 10th problem is: Give a mechanical procedure by which any Diophantine equation can be tested to see if solutions exist. In Hilbert's words: "Given a Diophantine equation with any number of unknown quantities and with rational integral numerical coefficients: to devise a process according to which it can be determined by a finite number of operations whether the equation is solvable in rational integers." Hilbert does not ask for a process to find the solutions but merely for a process to determine if the equation has solutions. The process should be a clear-cut formal procedure that could be programmed for a computing machine and that would be guaranteed to work in all cases. Such a process is known as an algorithm.

If Hilbert's problem is simply stated, Matyasevich' solution is even more simply stated: No such process can ever be devised; such an algorithm does not exist.

555

Worded in this way, the answer sounds disappointingly negative. Matyasevich' result, however, constitutes an important and useful addition to the understanding of properties of numbers.

Matyasevich' work extended a series of researches by three Americans: one of us (Davis), Julia Robinson and Hilary Putnam. Their work in turn was based on earlier investigations by several founders of modern logic and computability theory: Alan Turing, Emil Post, Alonzo Church, Stephen Kleene and the same Kurt Gödel who is famous for his work on the consistency of axiomatic systems (Hilbert's second problem) and on the continuum hypothesis of Cantor (Hilbert's first problem).

Let us start on Hilbert's 10th problem by looking at a few Diophantine equations. The term "Diophantine equation" is slightly misleading, because it is not so much the nature of the equation that is crucial as the nature of the admissible solutions. For example, the equation $x^2+y^2-2=0$ has infinitely many solutions if one does not think of it as a Diophantine equation. The solutions are represented by the graph of the equation, which is a circle in the plane formed by the x axis and the y axis. The center of the circle is at the coordinates $x=0$, $y=0$. That point is called the origin; it is abbreviated $(0,0)$. The radius of the circle is $\sqrt{2}$ [see Fig. 1]. The coordinates of any point on the circle satisfy the equation, and there are an infinite number of such points. If we consider the problem as a Diophantine equation, however, there are only four solutions: (1) $x=1$, $y=1$; (2) $x=-1$, $y=1$; (3) $x=1$, $y=-1$, and (4) $x=-1$, $y=-1$.

Suppose the equation is changed to $x^2+y^2-3=0$. There are still an infinite number of solutions if it is treated as an ordinary equation but no solutions at all if it is treated as a Diophantine equation. The reason is that now the graph is a circle with radius equal to $\sqrt{3}$, and no points on this curve have both coordinates simultaneously equal to whole numbers.

A famous family of Diophantine equations has the form $x^n+y^n=z^n$, where n may equal $2,3,4$ or any larger integer. If n is equal to 2, the equation is satisfied by the lengths of the sides of any right triangle and is called the Pythagorean theorem. One such solution is the set of numbers $x=3$, $y=4$, $z=5$. If n is equal to or greater than 3, the equation is what is known as Fermat's equation. The 17th-century French mathematician Pierre de Fermat thought he had proved that these equations have no positive whole-number solutions. In the margin of his copy of Diophantus' book he wrote that he had found a "marvelous proof" that was unfortunately too long to be written down in that space. The proof (if indeed Fermat had one) has never been found. Known as Fermat's last theorem, it is probably the oldest and most famous unsolved problem in mathematics. These examples show that Diophantine equations are easy to write down but hard to solve. They are hard to solve because we are so exclusive about the kind of numbers we accept as solutions.

For first-degree equations, that is, equations in which unknowns are not multiplied together and all exponents are equal to 1, such as $7x+4y-3z-99t+$

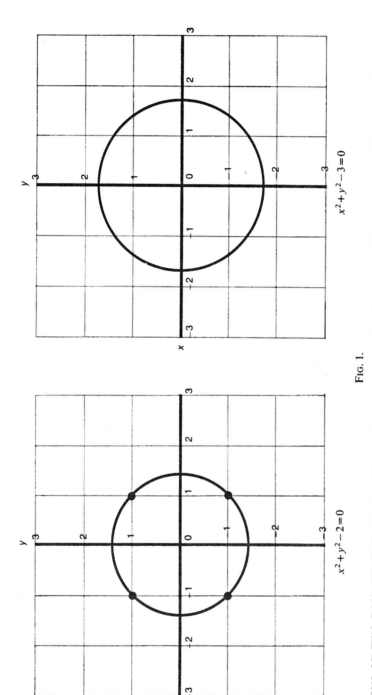

$x^2+y^2-2=0$

$x^2+y^2-3=0$

FIG. 1.

GRAPHS OF TWO EQUATIONS illustrate the difference between an ordinary equation and a Diophantine equation, for which one is interested only in whole-number solutions; this difference is central to Hilbert's 10th problem. The equations in point are $x^2+y^2-2=0$ (*left*) and $x^2+y^2-3=0$ (*right*); both are represented by circles with their center at the origin, that is, at the point with coordinates $x=0$, $y=0$. In the case of $x^2+y^2-2=0$ the circle has a radius of $\sqrt{2}$. If the equation is treated as an ordinary equation, there are infinitely many solutions. If, however, it is treated as a Diophantine equation, there are only four solutions: (1) $x=1$, $y=1$, (2) $x=-1$, $y=1$, (3) $x=1$, $y=-1$, and (4) $x=-1$, $y=-1$. These solutions are represented by dots where the graph crosses the four points with those coordinates on the Cartesian grid. In the case of $x^2+y^2-3=0$, the circle has a radius of $\sqrt{3}$. As an ordinary equation it has an infinite number of solutions; as a Diophantine equation, however, it has none at all.

557

$13u - 10 = 0$, the existence of solutions can be determined by a technique of division known since ancient times as Euclid's algorithm. For second-degree equations with two unknowns, such as $3x^2 - 5y^2 + 7 = 0$ or $x^2 - xy - y^2 = 1$, a theory developed early in the 19th century by the great Karl Friedrich Gauss enables one to determine whether there are any solutions. Recent work by the young British mathematician Alan Baker has shed considerable light on equations greater than the second degree that have two unknowns. For equations greater than the first degree that have more than two unknowns, there exist only some special cases that can be handled by special tricks, and a vast sea of ignorance.

Why is it so difficult to find a process such as the one Hilbert called for? The most direct approach would be to simply test all possible sets of values of the unknowns, one after another, until a solution is found. For example, if the equation has two unknowns, one could make a list of all pairs of integers. Then one would simply go through the list trying one pair after another to see if it satisfies the equation. This is certainly a clear-cut, mechanical procedure that a machine could carry out. What will be the result?

If the equation is the first one we mentioned, $x^2 + y^2 - 2 = 0$, one would test $(0,0)$, $(0,1)$, $(1,0)$, $(0, -1)$, $(-1,0)$ and reject them all. The next candidate, $(1,1)$, is a solution. We were lucky: only six pairs had to be considered. If, on the other hand, the equation were $x^2 + y^2 = 20,000$, one would have to test thousands of pairs of numbers before a solution was found. Still, it is clear that if a solution exists, it will be found in a finite number of steps.

On the other hand, what about the second equation: $x^2 + y^2 - 3 = 0$? One can try pairs of integers from now till eternity, and all that will ever be known is that a solution has not been found yet. One would never know whether or not the next pair tried would be a solution. For this particular example it is possible to prove there are no solutions. But the proof requires a new idea; it cannot be obtained merely by successively substituting integers into the equation.

A device that carries out a process of the kind suggested by Hilbert should accept as an input the coefficients of an arbitrary Diophantine equation. As an output it should turn on a green light if the equation has a solution and a red light if it has none. Such a machine might be called a Hilbert machine. By way of contrast a device that simply searches for solutions by successive trials *ad infinitum* could be described as a green-light machine. If the equation has a solution, the green light goes on after a finite number of steps. If the equation has no solution, the computation simply goes on forever; unlike the Hilbert machine, the green-light machine has no way of knowing when to give up.

It is easy to build a green-light machine for Diophantine equations. The question is, can we do better and build a Hilbert machine, that is, a green-light-red-light machine that will always stop after a finite number of steps and give a definite yes or no answer? What Matyasevich proved is that this can never be done. Even if we allow the machine unlimited memory storage and unlimited computing time, no program can ever be written and no machine can ever be built that will do what Hilbert wanted. A Hilbert machine does not exist.

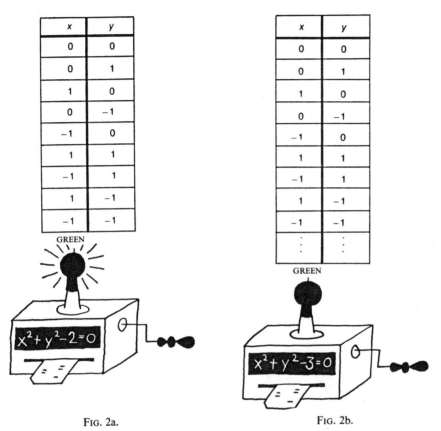

x	y
0	0
0	1
1	0
0	−1
−1	0
1	1
−1	1
1	−1
−1	−1

GREEN

$x^2+y^2-2=0$

x	y
0	0
0	1
1	0
0	−1
−1	0
1	1
−1	1
1	−1
−1	−1
⋮	⋮

GREEN

$x^2+y^2-3=0$

FIG. 2a.   FIG. 2b.

PAIRS OF INTEGERS can be individually tested by green-light machines to see if they are solutions to Diophantine equations. Trial and error comes up with a solution for the equation $x^2+y^2-2=0$ on the sixth try (Fig. 2a). Green-light machine testing equation $x^2+y^2-3=0$ has no way of knowing when to give up, however, because there are no whole-number solutions (Fig. 2b). All it knows is that it has found no solutions yet.

Hilbert continued in his address of 1900: "Occasionally it happens that we seek the solution under insufficient hypotheses or in an incorrect sense, and for this reason do not succeed. The problem then arises: to show the impossibility of the solution under the given hypotheses, or in the sense contemplated." That is exactly what has happened with the 10th problem.

In order to explain how we know that no Hilbert machine exists, we have to discuss some simple ideas about computability. Suppose S stands for a set of integers. S is "listable" if a green-light machine can be built that will do the following job: accept any integer as an input, and as an output turn on a green light after a finite number of steps if and only if the input (the integer) belongs to S. For example, the set of even numbers is listable. In this case the machine would divide the input by 2 and turn on a green light if the remainder is 0. In mathematical literature such sets are called recursively enumerable; the word "listable" is our informal equivalent.

The set S is "computable" if a green-light-red-light machine (similar to the Hilbert machine for Diophantine equations) can be built to do a more difficult job: accept any integer as input and, after a finite number of steps, turn on a green light if the integer is in S and a red light if the integer is not in S. For example, the set of even numbers is computable. The machine would divide the input by 2; if the remainder is 0, it turns on a green light, and if the remainder is 1, it turns on a red light [see Figs. 3 and 4].

There is a close connection between these two definitions. For the purposes of explanation, let $\bar{S}$ denote the complement of S, that is, the set of all integers that do not belong to S. If in the two examples S is the set of even integers, then $\bar{S}$ is the

FIG. 3.

GREEN-LIGHT-RED-LIGHT MACHINE is an imaginary device that tests numbers to determine if they are members of a given set. Hilbert's 10th problem asks if a green-light-red-light "Hilbert machine" can be built to test Diophantine equations to see whether or not they have solutions. In the case of testing numbers for membership in a set, green light goes on if the machine can determine in a finite number of steps that a given input is a member of the set. Say that S is the set of all even numbers. To test inputs one can devise an algorithm for dividing each input x by 2. If the remainder of the division is 0 (written Rem $x/2=0$), machine would turn on its green light, signifying that x is a member of S.

set of odd integers. We can prove that if S is computable, S and $\bar{S}$ are both listable. To put that statement another way: If a green-light-red-light machine exists for S, then there exists a green-light machine for S and a green-light machine for $\bar{S}$. The proof is simple. To build a green-light machine for S, just unscrew the red bulb of the green-light-red-light machine. To build a green-light machine for $\bar{S}$, unscrew the green bulb of the Hilbert machine and put it into the socket that held the red bulb. [See Fig. 5.]

The converse is also true: If S and $\bar{S}$ are listable, then S is computable. The equivalent of this statement is: If a green-light machine exists for each of S and $\bar{S}$, then a green-light-red-light machine can be built for S. This is easily done. In the green-light machine for $\bar{S}$, replace the green bulb with a red bulb. Then hook up the two machines in parallel, so that the input goes into both simultaneously. The result is clearly a green-light-red-light machine. [See Fig. 6.]

FIG. 4.

RED LIGHT GOES ON on the green-light-red-light machine if the machine can determine that the input is not a member of the set. Suppose the input x is the whole number 23; 2 goes into 23 with a remainder of 1, signifying that 23 is not a member of S. Complement of set S is $\bar{S}$, the set of odd numbers; 23 is a member of $\bar{S}$. Since a green-light-red-light machine can be built to sort members of S from members of $\bar{S}$, the set S is called computable.

FIG. 5.

GREEN-LIGHT-RED-LIGHT MACHINE FOR THE SET S can be transformed into a green-light machine for S (that is, a machine that simply lights up when the input is a member of S) plus a green-light machine for $\bar{S}$, the complement of S. The proof is simple. To build a green-light machine for S, unscrew the red lamp of the green-light-red-light machine. To build a green-light machine for $\bar{S}$, unscrew the green lamp of the green-light-red-light machine and put it into the socket that held the red lamp. This fact can be stated in another way. If a set (such as S) is computable, then both the set and its complement (such as $\bar{S}$) are listable, that is, the members of S (in this case the set of even numbers) can be listed separately and sorted from the members of $\bar{S}$ (the set of odd numbers).

562

Knowing all of this, we can now state one of the crucial facts in computability theory, one that plays a central role in the solution of Hilbert's 10th problem: There is a set K that is listable but not computable! That is, there exists a green-light machine for K, but it is impossible to build a green-light machine for $\bar{K}$, the complement of K.

To prove this seemingly strange fact, let each green-light machine be specified by a detailed "customer's manual" in the English language. The customer's manual describes exactly how the machine is constructed. The customer's manuals can be set in order and numbered sequentially 1, 2, 3 and so on. In that way all green-light machines are numbered; M_1 is the first machine, M_2 is the second and so on. There is a subtle point hidden here. Such an ordered list of customer's manuals would not be possible for green-light-red-light machines. The difficulty is that one cannot tell from the manual whether the red light or the green light will turn on for any input to the corresponding machine.

The set K is defined as the set of numbers n such that the nth machine lights up when it receives n itself as an input. In other words, the number 1 belongs to K if and only if M_1 turns on its green light when "1" is entered into its input. The number 2 belongs to K if and only if M_2 eventually lights up when "2" is entered into its input, and so on [see Fig. 7].

In order to build a green-light machine for K we need, along with the library of customer's manuals, a little man who can read them and carry out their instructions. He should perhaps be a wise old man, but he must be an obedient man who does exactly what he is told. We give the little man a number, say 3,781. The little man looks into customer's manual No. 3,781. Reading the manual, he is able to build the green-light machine $M_{3,781}$. Once this is done, he inserts the integer 3,781 as input into green-light machine $M_{3,781}$. If the green light goes on, the number 3,781 belongs to K. Thus we have a green-light machine for K.

What about $\bar{K}$? How can we be sure there is no green-light machine for it? Well, suppose there were such a machine. Then since $\bar{K}$ is the complement of K, this machine should light up for any input, say for 297, if and only if M_{297} does *not* light up for 297. (If M_{297} lit up, it would mean that the integer 297 belongs to K and not to $\bar{K}$.) Thus the machine for $\bar{K}$ certainly is not the same as M_{297} [see Fig. 8]. By the same token, however, it is not the same as M_n for any other value of n. The same argument would apply to any other number just as well as to 297, and it shows that no green-light machine for $\bar{K}$ appears anywhere in the library of customer's manuals. Since every possible green-light machine eventually turns up in our list, it follows that no green-light machine for $\bar{K}$ can possibly exist. That is to say, $\bar{K}$ is not listable.

The result is certainly remarkable. It deserves contemplation and appreciation. We know perfectly well what the set K is; in principle we can produce as much of it as we wish with a computer printout. Nevertheless, there can never be a formal procedure (an algorithm or a machine program) for sorting K from $\bar{K}$. Thus here is an example of a precisely stated problem that can never be solved by mechanical means.

FIG. 6a.

GREEN-LIGHT MACHINE FOR EACH OF S AND $\bar{S}$ can be used to construct a green-light-red-light machine for the set S. This statement is the converse of the one for Figure 5. In the green-light machine for $\bar{S}$ replace the green lamp with a red lamp. Then hook the machines in parallel so that the

This discussion has of course been informal and nonrigorous. It is possible, however, to reformulate all the ideas and arguments with precise mathematical definitions and proofs. In fact, they have been formulated in a branch of mathematical logic called recursive function theory, established in the 1930's by Gödel, Church, Post, Kleene and Turing.

Now, what has all this to do with Diophantine equations? Simply this. Matyasevich has proved that every listable set has a corresponding Diophantine equation. More precisely, if S is a listable set, then there is a corresponding polynomial P, with integer coefficients and variables $x, y_1, y_2, \ldots, y_n$, which is denoted by $P_S(x, y_1, y_2, \ldots, y_n)$. Any integer, such as 17, belongs to set S if and only if the Diophantine equation $P(17, y_1, y_2, \ldots, y_n) = 0$ has a solution.

It might be thought that for some sets we would have to resort to inconceivably complicated polynomials, but this is not the case. The degree of P need not exceed

... to build a green-light-red-light machine for S.

GREEN

RED

S

S̄

INPUT

FIG. 6b.

input goes into both simultaneously. The result is clearly a green-light-red-light machine. This assertion can be stated differently: If both a set and its complement are listable, then the set is computable.

the fourth power; the number of variables $y_1, y_2, \ldots, y_n$ need not exceed 14. (No one knows yet if both of these bounds can be achieved simultaneously.)

This result of Matyasevich' quickly leads to the conclusion that no Hilbert machine can exist. Recall the listable set K constructed a few paragraphs above. According to Matyasevich, there is a Diophantine equation, $P_K(x, y_1, y_2, \ldots, y_n) = 0$, associated with this set. If it were possible to build a Hilbert machine, that is, a green-light-red-light machine for testing Diophantine equations to see if they have solutions, then for any integer x we could determine whether or not there existed integers $y_1, y_2, \ldots, y_n$ such that the equation has a solution. In so determining, however, we would also be determining whether or not x belongs to K. In other words, a Hilbert machine applied to the Diophantine equation that describes K could be used as a green-light-red-light machine for K. We have proved, however, that K is not computable, so that no green-light-red-light machine can exist for K. The only way out of this dilemma is to conclude that there is no Hilbert machine. In other words, Hilbert's 10th problem is unsolvable!

FIG. 7.

THE SET K IS LISTABLE, that is, a green-light machine for K exists. Let all conceivable green-light machines be numbered: M_1 is the first machine, M_2 is the second machine, M_3 is the third machine and so forth up to the nth machine. K is defined as the set of numbers n such that the nth machine lights up when it receives n itself as an input. In the illustration a little man has entered the number 3,781 as an input to $M_{3,781}$ and the green light has turned on, indicating that the whole number 3,781 is a member of set K.

566

FIG. 8.

THE SET K IS NOT COMPUTABLE, that is, no green-light machine exists for $\overline{K}$, the complement of K. Suppose there was such a green-light machine for $\overline{K}$. Since $\overline{K}$ is the complement of K, this machine should light up for any input, say for 297, if and only if M_{297} does not light up for 297. Thus the machine for $\overline{K}$ is certainly not the same as M_{297}. By the same token, it is not the same as M_n for any other value of n. Thus no green-light machine exists for $\overline{K}$, meaning that K is not listable. A listable set whose complement is not listable is not computable; no green-light-red-light machine can be built for it. Thus there is no algorithm for sorting K from $\overline{K}$.

567

The fact that a Diophantine equation is associated with every listable set is a positive result that is of great interest in itself, quite aside from its application to Hilbert's 10th problem. A particularly important and interesting set of integers is the set of prime numbers. A prime number is one that is factorable (divisible) only by 1 and by itself. Some examples are 2, 3, 5, 7, 11, 13 and 17. That they are listable is rather obvious. An algorithm for listing them has come down from the Greeks with the name of "the sieve of Eratosthenes." Combining Matyasevich' result with a device developed by Putnam, we obtain a Diophantine equation $Q(y_1, y_2, \ldots, y_n)$ $= z$ such that a positive number z is a prime if and only if this equation has a positive integer solution $y_1, y_2, \ldots, y_n$. (The exact form of the polynomial Q is a bit too complicated to fully write out here.)

Another remarkable result can be proved by combining Matyasevich' theorem with Gödel's work on undecidability. If there is any system of axioms whatsoever from which information can be deduced about Diophantine equations, one can always obtain a particular Diophantine equation that has the following properties: (1) the equation has no positive integer solutions and (2) the fact that it has no positive integer solutions cannot be logically deduced from the given set of axioms. Of course, once the Diophantine equation is obtained we can make up a new set of axioms from which one can prove that the Diophantine equation has no solution. But then this new set of axioms will give rise to another Diophantine equation for which the same can be asserted.

What went into the proof of Matyasevich' theorem? In addition to the results from classical and even ancient number theory that we have already mentioned, there is a key result known as the Chinese remainder theorem. It will be helpful to illustrate the Chinese remainder theorem by a numerical example.

Suppose one wishes to find a number whose remainders, when divided by the numbers 10, 3, 7 and 11, are respectively 4, 2, 3 and 1 [see Fig. 9]. The Chinese remainder theorem assures us that there must be such a number. (In fact, in this case 584 is such a number.) All that is required for the Chinese remainder theorem to work is that no pair of the divisors used have any common factor (except, of course, 1). There can be any number of divisors, and the desired remainders can be any positive integers whatsoever.

In 1931 Gödel showed how to use the Chinese remainder theorem as a coding trick, in which an arbitrary finite sequence of numbers can be encoded as a single number. From the code number one recovers the sequence in the same way that 4, 2, 3 and 1 are obtained from 584 in the example—as remainders in successive divisions. The divisors can be chosen to be in arithmetic progression.

The first attempt to prove that a Hilbert machine cannot exist was made by one of us (Davis) in his doctoral dissertation in 1950. Gödel's technique of using the Chinese remainder theorem as a coding device was applied to associate a Diophantine equation, $P_S(k, x, z, y_1, y_2, \ldots, y_n) = 0$, with every listable set S. Unfortunately the relation between the set and the equation turned out to be more complicated than what was needed for Hilbert's 10th problem. Specifically, the relation was: A positive integer x belongs to the set S if and only if for some

PROBLEM: To find the smallest number n that has the remainders of 4, 2, 3 and 1 when it is divided by 10, 3, 7 and 11.

SOLUTION: Let x be the number sought. "Rem" will be the abbreviation for "The remainder of . . . ." The problem can then be rewritten:

$$\text{Rem}\left(\frac{x}{10}\right) = 4 \qquad \text{Rem}\left(\frac{x}{7}\right) = 3$$

$$\text{Rem}\left(\frac{x}{3}\right) = 2 \qquad \text{Rem}\left(\frac{x}{11}\right) = 1$$

In order to find x four auxiliary problems for new unknowns y_1, y_2, y_3 and y_4 must be solved. In each case the numerator is obtained by multiplying three of the divisors together and using the fourth as the denominator. For example, in the first equation with y_1 the numerator 231 is equal to $3 \times 7 \times 11$, and 10 is put in the denominator:

$$\text{Rem}\left(\frac{231y_1}{10}\right) = 4, y_1 < 10 \qquad \text{Rem}\left(\frac{330y_3}{7}\right) = 3, y_3 < 7$$

$$\text{Rem}\left(\frac{770y_2}{3}\right) = 2, y_2 < 3 \qquad \text{Rem}\left(\frac{210y_4}{11}\right) = 1, y_4 < 11$$

The set of smallest integers that are solutions to these auxiliary equations is $y_1 = 4$, $y_2 = 1$, $y_3 = 3$ and $y_4 = 1$.

To get x (the original number sought) the numerators of the four auxiliary equations are added together:

$$x = (231y_1) + (770y_2) + (330y_3) + (210y_4)$$

$$= (231 \times 4) + (770 \times 1) + (330 \times 3) + (210 \times 1)$$

$$= 924 + 770 + 990 + 210$$

$$= 2,894$$

Thus 2,894 is one value of x. A smaller number can be obtained if the product of all four divisors is subtracted from this solution:

$$2,894 - (10 \times 3 \times 7 \times 11) = 2,894 - 2,310 = 584.$$

Therefore 584 is the smallest solution to the problem.

FIG. 9.

CHINESE REMAINDER THEOREM is used in the solution to Hilbert's 10th problem. In this case the theorem is employed to find a number whose remainders, when divided by the numbers 10, 3, 7 and 11, are respectively 4, 2, 3 and 1. Integer 584 is the smallest solution.

positive integer value of z it is possible to find a solution for every one of the Diophantine equations obtained by substituting $k=1$, then $k=2$ and so on up to z into the equation $P_S(k,x,z,y_1,y_2,\ldots,y_n)=0$. Although the result seemed tantalizingly close to what was needed, it was only a beginning.

At about the same time Robinson began her own investigations of sets that can be defined by Diophantine equations. She developed various ingenious techniques for dealing with equations whose solutions behaved like exponentials (grew like a power). In 1960 she, Davis and Putnam collaborated in proving another result. They made use of both her work and Davis' result to show that to any listable set there corresponded a Diophantine equation of an "extended" kind, extended in the sense that variables in the equation were allowed to occur as exponents. An example of such an equation is $2^t + x^2 = z^3$. Davis, Robinson and Putnam combined their work with some of Robinson's earlier results and discovered the following: If even one Diophantine equation could be found whose solutions behaved exponentially in an appropriate sense, then it would be possible to describe every listable set by a Diophantine equation. This would in turn show that Hilbert's 10th problem is unsolvable.

1. 1
2. 1
3. $1 + 1 = 2$
4. $1 + 2 = 3$
5. $2 + 3 = 5$
6. $3 + 5 = 8$
7. $5 + 8 = 13$
8. $8 + 13 = 21$
9. $13 + 21 = 34$
10. $21 + 34 = 55$
11. $34 + 55 = 89$
12. $55 + 89 = 144$
13. $89 + 144 = 233$

.
.
.

$\underline{n}$ $\approx \frac{1}{\sqrt{5}} \left(\frac{1 + \sqrt{5}}{2} \right)^n$

FIG. 10.

FIBONACCI NUMBERS were discovered in A.D. 1202 by Leonardo of Pisa, known as Fibonacci. The sequence is obtained by starting with 1 and 1 and successively adding the last two numbers to get the next one. The sequence grows exponentially: the nth number in the sequence is approximately proportional to the nth power of the real number $\left[(1 + \sqrt{5})/2 \right]$.

It took a decade to find a Diophantine equation whose solutions grow exponentially in the appropriate sense. In 1970 Matyasevich found such an equation by using what are known as the Fibonacci numbers [see Fig. 10]. These celebrated numbers were discovered in A.D. 1202 by Leonardo of Pisa, who was also known as Fibonacci. He found them by computing the total number of pairs of descendants of one pair of rabbits if the original pair and each offspring pair reproduced itself once a month. The Fibonacci series is obtained by starting with 1 and 1 and successively adding the preceding two numbers to get the next: the first Fibonacci

I. $u + w - v - 2 = 0$

II. $l - 2v - 2a - 1 = 0$

III. $l^2 - lz - z^2 - 1 = 0$

IV. $g - bl^2 = 0$

V. $g^2 - gh - h^2 - 1 = 0$

VI. $m - c(2h + g) - 3 = 0$

VII. $m - fl - 2 = 0$

VIII. $x^2 - mxy + y^2 - 1 = 0$

IX. $(d - 1)l + u - x - 1 = 0$

X. $x - v - (2h + g)(e - 1) = 0$

FIG. 11.

MATYASEVICH' SOLUTION to Hilbert's 10th problem involves a Diophantine equation that is obtained by squaring each of these 10 equations and then adding them together and setting the resulting complicated polynomial equal to zero. In these equations the values u and v in the solutions are related in such a way that v is the $2u$th Fibonacci number. From the solution it followed that for every listable set there is an associated Diophantine equation. Since there exist listable sets whose complements are not listable, then not every listable set can have a green-light-red-light machine. Since having a green-light-red-light machine for a set is equivalent to having a Hilbert machine for Diophantine equations, Matyasevich' result means that no Hilbert machine can be built to test Diophantine equations.

number is 1, the second is 1, the third is $1 + 1 = 2$, the fourth is $1 + 2 = 3$, the fifth is $2 + 3 = 5$ and so on. The property that is important for Hilbert's 10th problem is that the Fibonacci numbers grow exponentially. That is, the nth Fibonacci number is approximately proportional to the nth power of a certain fixed real number.

If one could find a Diophantine equation whose solutions relate n to the nth Fibonacci number, it would be the desired example of a Diophantine equation whose solutions behave exponentially. The solution of Hilbert's 10th problem would follow from this example. What Matyasevich did was to construct such a Diophantine equation [see Fig. 11]. Once he had shown that the set of Fibonacci numbers is associated in this way with a Diophantine equation, it followed immediately from the theorem of Davis, Robinson and Putnam that for every listable set there is an associated Diophantine equation, including in particular the set K, which is not computable. And so ends the story of Hilbert's 10th problem.

24

LAWRENCE ZALCMAN

Lawrence Zalcman was born on June 9, 1943, in Kansas City, Missouri, and attended the public schools of that city. He attended Dartmouth College as a General Motors National Scholar, receiving his A.B. in 1964 *summa cum laude* with valedictory rank. He studied at M.I.T. on a National Science Foundation Graduate Fellowship and received his Ph.D. (under the direction of Kenneth Hoffman) in 1968. From 1968 to 1972 he was an assistant professor at Stanford University. In 1972 he joined the staff of the University of Maryland mathematics department, where he has been a professor since 1974.

Zalcman was a NATO Postdoctoral Fellow at the Hebrew University of Jerusalem and the Weizman Institute of Science in Rehovoth, Israel (1970–71). He has also been a visiting professor at the Technion (Israel Institute of Technology, Haifa), a member of the *Forschungsinstitut für Mathematik* of the *Eidgenössische Technische Hochschule* (Swiss Federal Institute of Technology, Zürich), and a visiting lecturer at a number of American Universities.

Zalcman is a member of the Editorial Board of the *Proceedings of the American Mathematical Society*. He is a past winner of the William Lowell Putman Competition and received the Association's Lester R. Ford Award in 1975 for the paper for which he later won the Chauvenet Prize.

Zalcman's research centers on complex analysis, especially on the interaction of function theory with related branches of mathematics (including approximation theory, Banach algebras, differential equations, and harmonic analysis). He also has strong interests in the history and philosophy of mathematics.

Professor Zalcman believes strongly in the values of good expository writing as exemplified by his statement: "If one believes, as I do, that a major component of mathematics (and perhaps *the* major function of a mathematical proof) is communication, good exposition acquires an almost overriding importance. Thus, it has always seemed to me that an excellent expository paper is genuinely more valuable than all but the most seminal research papers. I hope that the *Monthly* will continue to publish first-rate expositions and (what is anterior to that) that mathematicians will find the time to write them."

In his acceptance, Professor Zalcman expressed his delight and very great honor for having been awarded the Chauvenet Prize. The acceptance was read at the Annual Business Meeting by D. P. Roselle, Secretary of the Association.

REAL PROOFS OF COMPLEX THEOREMS (AND VICE VERSA)

LAWRENCE ZALCMAN

Introduction. It has become fashionable recently to argue that real and complex variables should be taught together as a unified curriculum in analysis. Now this is hardly a novel idea, as a quick perusal of Whittaker and Watson's *Course of Modern Analysis* or either Littlewood's or Titchmarsh's *Theory of Functions* (not to mention any number of *cours d'analyse* of the nineteenth or twentieth century) will indicate. And, while some persuasive arguments can be advanced in favor of this approach, it is by no means obvious that the advantages outweigh the disadvantages or, for that matter, that a unified treatment offers any substantial benefit to the student. What is obvious is that the two subjects do interact, and interact substantially, often in a surprising fashion. These points of tangency present an instructor the opportunity to pose (and answer) natural and important questions on basic material by applying real analysis to complex function theory, and vice versa. This article is devoted to several such applications.

My own experience in teaching suggests that the subject matter discussed below is particularly well-suited for presentation in a year-long first graduate course in complex analysis. While most of this material is (perhaps by definition) well known to the experts, it is not, unfortunately, a part of the common culture of professional mathematicians. In fact, several of the examples arose in response to questions from friends and colleagues. The mathematics involved is too pretty to be the private preserve of specialists. Publicizing it is the purpose of the present paper.

1. The Greening of Morera. One of the most useful theorems of basic complex analysis is the following result, first noted by Giacinto Morera.

MORERA'S THEOREM [**37**]. *Let $f(z)$ be a continuous function on the domain D. Suppose that*

$$(1) \qquad \int_\gamma f(z)dz = 0$$

for every rectifiable closed curve γ lying in D. Then f is holomorphic in D.

Morera's Theorem enables one to establish the analyticity of functions in situations where resort to the definition and the attendant calculation of difference quotients would lead to hopeless complications. Applications of this sort occur, for instance, in the proofs of the Schwarz Reflection Principle and other theorems on the extension of analytic functions. Nor is its usefulness limited to this circle of ideas; the important fact that the uniform limit of analytic functions is again analytic is an immediate consequence (observed already by Morera himself, as well as by Osgood [**39**], who had rediscovered Morera's theorem).

Perhaps surprisingly, the proofs of Morera's theorem found in complex analysis texts all follow a single pattern. The hypothesis on f insures the existence of a single-valued primitive F of f, defined by

$$(2) \qquad F(z) = \int_{z_0}^{z} f(\zeta)d\zeta.$$

Here z_0 is some fixed point in D and the integral is taken over any rectifiable curve joining z_0 to z. The function F is easily seen to be holomorphic in D, with $F'(z) = f(z)$; since the derivative of a holomorphic function is again holomorphic, we are done.

Several remarks are in order concerning the proof sketched above. First of all, the assumption that (1) holds for all rectifiable closed curves in D is much too strong. It is enough, for instance, to assume that (1) holds for all closed curves consisting of a finite number of straight line segments parallel to the coordinate axes; the integration in (2) is then effected over a (nonclosed) curve composed of such segments, and the proof proceeds much as before. Second, since analyticity is a local property, condition (1) need hold only for an arbitrary neighborhood of each point of D; that is, (1) need hold only for *small* curves. Finally, the proof requires the fact that the derivative of an analytic function is again analytic. While this is a trivial consequence of the Cauchy integral formula, it can be argued that that is an inappropriate tool for the problem at hand; on the other hand, a proof of this fact without complex integration is genuinely difficult and was, in fact, only discovered (after many years of effort) in 1961 [**44**], [**10**], [**46**].

There is an additional defect to the proof, and that is that *it does not generalize.* Thus, it was more than thirty years after Morera discovered his theorem that Torsten Carleman realized the result remains valid if (1) is assumed to hold only for all (small) *circles* in D. It is an extremely instructive exercise to try to prove Carleman's version of Morera's theorem by mimicking the proof given above. The argument fails because it cannot even be started: the very existence of a single-valued primitive is in doubt. This leads one to try a different (and more fruitful) approach, which avoids the use of primitives altogether.

Suppose for the moment that f is a smooth function, say continuously differentiable. Fix $z_0 \in D$ and suppose (1) holds for the circle $\Gamma_r(z_0)$ of radius r, centered at z_0. Then, by the complex form of Green's theorem

$$0 = \int_{\Gamma_r(z_0)} f(z)dz = 2i \iint_{\Delta_r(z_0)} \frac{\partial f}{\partial \bar z}\, dxdy,$$

where $\Delta_r(z_0)$ is the disc bounded by $\Gamma_r(z_0)$ and $\partial f/\partial \bar z = \frac{1}{2}(\partial f/\partial x + i\, \partial f/\partial y)$. (There's no cause for panic if the $\partial/\partial \bar z$ operator makes you uneasy or you are not familiar with the complex form of Green's theorem; just write $f(z) = u(z) + iv(z)$, $dz = dx + idy$, and apply the usual version of Green's theorem to the real and imaginary parts of the integral on the left.) Dividing by an appropriate factor, we have

$$\frac{1}{\pi r^2} \iint_{\Delta_r(z_0)} \frac{\partial f}{\partial \bar{z}} \, dxdy = 0;$$

i.e., the average of the continuous function $\partial f/\partial \bar{z}$ over the disc $\Delta_r(z_0)$ equals 0. Make $r \to 0$ to obtain $(\partial f/\partial \bar{z})(z_0) = 0$. Since this holds at each point $z_0 \in D$, $\partial f/\partial \bar{z} = 0$ identically in D. Writing this in real coordinates, we see that $u_x = v_y$, $u_y = -v_x$ in D; thus the Cauchy-Riemann equations are satisfied and f is analytic.

Notice that we did not need to assume that (1) holds for all circles in D or even all small circles; to pass to the limit it was enough to have, for each point of D, a sequence of circles shrinking to that point. Moreover, since f has been assumed to be continuously differentiable, it is sufficient to prove that $\partial f/\partial \bar{z}$ vanishes on a dense set. Finally, and most important, *the fact that our curves were circles was not used at all!* Squares, rectangles, pentagons, ovals could have been used just as well. To conclude that $(\partial f/\partial \bar{z})(z_0) = 0$, all we require is that (1) should hold for a sequence of simple closed curves γ that accumulate to z_0 (z_0 need not even lie inside or on the γ's) and that the curves involved allow application of Green's theorem. It is enough, for instance, to assume that the curves are piecewise continuously differentiable.

To summarize, we have shown that Green's theorem yields in a simple fashion a very general and particularly appealing version of Morera's theorem for C^1 functions. It may reasonably be asked at this point if the proof of Morera's theorem given above can be modified to work for functions which are assumed only to be continuous. That is the subject of the next section.

2. Smoothing. Let $\phi(z)$ be a real valued function defined on the entire complex plane which satisfies

(a) $\phi(z) \geqq 0$,

(b) $\iint \phi(z)dxdy = 1$,

(c) ϕ is continuously differentiable,

(d) $\phi(z) = 0$ for $|z| \geqq 1$.

It is trivial to construct such functions; we can even require ϕ to be infinitely differentiable and to depend only on $|z|$, but these properties will not be required in the sequel. Set, for $\varepsilon > 0$, $\phi_\varepsilon(z) = \varepsilon^{-2}\phi(z/\varepsilon)$. Then, clearly, ϕ_ε satisfies (a) through (c) above and ϕ_ε vanishes off $|z| < \varepsilon$. The family of functions $\{\phi_\varepsilon\}$ forms what is known in harmonic analysis as an approximate identity (a smooth approximation to the Dirac delta function); workers in the field of partial differential equations, where the smoothness properties of the ϕ_ε are emphasized, are accustomed to call similar functions (Friedrichs) mollifiers.

Suppose now that f is a continuous function on some domain D and set

(3) $$f_\varepsilon(z) = \iint f(z - \zeta)\phi_\varepsilon(\zeta)d\xi d\eta \qquad \zeta = \xi + i\eta,$$

where the integral is extended over the whole complex plane. This integral exists and defines a continuous function for all points z whose distance from the boundary of D is greater than ε. Moreover, $f_\varepsilon(z)$ is continuously differentiable for such points. Indeed, changing variable in (3), we have

$$f_\varepsilon(z) = \iint \phi_\varepsilon(z - \zeta)f(\zeta)d\xi d\eta$$

and the x and y derivatives can be brought inside the integral since we have chosen ϕ_ε to be continuously differentiable. Finally, we note that for any compact subset K of D, $f_\varepsilon(z)$ converges uniformly to $f(z)$ on K as $\varepsilon \to 0$. This expresses the delta-function-like behavior of the family $\{\phi_\varepsilon\}$. Here is the simple proof. By (b),

$$f(z) - f_\varepsilon(z) = \iint_{|\zeta| \leq \varepsilon} \{f(z) - f_\varepsilon(z - \zeta)\}\phi_\varepsilon(\zeta)d\xi d\eta,$$

whence by (a) and (b)

(4)
$$
\begin{aligned}
\left| f(z) - f_\varepsilon(z) \right| &\leq \iint_{|\zeta| \leq \varepsilon} \left| f(z) - f(z - \zeta) \right| \phi_\varepsilon(\zeta)d\xi d\eta \\
&\leq \sup_{|\zeta| \leq \varepsilon} \left| f(z) - f(z - \zeta) \right|.
\end{aligned}
$$

Since K is compact, f is uniformly continuous on K; so (4) shows that

$$\sup_{z \in K} \left| f(z) - f_\varepsilon(z) \right| \to 0 \qquad \text{as } \varepsilon \to 0.$$

The proof of Morera's theorem is now easily completed. Suppose, for instance, that f is continuous on D and that there exists a sequence of positive numbers $r_1 \geq r_2 \geq r_3 \geq \cdots \to 0$ such that

(5)
$$\int_{\Gamma_{r_n}(z)} f(w)dw = 0$$

for each $z \in D$ whenever the circle $\Gamma_{r_n}(z) = \{w : \left| w - z \right| = r_n\}$ lies in D. Fix a compact set $K \subset D$ and take $\varepsilon < \frac{1}{2}\text{dist}(K, \partial D)$. Then for $r = r_n < \frac{1}{2}\text{dist}(K, \partial D)$ and $z \in K$ we have

$$
\begin{aligned}
\int_{\Gamma_r(z)} f_\varepsilon(w)dw &= \int_{\Gamma_r(z)} \left\{ \iint f(w - \zeta)\phi_\varepsilon(\zeta)d\xi d\eta \right\} dw \\
&= \iint \left\{ \int_{\Gamma_r(z)} f(w - \zeta)dw \right\} \phi_\varepsilon(\zeta)d\xi d\eta \\
&= \iint \left\{ \int_{\Gamma_r(z-\zeta)} f(w)dw \right\} \phi_\varepsilon(\zeta)d\xi d\eta \\
&= 0.
\end{aligned}
$$

Since f_ε is continuously differentiable, it is analytic on the interior of K; and since f_ε converges to f uniformly on K, f must be analytic there. Finally, because K is arbitrary, f is analytic on all of D.

Again, there is nothing particularly sacred about circles: if $\{\gamma_n\}$ is a sequence of simple closed piecewise continuously differentiable curves which shrink to the origin and $\gamma_n(z)$ is the image of γ_n under the map $w \mapsto w + z$, we may replace (5) by

$$(6) \qquad \int_{\gamma_n(z)} f(w)\,dw = 0$$

and the rest of the argument remains unchanged. Similarly, it is enough to assume that (5) or (6) hold only for a dense set of $z \in D$, since the full condition then follows from the continuity of f.

The result can be extended even further. The requirement that f be continuous may be relaxed to the assumption that f is measurable and integrable with respect to Lebesgue area measure on compact subsets of D. Of course, the conclusion now reads that f agrees almost everywhere with a function analytic on D. For a complete treatment, together with an historical discussion, see [62]. The use of smoothing operators is a standard tool among workers in partial differential equations and approximation theory; for a systematic exposition of its use in this last subject, see [52].

The success of the smoothing technique in dealing with Morera's theorem suggests using it to prove Cauchy's theorem. This is a good idea, but one which, unfortunately, simply does not work. Here's the rub. Suppose $f(z)$ is analytic in the disc D. We know (by Green's theorem) that $\int_T f(z)\,dz = 0$ for every triangle T in D *if f is continuously differentiable*. Of course, in general, f is *not* known *a priori* to be continuously differentiable; but we may construct $f_\varepsilon(z)$, as in (3), which is. *However, it is not clear that $f_\varepsilon(z)$ is holomorphic.* The problem is that while $f'(z)$ is known to exist for each $z \in D$, and is easily proved to be measurable, it is *not* known to be integrable; we cannot, therefore, differentiate f inside the integral sign of (3). (A similar difficulty arises in the proof of Hartogs' theorem: *If a function of two complex variables $g(z_1, z_2)$ is analytic in each variable separately, then it is analytic as a function of the joint variables z_1, z_2*). The argument does work if f' is assumed to be area integrable, but this assumption is (of course) unnecessary, and it seems best to base the proof of Cauchy's theorem on Pringsheim's device [45] of subdividing triangles. This is the pattern followed in most modern texts.

3. In circles. All the versions of Morera's theorem discussed up to now have depended in an essential fashion on the fact that (1) holds for a certain class of contours containing arbitrarily small curves. The obvious question to ask is what happens if (1) holds for circles which do *not* shrink in radius. In this situation, it is natural to assume that the function in question is defined on the entire complex plane. A satisfying answer is provided by the following result, proved in 1970 [62].

THEOREM. *Let f be a continuous function on the complex plane and suppose that there exist numbers $r_1, r_2 > 0$ such that*

(7) $$\int_\Gamma f(z)dz = 0$$

for every circle having radius r_1 or r_2 (and arbitrary center). Then f is an entire function unless r_1/r_2 is a quotient of zeroes of the Bessel function $J_1(z)$.

The hypothesis on f may be relaxed to the assumption of local integrability, and (7) need hold only for 'almost all' circles. The restriction on the pair r_1, r_2 is, however, essential: in case it is not satisfied, f may fail to be holomorphic anywhere.

The proof is considerably more involved than (and of an altogether different character from) the sort of argument we have seen in the preceding sections; essential ingredients include the harmonic analysis of an appropriate space of distributions and the Delsarte-Schwartz theory of mean-periodic functions. See [62], where related results are discussed, for details. One can also show that if f is continuous on the plane and (7) holds for every square (of arbitrary center and orientation) having side of fixed length, then f is entire. Again, a reference is [62]. Further perspectives on results of this sort will be found in [63].

4. Reflections on reflection. According to the Schwarz Reflection Principle, if $f(z)$ is analytic in $\Delta = \{z : |z| < 1\}$ and continuously extendible to an open arc γ of $\Gamma = \{z : |z| = 1\}$, and if the values of f corresponding to points of γ lie on a circular, or, more generally, an analytic arc γ^*, then f may be extended by 'reflection' to a function analytic in a domain containing $\Delta \cup \gamma$. The usefulness of this technique can hardly be overestimated: it provides an essential tool in problems involving the extension of conformal mappings and plays a traditional role in the 'slick' proof [49, pp. 322–325] of Picard's little theorem. Another application yields what is surely the simplest proof that a nonzero function analytic in Δ cannot vanish identically on an arc of Γ.

The question thus naturally arises whether an analogous result holds if γ^* is no longer analytic but simply smooth, C^∞ say. A negative answer is immediate. Indeed, let Γ^* be an infinitely differentiable, nowhere analytic, simple closed Jordan curve and let f map Δ conformally onto the interior D of Γ^*. The univalent function f, extends to a homeomorphism of $\Delta \cup \Gamma$ onto $D \cup \Gamma^*$ and induces a one-one correspondence between the points of Γ and those of Γ^*. However, f cannot be continued analytically across any subarc of Γ, for then f would establish an analytic correspondence between a subarc γ of Γ and a subarc γ^* of Γ^*. Thus γ^* would be analytic, contrary to hypothesis. This example is really quite striking, providing, as it does, an example of a (univalent!) function analytic on Δ and of class C^∞ on $\Delta \cup \Gamma$ which cannot be extended analytically across any arc of Γ.

What is not generally realized is that the example can be worked backward to provide an example of an infinitely differentiable, yet nowhere analytic, Jordan curve. This approach avoids altogether reliance on the plausible (and true) but nonobvious facts concerning smoothness and univalence of the boundary function which we invoked so shamelessly above. The tools we need are two, the first of which is the following simple lemma.

LEMMA. *Let* $f(z) = z + a_2 z + a_3 z^3 + \cdots$ *be analytic in* Δ. *Suppose that* $\Sigma_{n=2}^{\infty} n|a_n| < 1$. *Then f is continuous on $\Delta \cup \Gamma$ and univalent there.*

Proof. Continuity is clear from the absolute convergence of the series. Let $z, \zeta \in \Delta \cup \Gamma$. Then

$$\frac{f(z) - f(\zeta)}{z - \zeta} = 1 + \sum_{n=2}^{\infty} a_n(z^{n-1} + z^{n-1}\zeta + \cdots + \zeta^{n-1}).$$

Thus

$$\left| \frac{f(z) - f(\zeta)}{z - \zeta} \right| \geqq 1 - \sum_{n=2}^{\infty} n|a_n| > 0,$$

so that f is univalent.

The second ingredient we need is the celebrated Hadamard gap theorem.

HADAMARD GAP THEOREM. *Let* $f(z) = \Sigma_{k=0}^{\infty} a_k z^{n_k}$ *have* Δ *as its disc of convergence. If* $n_{k+1}/n_k \geqq q$ *for some* $q > 1$ *and all large* k, *then f has Γ as its natural boundary; that is, f cannot be continued analytically across any subarc of Γ.*

The beautiful proof of this theorem due to L. J. Mordell ([36], cf. [54, p. 223]) should be standard fare in graduate courses in complex analysis.

The construction of the required function is now almost trivial. We choose the sequences $\{a_k\}$ and $\{n_k\}$ to satisfy

(a) $a_0 = n_0 = 1$,
(b) $\Sigma_{k=1}^{\infty} n_k |a_k| < 1$,
(c) $(a_k)^{1/n_k} \to 1$,
(d) $n_{k+1}/n_k \geqq 2$,
(e) $\Sigma_{k=0}^{\infty} n_k^j |a_k| < \infty \qquad j = 0, 1, 2, \cdots$.

A simple concrete example is provided by the function

$$f(z) = z + \sum_{n=5}^{\infty} z^{2^n}/n!.$$

By the lemma, f establishes a homeomorphism between Γ and a simple closed Jordan curve Γ^*. Since f satisfies the hypothesis of Hadamard's gap theorem, f cannot be extended analytically across any arc of Γ. Hence, Γ^* must be nowhere analytic since otherwise the Schwarz principle would apply. Finally, by (e), the series for $f^{(j)}(z)$ converges absolutely on $\{z: |z| \leqq 1\}$ for each j; thus f is infinitely differentiable on $\Delta \cup \Gamma$, so that Γ^* is a C^{∞} curve.

Interestingly enough, one can trace the basic ideas of this section back to before the turn of the century, (see Osgood [38]). In particular, the lemma, which is usually attributed to the American topologist J. W. Alexander [67], was known to Fredholm as early as 1897 ([38, p. 17]).

5. Extensions. The reflection principle enables one (in certain circumstances) to extend a holomorphic function across an analytic arc to a somewhat larger domain. As we have seen, it is in general impossible to relax the condition of analyticity; nevertheless, the much weaker hypothesis of rectifiability suffices in case a continuous extension analytic in an abutting domain is already known. The precise result may be stated (somewhat informally) as follows.

THEOREM. *Let D be a domain and let J be a simple rectifiable Jordan arc dividing D into disjoint domains D_1 and D_2. Suppose f_j $(j = 1, 2)$ is analytic in D_j and continuous on $D_j \cup J$ and that $f_1 = f_2$ on J. Then the function f obtained by setting $f(z) = f_j(z)$ for $z \in D_j \cup J$ is analytic in D.*

The proof is a standard application of Morera's theorem, with due care exercised in dealing with the assumption that J is merely rectifiable.

The precise nature of the hypothesis of rectifiability on J in the above theorem is by no means clear, and the proof (which we leave to the reader) does little to explicate it. My experience has been that students — especially good ones — generally guess that the result remains true if rectifiability is dispensed with. This, however, is *not* the case, as the following example shows.

Let K be a compact set of positive Lebesgue measure and set

$$(8) \qquad\qquad f(z) = \iint_K \frac{d\xi d\eta}{\zeta - z} \qquad\qquad \zeta = \xi + i\eta.$$

The function $f(z)$ is obviously analytic off K and satisfies $f(\infty) = 0$; moreover since $\lim_{z \to \infty} zf(z) = - \iint_K d\xi d\eta \neq 0$, f is nonconstant on the unbounded component of K. We claim f is actually continuous on the complex sphere. Indeed, formula (8) exhibits f explicitly as the convolution of the locally (area) integrable function $1/\zeta$ with the bounded measurable function of compact support $\chi_K(\zeta)$, the characteristic function of K. Such a convolution is well known (and easily proved) to be continuous (see, for instance, [5, p. 154]).

Suppose now that $K = J$, a simple closed Jordan curve. The existence of such curves having positive area was first proved by Osgood [41] in 1902. (This is one of the relatively few examples in mathematics that retains its original vigor unimpaired: students today — even those who know about Peano curves — are as baffled and surprised by this fact as mathematicians were 70 years ago. The construction is not too complicated for presentation in class, and the example itself instills a healthy respect for the Jordan curve theorem.) One can actually construct J to have the additional property that it has positive area everywhere, that is, if D is an open set and $D \cap J \neq \emptyset$ then $D \cap J$ has positive area. The function f defined by (8) with $J = K$ is continuous on $\hat{\mathbb{C}} = \mathbb{C} \cup \{\infty\}$ and analytic off J; thus, it is analytic in both components D_1, D_2 of $\hat{\mathbb{C}} \setminus J$. However, f is not analytic at any point of J. Indeed, suppose f analytic at $z_0 \in J$ and let D be a small open disc about z_0 lying in the domain of analyticity of f. Set $J_1 = D \cap J$. Then

$$f(z) = \int_{J \setminus J_1} \frac{d\xi d\eta}{\zeta - z} + \int_{J_1} \frac{d\xi d\eta}{\zeta - z} = g(z) + h(z)$$

for $z \notin J$, and $g(z)$ is clearly analytic in D. Thus $h(z)$ must be analytic in D as well. But h is obviously analytic off $\bar{D}$ and continuous on $\hat{\mathbb{C}}$. Thus, according to the theorem of the present section, h is analytic on all of $\hat{\mathbb{C}}$, hence a constant. But $J \cap D = J_1$ has positive area, so that $h(z)$ is nonconstant. We have reached the desired contradiction.

Thus f cannot be continued analytically across any arc of J. In particular, the restrictions f_1, f_2 of f to the components D_1, D_2 of $\hat{\mathbb{C}} \setminus J$ determine analytic functions *which are not analytic continuations of one another*; indeed, J forms a natural boundary for each of these functions.

Actually, the requirement that J have positive measure was used merely to insure the existence of nontrivial functions continuous on $\hat{\mathbb{C}}$ and analytic off J. The same result can be obtained (but with more work) if the set in question has positive Hausdorff $(1 + \varepsilon)$-measure for some $\varepsilon > 0$ [61]. Even this condition is not necessary; in fact, Denjoy [11] has constructed an arc which is the graph of a function and which has the required property.

6. Blowing up the boundary. Questions involving length and area arise in conformal mapping as well. A conformal map, being analytic, must map sets of zero area to sets of zero area; however, distortion at the boundary is an *a priori* possibility. Writing $\Delta \cup \Gamma = \{z : |z| \leq 1\}$ as before, let us assume that the univalent function $f(z)$ maps Δ conformally onto the Jordan region D. According to the Osgood-Taylor-Carathéodory theorem, f extends to a homeomorphism of $\Delta \cup \Gamma$ onto $D \cup \partial D$. (Proofs of this important result, announced by Osgood [65] and proved independently by Osgood and Taylor [66, p. 294], and Carathéodory [6], [7] are available in [9, pp. 46–49] and [24, p. 129–134]. The reader will find a comparison of the treatments in these references particularly instructive in the matters of style of exposition and attention to detail.) In case ∂D is rectifiable, a theorem of the Riesz brothers [47] insures that f and f^{-1} preserve sets of zero length (= Hausdorff one-dimensional measure). When ∂D fails to be rectifiable, however, all hell breaks loose. In particular, a subset of ∂D having positive area may correspond to a subset of Γ having zero Lebesgue (linear) measure! For the construction, we need an important result from plane topology.

MOORE-KLINE EMBEDDING THEOREM [35]. *A necessary* and *sufficient condition* *that a compact set $K \subset \mathbb{C}$ should lie on a simple Jordan arc is that each closed connected subset of K should be either a point or a simple Jordan arc with the property that $K - \gamma$ does not accumulate at any point of γ, except (perhaps) the endpoints.*

Now let K be a Cantor set having positive area; K may be realized, for instance as the product of two linear Cantor sets, each of which has positive linear measure. Construct countably many disjoint simple Jordan arcs $J_n \subset \mathbb{C} \backslash K$ such that the sequence $\{J_n\}$ accumulates at each point of K and at no other points of $\hat{\mathbb{C}}$ and with the additional property that if $z_0 \in \mathbb{C} \backslash (K \cup \{J_n\}) = R$ and $z \in K$, then any arc from z_0 to z which lies, except for its final endpoint, in R must have infinite length. By the Moore-Kline embedding theorem, we may pass a simple closed Jordan curve J through $K \cup \{J_n\}$. Let f be a conformal map from Δ to D, the domain bounded by J. Then f extends to a homeomorphism from Γ to J. Let $S = f^{-1}(K)$. That S has zero linear measure follows at once from the following theorem, due to Lavrentiev.

THEOREM. *Let f be a conformal homeomorphism of $\Delta \cup \Gamma$ onto the Jordan domain $D \cup J$. If $S \subset J$ is not rectifiably accessible from D then $f^{-1}(S) \subset \Gamma$ has zero measure.*

Proof. Since D is a bounded domain, its area, given by the expression $\iint_{\Delta} |f'(z)|^2 dxdy$, is finite. Thus

$$\int_0^{2\pi} \int_0^1 |f'(re^{i\theta})| \, rdrd\theta$$

$$\leq \left(\int_0^{2\pi} \int_0^1 rdrd\theta \right)^{1/2} \left(\int_0^{2\pi} \int_0^1 |f'(re^{i\theta})|^2 r \, drd\theta \right)^{1/2} < \infty.$$

It follows that $\int_0^1 |f'(re^{i\theta})| \, r \, dr < \infty$ for almost all θ or, what is the same, $l(\theta) = \int_0^1 |f'(re^{i\theta})| \, dr < \infty$ almost everywhere. But $l(\theta)$ is the length of the image of the radius from 0 to $e^{i\theta}$ under f. So almost every point of Γ corresponds to a rectifiably accessible point of J, and we are done.

Actually, much more is true. It follows from a result of Beurling [3] (cf. [9, p. 56]) that the set of points on J which are not rectifiably accessible from D must correspond to a set of logarithmic capacity 0 on the unit periphery. It would take us too far afield to enter into a detailed discussion of the capacity of plane sets here; for our purposes it is enough to know that sets of capacity zero are exceedingly small. For instance, such a set must have zero Hausdorff ε-measure for all $\varepsilon > 0$. The first person to show that a set of capacity zero on Γ could correspond under a conformal mapping to a set having positive area was Kikuji Matsumoto [33]. He actually proved (what is implicit in the above discussion) that for each totally disconnected compact subset K of the plane there exists a Jordan domain D with boundary $J \supset K$ such that K corresponds under conformal mapping to a set of capacity zero on Γ. The discussion here (in particular, the ingenious proof of the central result) is based on an idea of Walter Schneider [51].

The *compression* of the boundary of the unit disc presents greater difficulties. Lavrentiev, however, has shown that a set of positive measure on Γ may be mapped onto a set of zero length under a conformal mapping of Jordan domains [30]. A more recent construction is due to McMillan and Piranian [32].

7. Absolute convergence and uniform convergence. Conformal mapping techniques are also useful in constructing examples concerning the convergence of power series and Fourier series. Below we offer some simple but instructive examples.

The first example of a power series which converges uniformly but not absolutely on the closed unit disc was given by Fejér [15], cf. [25, vol. 1, p. 122]. The following geometric example, due to Gaier [68] and rediscovered by Piranian (see [1, pp. 289, 314]), is particularly appealing. Let D be the region of figure 1, a triangle from

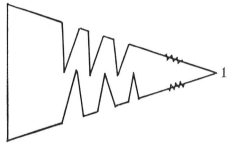

Fig. 1

which wedges have been removed in such a way that the vertex at $z = 1$ is not rectifiably accessible from the interior of D. Since D is a Jordan region, any conformal map of $\Delta = \{z : |z| < 1\}$ onto D extends to a homeomorphism of the closed regions. Suppose that $f(z) = \Sigma a_n z^n$ is such a homeomorphism satisfying $f(1) = 1$. Clearly,

$$(9) \qquad \int_0^1 |f'(r)| \, dr = \int_0^1 \left| \sum_{n=1}^{\infty} n a_n r^{n-1} \right| dr$$

$$\leq \int_0^1 \left(\sum_{n=1}^{\infty} n |a_n| r^{n-1} \right) dr = \sum_{n=1}^{\infty} |a_n|.$$

Since the length of the image of $[0,1]$ under f is infinite and is given by the extreme left member of (9), the series for f is not absolutely convergent. That the series is uniformly convergent on the closed disc follows from a result due to Fejér.

FEJÉR'S TAUBERIAN THEOREM [16], [55, p. 357]. *Let $f(z) = \Sigma_{n=0}^{\infty} a_n z^n$ and suppose $\Sigma_{n=1}^{\infty} n |a_n|^2 < \infty$. If $\lim_{r \to 1-} f(re^{i\theta}) = f(e^{i\theta})$ exists, then the sum $\Sigma_{n=0}^{\infty} a_n e^{in\theta}$ exists and is equal to $f(e^{i\theta})$. Moreover, if $\lim_{r \to 1-} f(re^{i\theta}) = f(e^{i\theta})$ uniformly for $\theta_1 \leq \theta \leq \theta_2$, then $\Sigma_{n=0}^{\infty} a_n e^{in\theta} = f(e^{i\theta})$ uniformly for $\theta_1 \leq \theta \leq \theta_2$.*

This is a typical theorem of Tauberian type, the Tauberian condition being, of course, $\sum_{n=1}^{\infty} n |a_n|^2 < \infty$.

Proof. Set $s_N(e^{i\theta}) = \sum_{n=0}^{N} a_n e^{in\theta}$. Then

$$\left| f(re^{i\theta}) - s_N(e^{i\theta}) \right| \leq \sum_{n=1}^{N} |a_n|(1 - r^n) + \sum_{n=N+1}^{\infty} |a_n| r^n = S_1 + S_2.$$

Now $1 - r^n \leq n(1 - r)$ (divide both sides by $1 - r$). Thus,

$$S_1 \leq (1 - r) \sum_{n=1}^{N} n |a_n| \leq (1 - r) \left(\sum_{n=1}^{N} n \right)^{1/2} \left(\sum_{n=1}^{N} n |a_n|^2 \right)^{1/2} \leq KN(1 - r),$$

where $K = (\sum_{n=1}^{\infty} n |a_n|^2)^{1/2}$. Here we have used the Cauchy-Schwarz inequality and the fact that $\sum_{n=1}^{N} = N(N + 1)/2 \leq N^2$. Applying Cauchy-Schwarz to S yields

$$S_2 = \sum_{n=N+1}^{\infty} \sqrt{n} |a_n| \frac{r^n}{\sqrt{n}} \leq \left(\sum_{n=N+1}^{\infty} \frac{r^{2n}}{n} \right)^{1/2} \left(\sum_{n=N+1}^{\infty} n |a_n|^2 \right)^{1/2}$$

$$\leq \left(\frac{1}{N(1 - r)} \sum_{n=N+1}^{\infty} n |a_n|^2 \right)^{1/2}$$

since

$$\sum_{n=N+1}^{\infty} r^{2n}/n \leq 1/N \sum_{n=0}^{\infty} r^n = 1/N(1 - r).$$

Having fixed N, we may, by the intermediate value theorem for continuous functions, choose $r = r_N$ such that $N(1 - r_N) = (\sum_{n=N+1}^{\infty} n |a_n|^2)^{1/2}$. Clearly, as $N \to \infty$, $r_N \to 1$. Thus

$$\left| f(r_N e^{i\theta}) - s_N(e^{i\theta}) \right| \leq K \left(\sum_{n=N+1}^{\infty} n |a_n|^2 \right)^{1/2} + \left(\sum_{n=N+1}^{\infty} n |a_n|^2 \right)^{1/4},$$

and the right hand side tends to 0 as $N \to \infty$ since $\sum_{n=1}^{\infty} n |a_n|^2 < \infty$. Since $f(re^{i\theta}) \to f(e^{i\theta})$, $s_N(e^{i\theta}) \to f(e^{i\theta})$; hence $\sum_{n=0}^{\infty} a_n e^{in\theta} = f(e^{i\theta})$. Finally, all our calculations are uniform in θ, so if $f(re^{i\theta}) \to f(e^{i\theta})$ uniformly on some arc, then $\sum_{n=0}^{\infty} a_n e^{in\theta} = f(e^{i\theta})$ uniformly on that arc.

To apply Fejér's theorem to the situation at hand, simply note that if f maps Δ conformally onto the Jordan region D then (by the Osgood-Taylor-Carathéodory theorem) f extends continuously to $\Delta \cup \Gamma$ so that $f(re^{i\theta}) \to f(e^{i\theta})$ uniformly for $0 \leq \theta \leq 2\pi$. Since

$$\pi \sum_{n=1}^{\infty} n |a_n|^2 = \int_0^1 \int_0^{2\pi} |f'(re^{i\theta})|^2 \, d\theta \, r \, dr = \text{area of } D < \infty,$$

the Taylor series for f converges uniformly on Γ.

We should observe that it is easy to modify the domain of figure 1 so that its boundary becomes analytic at every point except $z = 1$. The corresponding mapping function then extends (by Schwarz reflection) across $\Gamma \backslash \{1\}$ and yields a function univalent and analytic on a domain containing $(\Delta \cup \Gamma) \backslash \{1\}$ whose Taylor series converges uniformly but not absolutely on $\Delta \cup \Gamma$.

8. Fourier series. One of the loveliest applications of complex analysis to real variables occurs in the theory of Fourier series. The result in question is the so-called Pál-Bohr theorem, which may be stated as follows.

PÁL-BOHR THEOREM. *Let $f(e^{i\theta})$ be a continuous real-valued function on the unit circle Γ. There is a self-homeomorphism ϕ of Γ such that the Fourier series of $f \circ \phi$ converges uniformly.*

It is well known, of course, that the Fourier series of a continuous function may diverge on a dense subset of Γ [28, p. 58]; this gives the Pál-Bohr theorem added poignancy. On the other hand, a deep and famous result of Lennart Carleson [64] insures that the Fourier series of a continuous function converges *almost* everywhere in the sense of Lebesgue measure.

The Pál-Bohr theorem has an interesting history. It was first proved by Jules Pál in 1914 with the weaker conclusion that uniform convergence could be obtained on any *proper* closed subarc of Γ, however large. Bohr [4], in 1935, removed the restriction in Pál's theorem. Finally, in 1944, Salem [50] introduced a trick which yields the full strength of the result very quickly.

Proof of the Pál-Bohr Theorem. Regard f as a function on the interval $[-\pi, \pi]$ satisfying the periodicity condition $f(-\pi) = f(\pi)$. We rule out at the outset the trivial case in which f is identically constant. By adding, if necessary, a continuous periodic function of bounded variation, we may assume that $f(-\pi) = f(\pi) = f(x)$ for exactly one point $x \in (-\pi, \pi)$. (This is Salem's trick; see [50] for a complete verification.) Since the Fourier series of a continuous function of bounded variation converges uniformly, it is enough to prove the theorem under this additional assumption. Let g be a continuous periodic function on $[-\pi, \pi]$ which increases on $(-\pi, x)$ and decreases on (x, π). Then the image of $[-\pi, \pi]$ under the map $H(t) = g(t) + if(t)$ is a simple closed Jordan curve J in the plane.

Let $F(z) = \sum_{n=0}^{\infty} a_n z^n$ be a Riemann map of Δ onto the interior of J such that $F(-1) = H(-\pi)$. Then F extends to a homeomorphism of Γ onto J, and by the discussion following the proof of Fejér's theorem, the series $F(e^{i\theta}) = \sum_{n=0}^{\infty} a_n e^{in\theta}$ converges uniformly on Γ. The required homeomorphism of $[-\pi, \pi)$ is obtained by setting $\phi(t) = H^{-1}(F(e^{it}))$. Indeed, this is clearly a homeomorphism, and $f(\phi(t)) = f \circ H^{-1}(F(e^{it})) = \operatorname{Im} F(e^{it})$, which has a uniformly convergent Fourier series since $F(e^{it})$ does.

Perhaps surprisingly, the argument given above is (essentially) the *only* known proof of this theorem. Whether an analogous result holds for complex-valued functions remains an open question; of course, this is equivaient to the question of whether, given *two* real-valued continuous functions f, g on Γ, one can find a single homeomorphism ϕ for which $f \circ \phi$ and $g \circ \phi$ both have uniformly convergent Fourier series.

We learned of the Pál-Bohr theorem from the interesting survey article of Goffman and Waterman [20], and our treatment parallels the discussion given there. The decision to reproduce the proof in some detail was based on our feeling that this beautiful result deserves a wider public.

9. Harmonic conjugates. A somewhat different application of conformal mapping to problems involving Fourier series involves the construction of functions having certain prescribed bad boundary behavior. Thus, one may ask (and Prof. A. Devinatz did) for an explicit example of a function harmonic on Δ and continuous on $\Delta \cup \Gamma$ whose harmonic conjugate is discontinuous but bounded. Although the problem has been framed (for simplicity) in terms of harmonic functions, it is actually a pure real variable question concerning the lack of smoothness of a certain singular integral operator.

For the solution, consider the simply connected domain D, indicated in Figure 2, bounded by an (open) analytic curve J together with its asymptote the segment

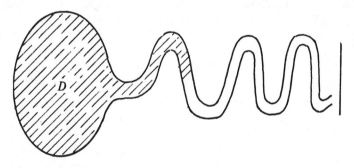

Fig. 2

$\{y: -1 \leq y \leq 1\}$ of the y-axis in the complex plane. Map Δ conformally onto D by the univalent function $f(z) = u(z) + iv(z)$. A standard result in conformal mapping [55, p. 353] insures that a single point of Γ, say 1, corresponds to the "bad" part of the boundary and that f establishes a homeomorphism between $\Gamma \setminus \{1\}$ and J. By the reflection principle, f actually extends analytically across $\Gamma \setminus \{1\}$. One proves that as $z \to 1$, $u(z) \to 0$; and it is now obvious that u is not only harmonic on Δ and harmonically extendible across $\Gamma \setminus \{1\}$ but also continuous on $\Delta \cup \Gamma$. On the other hand, the harmonic function v, which is clearly bounded, is *not* continuous at $z = 1$. The details of the proof will be easily supplied by anyone familiar with Carathéodory's important theory of prime ends [7], [55, pp. 352–355], [9]. An obvious modification yields a bounded continuous function whose conjugate is unbounded.

10. Tauberian theorems. Tauberian theorems, such as Fejér's, have an intrinsic interest quite independent of applications. Of these, the most celebrated is certainly that due to Littlewood, which states that if $\lim_{r \to 1-} \sum_{n=0}^{\infty} a_n r^n = L$ exists and $a_n = O(1/n)$, then $\sum_{n=0}^{\infty} a_n = L$. This result resisted considerable efforts at proof for several years before it was finally settled by Littlewood [31], whose argument required six pages of ingenious and delicate analysis. Much later, Karamata [27] introduced a new technique, based on approximation theory, resulting in an enormous simplification of the proof. Much less well-known is Wielandt's modification [59] of Karamata's proof, which yields a simple and transparent proof of the theorem in question. Below, we present Wielandt's proof of a strengthened version (due to Hardy and Littlewood [22]) of the Littlewood Tauberian theorem.

THEOREM. *Let* $f(z) = \sum_{n=0}^{\infty} a_n z^n$ *be analytic in* $|z| < 1$ *and suppose that the* a_n *are real and that* $na_n \leq K$ *for some* $K > 0$. *If* $\lim_{r \to 1-} f(r) = L$ *exists as* $r \to 1 -$, *then* $\sum_{n=0}^{\infty} a_n = L$.

The advantage of this result over the original Littlewood theorem lies, of course, in the fact that the order estimate on the coefficients is replaced by a *one-sided* bound.

Proof. Trivial normalizations allow us to assume that $L = 0$, $a_0 = 0$, $K = 1$. Consider the family $\mathscr{F}$ of real functions $\phi(x)$ on $(0, 1)$ which satisfy

   (a) $\sum_{n=1}^{\infty} a_n \phi(x^n)$ is convergent for $x \in (0, 1)$,
   (b) $\Phi(x) = \sum_{n=1}^{\infty} a_n \phi(x^n) \to 0$ as $x \to 1 -$.

Clearly, if $\phi(x) \in \mathscr{F}$, $\phi(x^k) \in \mathscr{F}$ $(k = 1, 2, \cdots)$ and $\mathscr{F}$ is closed under linear combinations. Since (by hypothesis) $x \in \mathscr{F}$, each polynomial vanishing at the origin belongs to $\mathscr{F}$. The proof depends on a simple lemma concerning the approximation of functions.

LEMMA. *Let* $\phi(x)$ *satisfy* (a). *Suppose that for each* $\varepsilon > 0$ *there exist polynomials* $p_1(x)$, $p_2(x)$ *such that* $p_i(0) = 0$, $p_i(1) = 1$ $(i = 1, 2)$ *and*

$$p_1(x) \leq \phi(x) \leq p_2(x) \qquad \frac{p_2(x) - p_1(x)}{x(1 - x)} = q(x) > 0,$$

where $\int_0^1 q(x)dx < \varepsilon$. *Then* $\phi(x)$ *satisfies* (b) *and hence belongs to* $\mathscr{F}$.

Proof of Lemma. Let $\Phi(x) = \sum_{n=1}^{\infty} a_n \phi(x^n)$, $q(x) = \sum_{k=0}^{r} b_k x^k$. Then

$$\Phi(x) - \sum_{n=1}^{\infty} a_n p_1(x^n) = \sum_{n=1}^{\infty} a_n(\phi(x^n) - p_1(x^n)) \leq \sum_{n=1}^{\infty} \frac{1}{n}(p_2(x^n) - p_1(x^n))$$

$$= \sum_{n=1}^{\infty} \frac{1}{n} (1 - x^n)x^n q(x^n) \leqq (1 - x) \sum_{n=1}^{\infty} x^n q(x^n)$$

$$= (1 - x) \sum_{k=0}^{r} b_k \sum_{n=1}^{\infty} x^{n(k+1)} = \sum_{k=0}^{r} \frac{b_k(1 - x)x^{k+1}}{1 - x^{k+1}} \rightarrow \sum_{k=0}^{r} \frac{b_k}{1 + k}$$

$$= \int_{0}^{1} q(x)dx < \varepsilon \text{ as } x \rightarrow 1 - .$$

Here we have used the fact that $1 - x^n \leqq n(1 - x)$ and that $(1 - x^n)/(1 - x) \rightarrow n$ as $x \rightarrow 1$. Since $p_1(x) \in \mathscr{F}$, $\sum_{n=1}^{\infty} a_n p_1(x^n) \rightarrow 0$ as $x \rightarrow 1 -$ so that $\Phi(x) < \varepsilon$ for x near 1. Consideration of $p_2(x) - \phi(x)$ shows similarly that $\Phi(x) > -\varepsilon$ if x is near enough to 1. Thus $\Phi(x) \rightarrow 0$ as $x \rightarrow 1$, so that $\phi \in \mathscr{F}$.

Continuing with the proof of the theorem, let

$$\phi^*(x) = \begin{cases} 0 & 0 \leqq x < \frac{1}{2} \\ 1 & \frac{1}{2} \leqq x \leqq 1 \end{cases}$$

so that $\Phi(x) = \sum_{n=1}^{\infty} a_n \phi^*(x^n) = \sum_{2x^n \geqq 1} a_n = \sum_{n=1}^{N} a_n = s_N$, where

$$N = \left[\log 2 / \log \frac{1}{x} \right].$$

It suffices to show that $\phi^*(x) \in \mathscr{F}$, for then $s_N \rightarrow 0$ as $N \rightarrow \infty$, whence $\sum_{n=0}^{\infty} a_n = 0$ as required. Now ϕ^* clearly satisfies (a), so it is enough to show that the conditions of the lemma are fulfilled. Since continuous functions are dense in the integrable functions, we can find continuous functions $g_1(x)$ and $g_2(x)$ such that

$$(10) \qquad g_1(x) < \frac{\phi^*(x) - x}{x(1 - x)} < g_2(x) \qquad \int_{0}^{1} [g_2(x) - g_1(x)]dx < \varepsilon.$$

The functions g_1 and g_2 may then be approximated uniformly by polynomials q_1 and q_2 in such a way that (10) still holds with the g_i's replaced by the q_i's. Putting $p_i(x) = x + x(1 - x)q_i(x)$, $q(x) = q_2(x) - q_1(x)$, we obtain polynomials satisfying the hypothesis of the lemma. This completes the proof.

The subject of Tauberian theorems extends far beyond questions concerning the convergence or divergence of a power series on its circle of convergence. One of the central results in the harmonic analysis of the real line is Wiener's Tauberian theorem, which states that if $f \in L^1(\mathbb{R})$ and the Fourier transform of f never vanishes, then linear combinations of translates of f are dense in $L^1(\mathbb{R})$. The relation between the theorems of Wiener and Littlewood is far from obvious, and it has become customary to deduce the latter from the former by way of explicating the Tauberian character of Wiener's theorem. This deduction is standard and may be found, for instance, in [60, pp. 104–106]. The proof involves the function $K(x) = e^{-x} \exp(-e^{-x})$ and uses the fact that the gamma function $\Gamma(z)$ has no zeroes on the line $\text{Re } z = 1$.

Unfortunately, the *deduction* of Littlewood's theorem from Wiener's is longer and significantly more complicated in both conception and detail than Wielandt's proof of the (more general!) Hardy-Littlewood theorem: it is a little like proving that the medians of a triangle are concurrent by invoking the fact that a nested sequence of compact sets has nonvoid intersection. Of course, Wiener's powerful methods have applications in many situations where the simple approximation theory argument we have given does not apply.

One such instance concerns the so-called high indices theorem.

HIGH INDICES THEOREM. *Let* $f(z) = \sum_{k=0}^{\infty} a_k z^{n_k}$ *be analytic in* $|z| < 1$ *and suppose that* $n_{k+1}/n_k \geqq q > 1$ *for all* k. *If* $\lim_{r \to 1^-} f(r) = L$ *exists, then* $\sum_{k=0}^{\infty} a_k = L$.

This theorem lies considerably deeper than Littlewood's theorem or its extension proved above; it was first proved, by Hardy and Littlewood, in 1925 [23], having been conjectured by Littlewood as early as 1910. The novelty of the result lies in the fact that the Tauberian condition (the lacunarity of the sequence of coefficients) involves no bound on the size of the coefficients. It is most instructive to try to apply the ideas used in proving the Tauberian theorems of Fejér and Hardy-Littlewood to the high indices theorem: they all fail miserably. In fact, I am aware of no really simple proof of this result. A particularly attractive argument, marked by considerable ingenuity in the use of such tools as the Phragmén-Lindelöf principle and Blaschke products, has been given by Halász [21], following some ideas of the German mathematician Dieter Gaier.

In concluding this section we should like to mention an amusing sidelight. Wielandt's proof of the Hardy-Littlewood theorem shares, with Mordell's proof of the Hadamard gap theorem, the property of being a gem of complex analysis mined by a mathematician whose central interests lay altogether outside analysis. The late Professor Mordell was, of course, one of the world's leading number theorists; Professor Wielandt is a group theorist of international repute. Is there a moral to be drawn here?

11. Category. The usual theorems on convergence of sequences of analytic functions, such as Vitali's convergence theorem [54, p. 168], require the uniform boundedness of the sequence in question on compact subsets of the domain. There is, however, a sometimes useful result, due to Osgood, which avoids altogether hypotheses other than simple pointwise convergence.

OSGOOD'S THEOREM [40]. *Let* D *be a domain and let* $\{f_n\}$ *be a sequence of functions analytic in* D. *Suppose* $f_n(z) \to f(z)$ *for each* $z \in D$. *Then* f *is analytic in an open set* $D_1 \subset D$ *which is dense in* D, *and convergence is uniform on compact subsets of* D_1.

This result has been rediscovered countless times and has on innumerable other occasions brought the experts to grief. Indeed, the question as to whether f must be analytic *anywhere*, appears (happily, with a correct solution) in the problem section of a recent symposium [**29**, p. 543]. The present formulation suggests — correctly — the use of the Baire category theorem.

Proof of Osgood's Theorem. Let $F_m = \{z\colon |f_n(z)| \leq m, n = 1, 2, 3, \cdots\}$. The F_m are clearly relatively closed in D and $\cup F_m = D$. By the Baire category theorem, some F_m must have interior. For this m, the sequence $\{f_n\}$ is uniformly bounded on F_m^0 hence by Vitali's theorem converges uniformly on compact subsets of F_m^0 to an analytic function. Thus f is analytic on F_m^0. Since the argument can be applied to any subdomain R of D — in particular, to an arbitrary disc — it follows that f must be analytic on a dense open subset D_1 of D. That convergence is uniform on compacta contained in D_1 is a standard argument, which we suppress.

A comment is perhaps in order on our use of the Baire category theorem, which states that a complete metric space is not the countable union of closed nowhere dense sets. Obviously, D is *not* complete in the Euclidean metric. However, it is easy to see that D can be given a new metric which induces the same (Euclidean) topology, under which D is complete. Alternatively, one may replace D by a slightly smaller *compact* set K and relativize the argument to K. We should also mention that category arguments appear elsewhere in complex analysis as well. A notable example is the proof of Hartogs' theorem, mentioned earlier in Section 2.

A nice complement to Osgood's theorem is provided by an example of a sequence of entire functions $f_n(z)$ with the property that

$$(11) \qquad\qquad \lim_{n \to \infty} f_n(z) = \begin{cases} 0 & z \neq 0 \\ 1 & z = 0. \end{cases}$$

There are (at least) two essentially distinct ways of constructing such a sequence. One method is to construct an entire function $F(z)$ such that $F(0) = 1$ and $F(z) \to 0$ as $|z| \to \infty$ on each ray through the origin. Such functions were first ex-

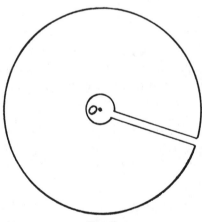

Fɪɢ. 3

hibited by Mittag-Leffler; see [34] for a detailed discussion and some surprising extensions. The construction of F is, with appropriate hints, a nice and doable exercise and occurs as such in Rudin's text [49, pp. 326–327]. Once F has been obtained, one observes that the sequence $f_n(z) = F(nz)$ satisfies (11).

Alternatively, one can apply Runge's theorem to "notched annuli" to construct polynomials satisfying (11). To be explicit, consider the set indicated in figure 3. A moment's reflection will reveal that one can choose a sequence K_n of such sets, with the property that for each $z \in \mathbb{C} \setminus \{0\}$ there exists an integer N such that $z \in K_n$ for all $n \geq N$. (The inner circle contracts, the outer circle expands, and the notch gets thinner and rotates, tending toward but never reaching its limiting line.) The function

$$g_n(z) = \begin{cases} 0 & z \in K_n \\ 1 & z = 0 \end{cases}$$

clearly extends to a function analytic in a neighborhood of the (disconnected) set $K_n \cup \{0\}$. Since this set does not separate the plane, g may be approximated uniformly (to within $1/n$, say) on $K_n \cup \{0\}$ by a polynomial p_n. These polynomials clearly satisfy (11).

12. Miscellany. The interactions between real and complex analysis are by no means limited to the areas mentioned above. To keep the discussion within manageable limits, we have restricted ourselves to (a subset of) those applications, examples, and aspects of the theory that have not found sustained treatment in the "popular" literature of texts and survey articles. Subjects which are treated elsewhere at adequate length but which deserve mention here by virtue of their interdisciplinary nature include the following:

(a) *The evaluation of real integrals and sums by residue techniques.* This is surely one of the most striking applications of complex function theory to real analysis. Fortunately, any good text on complex analysis will contain a fairly detailed discussion.

(b) *Complex methods in harmonic analysis.* This is a substantial area, which includes topics as diverse as interpolation theorems (see, for instance, [28, pp. 93–98]) and theorems of Paley-Wiener type [43]. Two of the most attractive recent texts in harmonic analysis [13], [28] devote whole chapters to this aspect of the theory. Further developments are discussed in the survey article of Weiss [57].

(c) *Functional analysis.* Complex variable methods appear here perhaps most notably in the construction of functional calculi for operators on Hilbert space or Banach space. The applications to commutative Banach algebras are particularly substantial; indeed, parts of this last-named subject are virtually coextensive with certain aspects of several complex variable theory. For further references, see [17] and [58]. In the opposite direction, techniques of functional analysis can be used to establish many results in function theory; this is the programme of [48]. Finally, an honest partnership between complex variables and functional analysis occurs in the study of certain Banach spaces of analytic functions, especially H^p spaces [12], [26].

(d) *Function theoretic methods in differential equations.* Complex methods occur rather naturally in the study of ordinary differential equations [8]. Their appearance in the study of partial differential equations is perhaps more surprising. Yet there are substantial applications, and more than one book [2], [19] has been devoted to this area. Further applications of function theory to problems in partial differential equations will be found in [18]. In a rather different direction, the theory of linear partial differential equations with constant coefficients is intimately connected with the study of certain spaces of entire functions of several complex variables; see [14] for an exhaustive treatment.

13. Monodromy. No excursion onto the bypaths of complex analysis would be complete without some mention of the monodromy theorem.

MONODROMY THEOREM. *Let D be a simply connected domain and let $f(z)$ be analytic in a neighborhood of $z_0 \in D$. Then if $f(z)$ can be continued analytically from z_0 along every path lying in D, the continuation gives rise to a single-valued function analytic on all of D.*

A more general version states that analytic continuation along paths is a homotopy invariant; see, for instance, [53]. Like the reflection principle, the monodromy theorem is an essential ingredient in the short proof of Picard's little theorem; in its extended form, it is the central result in the subject of analytic continuation. Yet no theorem of basic complex analysis is more abused or less understood. Indeed, it has been misapplied more than once even by mathematicians of the first rank (and specialists in complex analysis, at that!). One may speculate that a source of at least some of the confusion surrounding this result is the essentially topological, rather than function-theoretic, nature of the theorem.

The sort of error into which one may lapse is best indicated by an explicit example. Let D be a simply connected domain and f a function analytic in D which satisfies $f'(z) \neq 0$ on D. Suppose $R = f(D)$ is also simply connected. Question: Must f be univalent (one-one)? An affirmative answer may be found in [56, p. 243] and in other references as well. The argument is as follows. At each point $w_0 \in R$ one may define a local inverse $f_{w_0}^{-1}(w)$ of f, analytic in a neighborhood of w_0. Since R is simply connected, the totality of these functions defines a single-valued analytic function f^{-1} on R, which is a global inverse for f. Thus f must be univalent. Note further that the simple-connectivity of D is quite extraneous to the demonstration.

Unfortunately, the argument given above is altogether incorrect, since the essential hypothesis of the monodromy theorem, that analytic continuation be possible along every path in R, has not been verified. Can the proof be salvaged?

The answer is no. In fact, consider the function $f(z) = \int_0^z e^{\zeta^2} d\zeta$. This f is analytic (entire) on the simply connected domain $\mathbb{C}$ and $f'(z) = e^{z^2}$ is nowhere zero. Clearly, f is not univalent. So it suffices to prove that $f(\mathbb{C})$ is simply connected. We claim $f(\mathbb{C}) = \mathbb{C}$. Indeed, suppose $f(z) \neq w$. Since e^{z^2} is an even function, $f(z)$ is odd, so that if f fails to take on the value w it also misses the value $-w$. If $w \neq 0$, this contradicts Picard's (small) theorem. Since $f(0) = 0$, f takes on every value in the complex plane.

For $D = \mathbb{C}$, any function which satisfies $f'(z) \neq 0$ must be transcendental and hence must (by Picard's theorem) take on most values infinitely often. One can, however, construct a non-univalent, locally univalent function mapping the disc $\Delta = \{z : |z| < 1\}$ onto itself, which takes on no value more than three times. The extremely elegant example given above is due to D. S. Greenstein and appears as a solution to MONTHLY Problem 4740. It is an appropriate note on which to end this survey.

Preparation of this paper was supported in part by NSF GP 28970.

References

1. László Alpár, Sur certaines transformées des séries de puissance absolument convergentes sur la frontière de leur cercle de convergence, Magyar Tud. Akad. Mat. Kutató Int. Közl., 7 (1962) 287–316.

2. Stefan Bergman, Integral Operators in the Theory of Linear Partial Differential Equations, Springer-Verlag, 1969.

3. Arne Beurling, Ensembles exceptionnels, Acta Math., 72 (1940) 1–13.

4. Harold Bohr, Über einen Satz von J. Pál, Acta Sci. Math. (Szeged), 7 (1935) 129–135.

5. Andrew Browder, Introduction to Function Algebras, Benjamin, New York, 1969.

6. C. Carathéodory, Über die gegenseitige Beziehung der Ränder bei der konformer Abbildung des Inneren einer Jordanschen Kurve auf einen Kreis, Math. Ann., 73 (1913) 305–320.

7. ———, Über die Begrenzung einfach zusammenhängender Gebiete, Math. Ann., 73 (1913) 323–370.

8. Earl A. Coddington and Norman Levinson, Theory of Ordinary Differential Equations, Mc-Graw-Hill, New York, 1955.

9. E. F. Collingwood and A. J. Lohwater, The Theory of Cluster Sets, Cambridge University Press, New York, 1966.

10. E. H. Connell, On properties of analytic functions, Duke Math. J., 28 (1961) 73–81.

11. Arnaud Denjoy, Sur la continuité des fonctions analytiques singulières, Bull. Soc. Math. France, 60 (1932) 27–105.

12. Peter L. Duren, Theory of H^p Spaces, Academic Press, New York, 1970.

13. H. Dym and H. P. McKean, Fourier Series and Integrals, Academic Press, New York, 1972.

14. Leon Ehrenpreis, Fourier Analysis in Several Complex Variables, Wiley-Interscience, New York, 1970.

15. L. Fejér, Über gewisse Potenzreihen an der Konvergenzgrenze, Sitzungsber. Bayer. Akad. Wiss. Math. -phys. Kl., no. 3 (1910) 1–17.

16. ———, La convergence sur son cercle de convergence d'une série de puissance effectuant une représentation conforme du cercle sur le plan simple, C. R. Acad. Paris, 156 (1913), 46–49.

17. Theodore W. Gamelin, Uniform Algebras, Prentice-Hall, Englewood Cliffs, N.J., 1969.

594 LAWRENCE ZALCMAN

18. P. R. Garabedian, Partial Differential Equations, Wiley, New York, 1964.

19. Robert P. Gilbert, Function Theoretic Methods in Partial Differential Equations, Academic Press, New York, 1969.

20. Casper Goffman and Daniel Waterman, Some aspects of Fourier series, this MONTHLY, 77 (1970) 119–133.

21. Gábor Halász, Remarks on a paper of D. Gaier on gap theorems, Acta Sci. Math. (Szeged), 28 (1967) 311–322.

22. G. H. Hardy and J. E. Littlewood, Tauberian theorems concerning power series and Dirichlet's series, whose coefficients are positive, Proc. London Math. Soc., 13 (1914) 174–191.

23. ――――, and ――――, A further note on the converse of Abel's theorem, Proc. London Math. Soc., 25 (1926) 219–236.

24. Maurice Heins, Selected Topics in the Classical Theory of Functions of a Complex Variable, Holt, Rinehart, and Winston, New York, 1962.

25. Einar Hille, Analytic Function Theory vols. 1 and 2, Ginn, New York, 1959 and 1962.

26. Kenneth Hoffman, Banach Spaces of Analytic Functions, Prentice-Hall, Englewood Cliffs, N. J., 1962.

27. J. Karamata, Über die Hardy-Littlewoodschen Umkehrungen des Abelschen Stetigkeitssatzes, Math. Zeitschr., 32 (1930) 319–320.

28. Yitzhak Katznelson, An Introduction to Harmonic Analysis, Wiley, New York, 1968.

29. Jacob Korevaar, ed., Entire Functions and Related Parts of Analysis, Proc. Symp. Pure Math. vol. 16, Amer. Math. Soc., 1968.

30. M. Lavrentieff, Sur quelques problèmes concernant les fonctions univalentes sur la frontière, Mat. Sb. N. S. 1, (43) (1936) 816–846. (Russian)

31. J. E. Littlewood, The converse of Abel's theorem on power series, Proc. London Math. Soc., 9 (1911) 434–448.

32. J. E. McMillan and George Piranian, The compression of the boundary, to appear.

33. Kikuji Matsumoto, On some boundary problems in the theory of conformal mappings of Jordan domains, Nagoya Math. J., 24 (1964) 129–141.

34. G. Mittag-Leffler, Sur la représentation analytique d'une branche uniforme d'une fonction monogène VI, Acta Math., 42 (1920) 285–308.

35. R. L. Moore and J. R. Kline, On the most general plane closed set through which it is possible to pass a simple continuous arc, Ann. Math., (2) 20 (1919) 218–223.

36. L. J. Mordell, On power series with circle of convergence as a line of essential singularities, J. London Math. Soc., 2 (1927) 146–148.

37. G. Morera, Un teorema fondamentale nella teoria delle funzioni di una variabile complessa, Rend. del R. Instituto Lombardo di Scienze e Lettere, (2) 19 (1886) 304–307.

38. W. F. Osgood, Example of a single-valued function with a natural boundary, whose inverse is also single-valued, Bull. Amer. Math. Soc., 4 (1898) 417–424; Supplementary note, *ibid.* 5 (1898) 17–18.

39. ――――, Some points in the elements of the theory of functions, Bull. Amer. Math. Soc., 2 (1896) 296–302.

40. ――――, Note on the functions defined by infinite series whose terms are analytic functions of a complex variable, with corresponding theorems for definite integrals, Ann. Math., (2) 3 (1901) 25–34.

41. ――――, A Jordan curve of positive area, Trans. Amer. Math. Soc., 4 (1903) 107–112.

42. Jules Pál, Sur des transformations de fonctions qui font converger leurs séries de Fourier, C. R. Acad. Sci. Paris, 158 (1914) 101–103.

43. Raymond E. A. C. Paley and Norbert Wiener, Fourier transforms in the complex domain, Colloq. Publ. no. 19, Amer. Math. Soc., 1934.

44. P. Porcelli and E. H. Connell, A proof of the power series expansion without Cauchy's formula, Bull. Amer. Math. Soc., 67 (1961) 177–181.

45. Alfred Pringsheim, Ueber den Goursat'schen Beweis des Cauchy'schen Integralsatzes, Trans. Amer. Math. Soc., 2 (1901) 413–421.

46. A. H. Read, Higher derivatives of analytic functions from the standpoint of topological analysis, J. London Math. Soc., 36 (1961) 345–352.

47. F. Riesz and M. Riesz, Über die Randwerte einer analytischen Funktion, C. R. du Congrès des Math. Scandinaves, (1916) 27–44.

48. L. A. Rubel and B. A. Taylor, Functional analysis proofs of some theorems in function theory, this MONTHLY, 76 (1969) 483–489.

49. Walter Rudin, Real and Complex Analysis, McGraw-Hill, New York, 1966.

50. R. Salem, On a theorem of Bohr and Pál, Bull. Amer. Math. Soc., 50 (1944) 579–580.

51. Walter J. Schneider, A short proof of a conformal mapping theorem of Matsumoto, Notices Amer. Math. Soc., 16 (1969) 549.

52. Harold S. Shapiro, Smoothing and Approximation of Functions, Van Nostrand, Princeton, N.J., 1969.

53. George Springer, Introduction to Riemann Surfaces, Addison-Wesley, Reading, Mass., 1957.

54. E. C. Titchmarsh, The Theory of Functions, Oxford University Press, New York, 1960.

55. M. Tsuji, Potential Theory in Modern Function Theory, Maruzen, 1959.

56. Hans F. Weinberger, A First Course in Partial Differential Equations with Complex Variables and Transform Methods, Blaisdell, Waltham, Mass., 1965, p. 243.

57. Guido Weiss, Complex methods in harmonic analysis, this MONTHLY, 77 (1970) 465–474.

58. John Wermer, Banach Algebras and Several Complex Variables, Markham, 1971.

59. Helmut Wielandt, Zur Umkehrung des Abelschen Stetigkeitssatzes, Math. Zeitschr., 56 (1952) 206–207.

60. Norbert Wiener, The Fourier Integral and Certain of its Applications, Dover, New York, 1958.

61. Lawrence Zalcman, Null sets for a class of analytic functions, this MONTHLY, 75 (1968) 462–470.

62. ———, Analyticity and the Pompeiu problem, Arch. Rat. Mech. Anal., 47 (1972) 237–254.

63. ———, Mean values and differential equations, Israel J. Math., 14 (1973) 339–352.

64. Lennart Carleson, On convergence and growth of partial sums of Fourier series, Acta Math., 116 (1966) 135–157.

65. W. F. Osgood, On the transformation of the boundary in the case of conformal mapping, Bull. Amer. Math. Soc., 9 (1903) 233–235.

66. William F. Osgood and Edson H. Taylor, Conformal transformations on the boundaries of their regions of definition, Trans. Amer. Math. Soc., 14 (1913) 277–298.

67. J. W. Alexander, II, Functions which map the interior of the unit circle upon simple regions, Ann. Math., (2) 17 (1915) 12–22.

68. Dieter Gaier, Schlichte Potenzreihen, die auf $|z| = 1$ gleichmässig, aber nicht absolut konvergieren, Math. Zeitschr., 57 (1952) 349–350.

DEPARTMENT OF MATHEMATICS, UNIVERSITY OF MARYLAND, COLLEGE PARK, MD 20742, U.S.A.

INDEX

Pages 1–312 refer to Volume I; pages 313–595 refer to Volume II.